Poplar Culture in North America

NRC Monograph Publishing Program

Inquiries: Monograph Publishing Program, NRC Research Press, National Research Council of Canada, Ottawa, Ontario K1A 0R6, Canada.
Web site: www.monographs.nrc.ca

Correct citation for this publication: Dickmann, D.I., Isebrands, J.G., Eckenwalder, J.E., Richardson, J. (*Editors*). 2001. *Poplar culture in North America*. NRC Research Press, Ottawa, Ontario, Canada. 397 pp.

A PUBLICATION OF THE NATIONAL RESEARCH COUNCIL OF CANADA
MONOGRAPH PUBLISHING PROGRAM

Poplar Culture in North America

Edited by

Donald I. Dickmann
Michigan State University, Department of Forestry, East Lansing, Michigan, U.S.A.

J.G. Isebrands
USDA Forest Service, North Central Research Station, Rhinelander, Wisconsin, U.S.A.

James E. Eckenwalder
University of Toronto, Department of Botany, Toronto, Ontario, Canada

Jim Richardson
Poplar Council of Canada, Ottawa, Ontario, Canada

Published on the occasion of the 21st session of the
International Poplar Commission by
National Research Council of Canada
in association with

Poplar Council of Canada   Poplar Council of the United States

This publication has been made possible by the generous financial support of the
Canadian Forest Service, Natural Resources Canada

NRC·CNRC
NRC Research Press
Ottawa 2001

Printed in Canada on acid-free paper.

ISBN 0-660-18145-2
NRC No. 43259

Canadian Cataloguing in Publication Data

Poplar culture in North America

Published on the occasion of the 21st session of the International Poplar Commission by National Research Council of Canada in association with Poplar Council of Canada and Poplar Council of the United States.

Includes bibliographical references.
Includes an abstract in French.
ISBN 0-660-18145-2

1. Poplar — North America — Congresses.
I. Dickmann, Donald I.
II. Poplar Council of Canada.
III. Poplar Council of the United States.
IV. National Research Council Canada. IV. Series.

SH397.P85D52 2001 634.9'725'097 C00-980189-8

Dedication

Louis Zsuffa has devoted his entire professional life to the study, cultivation, and breeding of poplars and willows. He was born in Sombor, Yugoslavia. He received his forestry education at the University of Zagreb, where he also obtained his Ph.D. in Forest Genetics in 1964. On an American Fellowship, he studied poplar breeding and cultivation in Italy and France. His scientific work on poplar genetics and breeding started at the Poplar Research Institute in Novi Sad, Yugoslavia, and was later continued with a Post-Doctoral Fellowship at the Faculty of Forestry, University of Toronto (1966–1967). In 1967, he joined the Ontario Ministry of Natural Resources as a Research Scientist, and was instrumental in setting up an active poplar research and breeding program in the Province of Ontario.

Louis returned to the Faculty of Forestry, University of Toronto, as a Professor in 1984. The University provided a unique forum where he was able to link his outstanding scientific experience with the advanced education and research of a new generation of young professionals and scientists. There he supervised the research of three Post-Doctoral Fellows, 13 Ph.D. and 14 Masters students. During his career, Louis has authored over 150 publications. His enthusiasm for his work on poplars and willows has always been infectious.

Louis Zsuffa's expertise has been sought by international development agencies for developing countries. Thus, he has given special lectures and participated in collaborative projects in China, India, Costa Rica, Malaysia, and Nepal. He is a founding member of the Poplar Council of Canada. He has served in executive positions on the International Poplar Commission of FAO, the Biomass Energy Agreement of the International Energy Agency, the Poplar Council of Canada, and the International Union of Forest Research Organizations. In recognition of his scientific achievements and international collaboration, Louis has received numerous honours and awards, among them, the Gold Medal for Scientific Achievement from the Canadian Institute of Forestry in 1982. He was a nominee for the prestigious Marcus Wallenberg Prize in Sweden.

Louis Zsuffa has dedicated his professional life to poplar and willow culture. As an individual, Louis has always been a kind, gentle, and generous person, showing support and understanding to students and colleagues alike. He continues to enjoy the support and companionship of his wife Mara. We dedicate this book to Louis, on behalf of his many colleagues and friends.

Contents

Preface

Since ancient times, poplars have fascinated humanity. Throughout the ages, people have capitalized on the diverse characteristics of the many species and clones that make up the genus *Populus* — their rapid growth, ease of reproduction by vegetative means, and widespread natural occurrence. These characteristics have led to the many valuable economic and environmental uses that have been found for poplar trees and poplar wood — pulp and paper, panel products, lumber, shelter for farm crops, streamside protection and erosion control, and, most recently, phytoremediation and treatment of wastes. Thus, it is not surprising that poplars were among the first trees to be cultivated. Poplar culture began in Asia but is practiced now wherever poplars are found naturally or will grow, on all six continents.

Like any other plant or animal that has been found useful, humanity has tried to improve the utility of poplars through breeding. Most poplars are particularly amenable to breeding and genetic improvement efforts due to their characteristics of early sexual maturity, relative ease of crossing, and ease of reproduction by vegetative means. In recent years, *Populus* has also become a model genus for research on plant molecular genetics and the implementation of advances in woody plant biotechnology.

North America is blessed with extensive natural resources of trembling aspen (*Populus tremuloides*). There are also a number of other native poplars which are easily cultivated and have been extensively hybridized amongst themselves and with related poplars from Europe and Asia. Poplar culture has been firmly established in North America for more than 100 years. Some of the front-line technological advances in poplar breeding, improvement, and utilization are still taking place in the United States and Canada.

However, poplar culture is a specialized science and art, requiring careful understanding and attention to clonal characteristics, site conditions, and specific cultural practices if the desired growth and benefits are to be achieved. Diseases and insect pests are a constant threat that can quickly reverse the gains of genetic improvement. Identification and control of the numerous poplar clones can be a poplar grower's nightmare.

This book is intended to gather under one cover important source information about North American poplars for the benefit of poplar growers and scientists. It provides perspectives on the current state and future prospects of all aspects of poplar culture and use in North America. As such, it updates and considerably expands the scope of the 1983 Michigan State University publication, *The culture of poplars in eastern North America* by Donald I. Dickmann and Katherine W. Stuart. The core of that publication and of the present volume is an introduction to the genus *Populus*, poplar breeding strategies, and the ecology and silviculture of poplar plantations. Included in the expanded scope is discussion of the physiological ecology of poplars, the ecology and silviculture of natural stands of poplars, particularly trembling aspen, the environmental benefits of using poplars for windbreaks, phytoremediation, and urban planting, and the properties and utilization of poplar wood. Expanded treatment is given to insect pests and poplar diseases. Nearly 50 descriptions of commercial poplar clones and cultivars are included as well as a key to *Populus* species and major hybrids.

The present volume complements the 1996 National Research Council of Canada publication, *Biology of* Populus *and its implications for management and conservation*, edited by R.F. Stettler, H.D. Bradshaw, Jr., P.E. Heilman, and T.M. Hinckley. That book presents recent advances in our fundamental understanding of the biology of poplars and interprets the relevance of the research findings to practical aspects of poplar culture and conservation. In the present volume, the editors have focused on the practical side, in an effort to offer relevant information on how to grow and use poplars, with sufficient background on the supporting science that poplar growers and scientists alike can benefit.

The 21st Session of the International Poplar Commission (IPC) — a subsidiary body of the Food and Agriculture Organization (FAO) of the United Nations — was held in western North America in September 2000, with more than 250 participants from 31 countries. The IPC was established in 1947 to promote and coordinate the breeding, cultivation, management, and utilization of poplars and willows. The theme of the 21st Session (IPC 2000), 'Poplar and willow culture: Meeting the needs of society and the environment,' inspired the publication of a volume that would showcase the current status of the poplar sector on this continent and document succinctly the scientific and technological advances that have been made.

The editors are deeply indebted to the generosity of the Canadian Forest Service, Natural Resources Canada, for providing the funding to make this publication possible, and to the United States Forest Service for its strong support of the project. The Poplar Council of the United States and the Poplar Council of Canada / Conseil du peuplier du Canada, both voluntary, nonprofit organizations dedicated to the goals of IPC within their respective countries, have given valuable support and encouragement.

Many colleagues provided constructive criticism of drafts of individual chapters. The authors and editors acknowledge with gratitude the help of Bill Berguson, Perry Boskart, Toby Bradshaw, Andy David, John Davis, Jake Eaton, Rob Farmer, Dan Herms, Gary Hogan, Jon Johnson, Richard Kabzems, Chuck Kaiser,

Gary Kuhn, Graeme Lockerby, Peter McAuliffe, H.S. McNabb Jr., Alex Mosseler, Neil Nelson, Peggy Payne, Pierre Périnet, John Phelps, Ken Raffa, Don Riemenschneider, Randy Rousseau, Mike Roy, Bill Schroeder, Wayne Shepperd, Brian Stanton, Reini Stettler, Jerry Tuskan, Robert van den Driessche, Joanne van Oosten, Tim Volk, Cathy Wendt, Lynne Westphal, Lisa Zabek, and Louis Zsuffa. We greatly appreciate the patience, understanding, and cheerful help of Gerry Neville, Diane Candler, and editorial staff of NRC Research Press.

Donald I. Dickmann

J.G. Isebrands

James E. Eckenwalder

Jim Richardson

July 2001

Abstract

This book describes the status of culture and utilization of poplars in North America and documents succinctly recent scientific and technological advances. Gathering under one cover important source information about North American poplars for the benefit of poplar growers and scientists, the book provides perspectives on the current status and future prospects of all aspects of the poplar sector. The scope of the work includes all *Populus* species native to Canada and the United States, naturally-occurring hybrids and varieties, as well as cultivars in current regular use.

The first part of the book, with ten chapters prepared by 28 U.S. and Canadian authors, summarizes practical knowledge on growing and using poplars. Included are chapters describing natural species, varieties, and provenances; poplar breeding techniques and strategies; the physiological ecology of poplars; the silviculture and ecology of natural stands; the silviculture and ecology of plantations; the use of poplars in urban plantings, windbreaks, and phytoremediation; poplar insects; poplar diseases; and wood properties and utilization of poplars.

The second part, with four chapters prepared by five authors, describes the characteristics of the principal poplar clones and cultivars in use in North America. Included are discussions of clonal morphology, phenology, sensitivity to insects and pathogens, growth, site relations, wood properties, importance, and regional suitability. The second part also reviews the breeding strategies that may be used for poplars in the 21st century.

The book complements a previous NRC Research Press publication: Stettler, R.F., Bradshaw, H.D., Jr., Heilman, P.E., and Hinckley, T.M. 1996. Biology of *Populus* and its implications for management and conservation. NRC Research Press, Ottawa, Ontario, Canada. 539 pp.

Résumé

Ce livre fait le point sur la culture et l'utilisation du peuplier en Amérique du Nord et fait état succinctement des progrès scientifiques et technologiques accomplis ces dernières années dans ce domaine. Il réunit sous une même couverture une information de base importante sur les peupliers d'Amérique du Nord pour le bénéfice des planteurs et des scientifiques tout en ouvrant des perspectives sur l'état actuel et sur l'avenir de tous les aspects du secteur peuplier. L'ouvrage porte sur toutes les essences indigènes du *Populus* au Canada et aux États-Unis, sur les hybrides et les variétés naturelles, ainsi que sur les cultivars d'utilisation courante.

La première partie de l'ouvrage comprend dix chapitres rédigés par 28 auteurs des États-Unis et du Canada et présente sommairement les connaissances pratiques sur la culture et l'utilisation des peupliers. On y retrouve entre autres des chapitres sur les essences et les variétés naturelles de peupliers et leur provenance; les techniques et les stratégies d'hybridation; la physiologie et l'écologie des peupliers; la sylviculture et l'écologie des peuplements naturels; la sylviculture et l'écologie des plantations; l'utilisation des peupliers en plantations urbaines, en plantations de brise-vent et en phytoremédiation; les insectes du peuplier; les maladies du peuplier; et enfin les propriétés du bois et l'utilisation du peuplier.

La deuxième partie du livre comprend quatre chapitres rédigés par cinq auteurs et décrit les caractéristiques des principaux clones et cultivars de peuplier utilisés en Amérique du Nord. On y trouve des informations sur la morphologie et la phénologie clonale, la sensibilité des clones aux insectes et aux agents pathogènes, la croissance, les relations avec les sites, les propriétés du bois ainsi que l'importance et l'adaptabilité régionale des clones. Dans la deuxième partie, les auteurs se penchent sur les stratégies d'hybridation que l'on prévoit utiliser au 21e siècle.

Le livre constitue un complément à une publication antérieure des Presses scientifiques du CNRC, soit Stettler, R.F., Bradshaw, H.D., Jr., Heilman, P.E. et Hinckley, T.M. 1996. Biology of *Populus* and its implications for management and conservation. Presses scientifiques du CNRC, Ottawa (Ontario) Canada. 539 pp.

CHAPTER 1
An overview of the genus *Populus*

Donald I. Dickmann

Introduction

The subject of this book — the genus *Populus*, collectively known as the poplars — comprises a singular group of trees. Because they are ubiquitous in the Northern Hemisphere, extremely useful, and possess unique characteristics unmatched by any other temperate tree genus, almost everyone has some knowledge or opinion about poplars. Unfortunately, certain poplar attributes are considered less than stellar. Therefore, these trees invoke a range of sentiments among people that reaches from the sacred to the profane.

On the one hand, poplars have been treated with reverence and esteem. The white poplar (*P. alba*) was linked with the Greek mythological deities Hercules and Persephone. A poplar tree cured Hercules of his serpent bite. The indigenous people of North America regarded poplars as powerful and sacred medicine trees (Altman 1994; Moerman 1998). Because of their attractive looks or striking stature, poplars have been planted for millennia throughout the world as decorative ornamentals or as line plantings. They have been depicted in ink or impressionistic pastels by great artists such as van Gogh, Monet, and Cézanne (Fig. 1). Poets have immortalized them in romantic verse. Scientifically, *Populus* is among the most-studied and most-written-about tree genera. The wood of poplar is highly versatile and prized by the wood-products industry. The exceptionally fast growth rate of poplars amazes people, and this trait, among others, has led to their extensive use in wood-producing plantations and windbreaks.

On the other hand, poplars have a defective, even sinister, side. In Christian legend, aspen (*P. tremula*) is the tree from which the cross of Jesus was hewn. Then, to make matters worse, Judas Iscariot allegedly hanged himself on an aspen. This considerable disgrace cursed the tree to shudder and tremble ever afterward. The bite of this curse is lessened, however, because several other tree species also are implicated in these infamous events. Whatever the truth, this superstition was so powerful that 19th-century lumberjacks in the Great Lakes Region refused to

D.I. Dickmann. Department of Forestry, Michigan State University, East Lansing, MI 48824-1222, U.S.A.
Correct citation: Dickmann, D.I. 2001. An overview of the genus *Populus*. *In* Poplar Culture in North America. Part A, Chapter 1. *Edited by* D.I. Dickmann, J.G. Isebrands, J.E. Eckenwalder, and J. Richardson. NRC Research Press, National Research Council of Canada, Ottawa, ON K1A 0R6, Canada. pp. 1–42.

Fig. 1. Vincent van Gogh's pen-and-ink depiction of a lane of poplars, from 1884 (van Gogh Museum, Amsterdam, The Netherlands).

sleep in a cabin made from aspen logs (Graham et al. 1963). The Victorians added to the ignominy; in their "language of flowers" aspen symbolized scandal, lamentation, and fear (Rupp 1990).

The situation has not improved in modern times. Poplars — especially hybrids — are often regarded as "junk trees" because of their short life span, susceptibility to breakage, and predisposition to diseases and insect damage. Hucksters of "bargain" nursery plants sell unfit hybrid poplar varieties to the public via gaudy Sunday-supplement ads, reinforcing the junk-tree image. The handbook *Weeds of the Northeast* lists cottonwood (*P. deltoides*) among the other noxious plant competitors of the region (Uva et al. 1997). Timber companies spray poplars that volunteer in their conifer plantations with herbicides. Because of their messy habits, many municipalities have passed ordinances banning the planting of poplar trees. One author refers to them as "plebeian" (Rogers 1917).

So there is a yin to balance each yang. In this book, however, the authors embrace the yang, extolling the positive side of poplars and emphasizing their multifarious benefits. We are unashamedly enthusiastic about poplars, and we strongly feel that if their limitations are recognized, their many virtues can be employed to wonderful effect.

What's in a name?

All true poplars are members of the genus *Populus*. The origin of this generic Latin name is an interesting study in linguistics and the blurred roots of plant nomenclature. *Populus* is said to be derived from the early Roman expression *arbor populi*, meaning "the people's tree," because poplars were frequently planted in public places and meetings were held beneath them (Rupp 1990). In fact, the Latin word *populus* is defined as "a multitude, host, crowd, throng, a great number of persons or things; the people." Other versions of the root of *Populus* have been proposed based on ancient Greek or Latin usage (Rupp 1990; Edlin 1963), but they are more fanciful than etymologically correct.

The common English name "poplar," a derivative of its Latin appellation, seems to be one of the few common tree names to be shared, in different forms, by several other modern European tongues. It is *peuplier* in French, *populier* in Dutch, *Pappel* in German, *poppel* in Danish, *poppeli* in Finnish, *poplys* in Welsh, and *pioppo* in Italian. In some parts of North America, the related "popple" is colloquially used for aspen, perhaps because of the tendency of the wood to pop and spark when burned. In the southeastern U.S., however, the name poplar is equivocal because there it usually refers to yellow or tulip poplar (*Liriodendron tulipifera*). Furthermore, if you buy "poplar" wood at a lumber store anywhere, yellow poplar is what you will get. True poplar wood usually is marketed under the name "aspen" or "cottonwood."

The idea of constant motion, referring to the well-known tendency of leaves to noisily flutter in the slightest breeze, is a trait most people associate with poplars. Besides the normal common name, various colloquialisms applied to poplars reflect this trait (Harlow 1957). For example, one Gaelic name for European aspen is *crann critheach*, or "the shaking tree." Aspen's Welsh folkname — *coed tafod merched* — means "tree of the woman's tongue." The Greeks also shared this simile for a poplar leaf. New World people were not to be outdone in this regard; the Acadian refugees in Louisiana called their local poplar *langues de femmes*. Finally, the Onondaga Indian name for the North-American quaking aspen is *Nut-Ki-e*, meaning "noisy leaf."

This characteristic leafy motion, an effect produced by the airfoil-like flattened petiole displayed by most members of the genus, has not been lost on poets seeking inspiration. The sight and sound of the wind rippling a poplar canopy can conjure powerful romantic images in the mind of those so inclined. A few examples will make the point. In Act II of the ever-popular comic operetta *The Pirates of Penzance*, librettist William S. Gilbert conveyed a romantic reversal of fortune using a poplar metaphor. Composer Arthur Sullivan set the scene to music.

> And the brook, in rippling measure,
> Laughs for very love,
> While the poplars, in their pleasure,
> Wave their arms above.

But a few stanzas later the mood has changed:

> Pretty brook, thy dream is over,
> For thy love is but a rover!
> Sad the lot of poplar trees,
> Courted by the fickle breeze!

Poplars figure prominently among the trees in *A Shropshire Lad*, the classic collection of verses published in 1896 by Alfred Edward Housman. For example, these stanzas from poem LII allude to the poet's roots in the English countryside:

> Far in a western brookland
> That bred me long ago
> The poplars stand and tremble
> By pools I used to know.

A far more grim but no-less romantic take on people and poplars is conveyed in poem XXVI, which begins with these lines:

> Along the fields as we came by
> A year ago, my love and I,
> The aspen over stile and stone
> Was talking to itself alone.

Then again in the second stanza:

> And sure enough beneath the tree
> There walks another love with me,
> And overhead the aspen heaves
> Its rainy-sounding silver leaves;

Contemporary poets do not slight poplars. Seamus Heaney's 1996 collection *Spirit Level* included these lines from a short poem entitled "The Poplar":

> Wind shakes the big poplar, quicksilvering
> The whole tree in a single sweep.

Now we must make the transition from rhyme to reality. The two sections that follow will discuss poplars in general, the current classification of the genus, and each currently recognized species. Several books, monographs, and other general references were consulted in preparing these discussions: Burns and Honkala (1990), DeByle and Winokur (1985), Dickmann and Stuart (1983), Elias (1987), FAO (1958, 1980), Farrar (1995), Graham et al. (1963), Harlow et al. (1996), Little (1979), and Rehder (1940). These references can be consulted for help in identifying poplars or for more information. To avoid clutter, individual attributions to these references will not be given in the text. Scientific papers, articles, or books that present something new or from which data are taken, however, will be cited individually.

General characteristics of poplars

Poplars are all deciduous or (rarely) semi-evergreen forest trees with a wide distribution in the Northern Hemisphere, from the tropics to the northern latitudinal limits of tree growth. Stem form is characteristically tall and straight (excurrent), but even within a species stems of individuals or clones can be multiply branched (deliquescent) or twisted and convoluted (Fig. 2). Poplars are short-lived compared to other trees such as white pine, oaks, or Douglas fir, in large measure because they are host to many diseases and insect pests. Nonetheless, the fast growth rate of poplars often enables them to reach large size. The characteristic cottonwoods of eastern North America (*P. deltoides*) and the Pacific Northwest (*P. trichocarpa*), for example, can become enormous trees.

The cell nucleus of poplars contains two sets of 19 ($2n = 38$) chromosomes; rarely, triploid plants with three sets of chromosomes ($3n = 57$) are found. The physical size of the *Populus* genome is remarkably small — 6 times smaller than maize and 40 times smaller than loblolly pine. In addition, the favorable ratio between genetic length and physical length in *Populus* chromosomes makes the genus an attractive choice for genetic mapping and cloning of genes of special importance to forest trees (Bradshaw 2000).

All of the taxa in the genus are dioecious. Individual trees bear either male (pollen-bearing) or female (seed-producing) flowers, both of which are borne on pendant catkins (Fig. 3). Occasionally, some trees will produce a small number of bisexual catkins. In most populations of poplar, male and female trees tend to be approximately equal in number, but there are exceptions. In the Rocky Mountains of Colorado, for example, female aspen trees predominate at low elevation, but at high elevations nearly all trees are male (Mitton and Grant 1996). Generally, male trees tend to be more precocious, producing more catkins per tree, but — at least in aspen — female trees grow faster than males. Emergence of the male and female catkins precedes the flush of leaves in the spring; in aspens, flowers appear at the very first hint of spring (Fig. 3A). After wind pollination, the fruit — an elongated cluster of capsules sometimes likened to a necklace — quickly matures. Ripe capsules split into two, three, or four parts, and the tiny, cottony seeds take to the air (Fig. 3B). Old trees can produce over 50 million seeds in a single season. Because the cotton is regarded as a nuisance, male trees are preferred for

Fig. 2. The form of mature poplar trees varies, depending upon the genetic makeup of a tree and the environment in which it is growing. (A) An eastern cottonwood showing the deliquescent form of trees grown in the open. (B) The typical excurrent, narrow-crowned form of a forest-grown tree, here a hybrid cottonwood.

horticultural and silvicultural plantings; their pollen-bearing catkins quickly rot away when they fall to the ground.

Variation in virtually every trait is a hallmark of the genus, and leaf shape is no exception. There is no poplar leaf archetype. Leaves always are simple, never compound, but they may be narrow or lanceolate, maple-like, round, oblong, deltoid, heart-shaped, or rhombic; they may be longer than wide, wider than long, or equal in both dimensions (Fig. 4). Even on the same tree, leaves may differ considerably in shape; early-season leaves that are preformed in the bud often are distinctly different than leaves initiated during the growing season. Petioles vary in length from less than 1 cm to nearly 10 cm. Autumn coloration of poplar leaves generally is yellow or pale gold, although some western clones of *P. tremuloides* turn bright orange. In a mountainous setting, this marvelous though short-lived splash of color is a major scenic attraction.

Fig. 2 (*concluded*).

Poplar trees may produce few coarse branches, or a myriad of fine branches and twigs. Large branches tend to be brittle and are easily broken off by strong winds, wet snow, or ice. In many species, small branches are abscised in a process not dissimilar to autumn leaf abscission, a trait unique to poplars. Poplar bark, which can be creamy white, various shades of gray, olive green, orange–brown, or bronze in color, often remains smooth for many years, especially in the aspens. Lenticels are prominent on the young bark (Fig. 5A). On older trees, the lower bark breaks up into coarse, corky ridges. In cottonwoods, these ridges may extend well up into the crown (Fig. 5B).

Poplars are major invaders of disturbed sites, but the ecological habitats occupied by the various taxa in the genus break rather cleanly into two categories. Many poplar species typically grow in riparian or wet habitats, ranging from the far northern boreal latitudes to the tropics (Fig. 6A). In wetlands, they are adapted to seasonal flooding or high water table conditions, and their seeds find a favorable environment for germination on the fresh silt or sand left when the water recedes. In contrast, upland habitats are the major province of the aspen and white poplars

Fig. 3. Sexual reproduction in poplars. (A) Typical male or pollen catkins appear early in spring. (B) By early summer ripe capsules release their cottony seed.

(A)

(B)

Fig. 4. While all leaves in the genus *Populus* are simple — not compound — the shape, prominence of teeth on the margins, and petiole length vary considerably among pure species and hybrid taxa.

(Fig. 6B). One of aspen's major ecological roles is the colonization of upland areas burned by intense, stand-replacing fires, and its seeds germinate readily on ash-covered soil. In the Lake States, for example, large areas were colonized by aspen, following the extensive fires of the late 19th and early 20th centuries, moving aspen into a prominence that it did not have in the pre-European settlement forests of the region. Aspens also will grow in wet depressions or on swampy margins, provided some kind of disturbance has allowed them to become established.

Another distinctive feature of the genus *Populus* is its predisposition to form clones by vegetative propagation. "Clone" is derived from the Greek word meaning a twig or slip. The major characteristic of clonal offspring is that they are exact genetic copies of their parent — unless mutations occur to either parent or progeny. In technical jargon, "ortet" refers to the original seed-derived donor plant of a clone; "ramets" are the individual clonal offspring, which can be derived from the original ortet or other ramets propagated from it. In certain naturally growing poplars, the ortet may have originated millennia ago, and its clonal offspring have gone through thousands of self-replicating generations. Clones of planted poplar cultivars may go back hundreds of years and many generations to the ortet. Although exact genetic copies of the ortet, clonal ramets may not necessarily look exactly like their parent or like one another. Local environmental

Fig. 5. Bark characteristics markedly change as trees become older. (A) Smooth bark of a young Euramerican hybrid poplar tree; note the prominent lenticels. (B) Ridged bark of an old narrowleaf cottonwood.

conditions, the age of the donor plant (cyclophysis), or the part of the donor plant where the clonal offspring originated (topophysis) may influence the way in which a ramet grows and develops.

The practice of "cloning" recently has been given a sinister connotation by environmental radicals because it is linked to controversial biotechnology practices that produce genetically modified organisms. Unfortunately, neither the public at large nor fanatical "eco-terrorists" realize that clones are as common in the natural world of plants as are flowers. This fact applies to poplars more than any other tree group. There is nothing inherently sinister in a clone, and anyone who thinks so is ignorant of basic botany. Vegetative propagation of clones enables many plant species to successfully compete and reproduce in the ecological habitats they occupy. Cloning also can be a very effective strategy for invading new habitats. In horticulture and forestry, where plant clones have been used since antiquity, the desirable attributes of a plant variety or an unusual form can be captured

Fig. 5 (*concluded*).

and propagated by cloning. Virtually everyone who grows ornamentals or plants a garden uses clones. On the other hand, there are social and environmental issues concerning the use of genetically modified organisms — and the clones that may arise from them — that need to be dealt with by society. But eco-terrorism is not the way.

Certain types of vegetative propagation or cloning occur after a poplar tree has been cut down or killed suddenly; e.g., by a fire. Most poplars will produce copious sprouts from a cut stump or the root collar of a tree that has been killed, although this ability declines as the tree matures. Eventually, a few large, surviving stump sprouts may take on the appearance of individual trees. Suckers are the second way that poplars vegetatively reproduce under natural conditions. These shoots are produced abundantly from shallow, horizontal roots, especially in the aspens, and in this way a single parent tree may produce a forest of clonal offspring (Fig. 7). Suckering occurs principally after a tree is killed, but living trees also will send up suckers from roots that have invaded adjacent open areas. Certain riparian poplars also reproduce vegetatively by a process called cladoptosis.

Fig. 6. Members of the genus *Populus* are adapted to a variety of habitats. (A) Black cottonwoods growing in a riparian corridor in California. (B) Bigtooth aspen on a sandy, highly disturbed upland site in the Lake States; note the successional understory of white pine.

(A)

(B)

Fig. 7. Nothing about the aspens is more remarkable than their ability to produce tens of thousands of root suckers per hectare after mature trees are suddenly killed or harvested.

Whole lateral twigs, often with leaves attached, can be abscised from large trees, and they may root and form clonal plants if they lodge in moist soil. Larger branches that break off from trees during the dormant season also will root and form new trees if they become covered with silt or sand. These last two modes of vegetative reproduction represent minor but ecologically significant means by which riparian poplars become established along the banks of streams or on sandbars when high water recedes.

Most important for silvicultural and horticultural applications is the establishment of clonal poplar plants with hardwood stem cuttings. This strictly anthropogenic form of vegetative reproduction employs 20- to 30-cm long (sometimes shorter or longer) sections of dormant, 1-year-old woody shoots as planting stock. Amazingly, if these "sticks" are planted in the spring they will quickly produce roots from existing primordia in the inner bark and new shoots from the buds (Fig. 8). The resultant clonal plants often grow several meters tall in the first growing season. This trait alone has allowed the widespread and successful planting of cottonwood and balsam poplars. Unfortunately, the aspens cannot be reproduced from hardwood stem cuttings, although they will propagate — albeit with difficulty — from root cuttings. Leafy "softwood" cuttings from aspen shoots also will root, an expensive process that requires greenhouse misting facilities. Thus, aspen plantations are relatively rare.

The wood of poplar is very versatile and widely used by the forest products industry. It is light in weight, soft, light in color (except for a dark-colored heartwood

Fig. 8. Clones of most poplars — the aspens excluded — are propagated using dormant hardwood stem cuttings. These Euramerican hybrid cuttings were planted 1 month earlier in southern Michigan; in September, 4 months later, they had grown to a height of ca. 2.5 m.

or wetwood core), straight-grained, and, because it is diffuse porous, uniform in texture. The wood is used for pulp and paper, veneer, excelsior, composition boards (especially oriented-strandboard, also known as OSB), lumber, and energy. Among the North American poplars the wood of aspen is most highly prized, and because of its abundance and wide distribution it is one of the most important wood raw materials in Canada and the Lakes States.

Taxonomy

The genus *Populus* is one of two members of the family Salicaceae, the other being the willows (*Salix*). Because they are so similar biologically, poplars and willows often are treated similarly in the horticultural and silvicultural trades. The International Poplar Commission, for example, includes willows under its umbrella. *Populus* comprises six taxonomically distinct sections, consisting of nearly 30 species of worldwide natural distribution in the Northern Hemisphere (Table 1). Twelve species are indigenous to North America, with five or six currently of commercial importance.

Because the genus *Populus* is a complex amalgam of tree taxa, it makes a fascinating — if equivocal — study of taxonomic classification. The thorny problems associated with the sorting and orderly description of tree species are further complicated in the genus for two reasons. First, because many poplars have a

Table 1. Proposed taxonomic classification of the genus *Populus* (Eckenwalder 1996).

Section	Species	NA[a]	Notes and common name
Abaso	*P. mexicana* Wesm.	Yes	Mexican poplar
Turanga	*P. euphratica* Oliv.		Euphrates poplar
(Afro-Asian poplars)	*P. ilicifolia* (Engler) Roul.		Kenyan poplar
	P. pruinosa Schr.		
Leucoides	*P. glauca* Haines		Formerly *P. wilsonii*
(Swamp poplars)	*P. heterophylla* L.	Yes	Swamp cottonwood
	P. lasiocarpa Oliver		
Aigeiros (Cottonwoods and	*P. deltoides* Marsh.	Yes	Eastern cottonwood; includes *P. sargentii* and *P. wislizenii*
black poplar)	*P. fremontii* S. Wats.	Yes	Fremont cottonwood
	P. nigra L.		Black poplar
Tacamahaca	*P. angustifolia* James	Yes	Narrowleaf cottonwood
(Balsam poplars)	*P. balsamifera* L.	Yes	Balsam poplar
	P. ciliata Royle		Himalayan poplar; heretofore placed in section *Leucoides*
	P. laurifolia Ledeb.		Laurel poplar
	P. simonii Carr.		Simon poplar
	P. suaveolens Fisch.		Asian poplar; includes *P. cathayana*, *P. koreana*, and *P. maximowiczii*
	P. szechuanica Schn.		Szechuan poplar
	P. trichocarpa Torr. & Gray	Yes	Black cottonwood
	P. yunnanensis Dode		Yunnan poplar
Populus[b]	*P. adenopoda* Maxim.		Chinese aspen
(Aspens and	*P. alba* L.		White poplar
white poplars)	*P. gamblei* Haines		Himalayan aspen
	P. grandidentata Mich.	Yes	Bigtooth aspen
	P. guzmanantlensis Vazq. & Cuevas	Yes	Manantlán white poplar
	P. monticola Brand.	Yes	Baja white poplar
	P. sieboldii Miquel		Japanese aspen
	P. simaroa Rzedo.	Yes	Balsas white poplar
	P. tremula L.		
	P. tremuloides Mich.	Yes	Quaking (trembling) aspen

[a]NA, native to North America.

[b]Formerly section *Leuce*.

large natural range, they often segregate into geographically and morphologically distinct subspecies and varieties (Table 2). Second, poplars are outcrossing, wind-pollinated, and show few inherent barriers to crossbreeding, so they readily hybridize, both naturally and through controlled pollinations. Interspecific hybrids among species in sections *Aigeiros*, *Tacamahaca*, and *Populus* are commonplace, as are intersectional hybrids between members of *Aigeiros* and *Tacamahaca* (Table 3). Many hybrid combinations, however, are facile only in one direction.[1] For example, the well-known hybrid *P. trichocarpa* × *P. deltoides* almost always requires embryo rescue because of premature dehiscence of the capsule, whereas the reciprocal — *P. deltoides* × *P. trichocarpa* — usually produces viable seed (Stettler et al. 1996). Because most hybrids are fertile, they can generate backcrosses to either of the parent species or combine with one another to form advanced generation hybrids. The result of all this crossbreeding within *Populus* is a multitude of intermediate forms, creating taxonomic havoc.

Based on their manner of dealing with this genetic complexity, poplar taxonomists segregate into two distinct camps (Eckenwalder 1996). Splitters, exemplified by Russian and Chinese practitioners, regale in naming new poplar taxa. Their tally of species can reach as high as 85, with 53 species in China alone! On the other hand, Western and North American taxonomists tend to be lumpers, adopting rigid standards for acceptance of new species. Lumpers view many of the species proposed by the splitter camp either as hybrids that do not breed true or subspecies of a single diverse taxon. Their lumped list comprises as few as 22 distinct species. Whatever their camp, no two authorities seem to agree on the exact number of species. Being inclined to lumping, I have adopted here the conservative view of poplar speciation proposed by North American poplar taxonomist James Eckenwalder of the University of Toronto (Table 1). (See also Chaps. 12 and 13 in Part B.)

Only recognized species and hybrids warrant a distinct Latin binomial. The full Latin name, which always is italicized, also includes the name of the authority who first described the taxon (usually abbreviated but not italicized). Examples of a full Latin appellation are *Populus heterophylla* L. (this common abbreviation refers to the authority Karl von Linne or Linnaeus, who first proposed the system of binomial nomenclature) or *Populus guzmanantlensis* Vazq. & Cuevas. Often the authority is omitted in the interest of brevity. Recognized subspecies and varieties also are Latinized and italicized; e.g., *Populus fremontii* var. *mesetae*. The complete Latin name of a taxon always should be used, with or without authority, at first reference in a publication or an oral presentation (see Tables 1–3).

Clonal or cultivar names (also called epithets) are not italicized and are placed after the Latin designation, bracketed by single quotes. Therefore, the correct designation for the cultivar of eastern cottonwood named by the Southern Forest

[1] In hybrid nomenclature, the maternal (seed-producing) parent is listed first followed by the paternal (pollen-producing) parent; e.g., *P. trichocarpa* (♀) × *P. deltoides* (♂).

Table 2. Some natural subspecies and varieties in the genus *Populus*.[a]

Section/species	Subspecies or variety	Remarks
Abaso		
P. mexicana	*dimorpha* (Brand.) Ecken.	Western coast of Mexico
	mexicana	Eastern coast of Mexico; southernmost North American poplar
Aigeiros		
P. deltoides	*deltoides*	Southern form; synonyms var. *angulata*, var. *missouriensis*, or *P. virginiana*
	monilifera (Ait.) Ecken.	Northern form, including plains cottonwood (synonyms var. *occidentalis* or *P. sargentii*)
	wislizeni (Wats.) Ecken.	Rio Grande cottonwood
P. fremontii	*fremontii*	Occurs west of the Continental Divide
	mesetae Ecken.	Mexico, Texas, and New Mexico; synonym *P. arizonica*
P. nigra	*betulifolia* Torrey	Young leaves and twigs downy
	caudina Ten.	Leaves and shoots downy; angled stems; synonym *pubescens* Parl.
	italica Duroi	Lombardy poplar; fastigiate form, male
	neapolitana Ten.	Less erect and less hairy than variety below
	plantierensis Schneid.	Fastigiate, reddish petioles; hairy twigs; synonym *charkowiensis*
	sinensis Carr.	Yellow branches; probably from China
	thevestina Dode	Similar to Lombardy poplar but with light bark; female or sometimes hermaphroditic
Tacamahaca		
P. szechuanica	*tibetica* Schn.	
P. simonii	*fastigiata* C.S.	Pyramidal form
P. trichocarpa	*hastata* Henry	Thick, narrow leaves; smooth, long fruits
Populus		
P. alba	*globosa* Spaeth.	Shrub or tree; dense foliage
	hickeliana Dode	Mediterranean distribution
	microphylla Maire	Mediterranean distribution
	nivea Ait.	Leaves silvery and hairy beneath, lobed
	pendula Loud.	With pendulous branches
	pyramidalis Bge.	Bolleana poplar; columnar form
	richardii Henry	Leaves yellow above
	subintegerrima Lane	Mediterranean distribution

Table 2 (*concluded*).

Section/species	Subspecies or variety	Remarks
P. grandidentata	*meridionalis* Tidestr.	East coast form
	pendula Hort.	With pendulous branches
P. tremula	*villosa* (Lang.) Wesm.	Lowland type
	glandulosa	A cultivated variety in Korea; formerly included within *P. davidiana*
P. tremuloides	*aurea* (Tidestr.) Daniels	Rocky Mountain form
	pendula Jaeger & Beissner	Eastern form
	magnifica Vict.	Eastern form
	reniformis Tidestr.	Eastern form
	vancouveriana (Trel) Sarg.	Vancouver Island form

[a]This listing does not include all proposed varieties or subspecies in the genus, and it is not intended as the taxonomic last word on the subject. These names may be encountered in the poplar literature, and this listing is presented to help interested readers sort out taxonomic origins.

Experiment Station in Stoneville, MS, is *Populus deltoides* 'Bolivar Belle.' The venerable Lombardy poplar is *Populus nigra* var. *italica* 'Lombardy.' A hybrid cultivar would be similarly designated, e.g., the natural hybrid from Iowa *Populus alba* × *P. grandidentata* 'Hansen.' If the hybrid has been assigned its own Latin binomial — in the case of the previous example *P.* ×*rouleauiana* (Table 3) — it can be used in place of the two parent species; i.e., *P.* ×*rouleauiana* 'Hansen.' If known varieties, clones, cultivars, or hybrids were parents in a hybrid combination, they should be identified in the full name; e.g., *Populus nigra* var. *betulifolia* × *P. trichocarpa* 'NE-12.' It is essential to always use the times (×) symbol when designating a hybrid, otherwise an uninformed reader may take it for an actual species.

A general description of the members in each section in the genus *Populus* follows, along with known varieties and hybrids. Emphasis is on North American taxa.

Section Abaso

Abaso is a monotypic section with *P. mexicana* as the single member (Table 1). This species had been previously placed in synonymy with *P. fremontii* because it superficially resembles the southwestern *Aigeiros* cottonwoods. But Eckenwalder (1977) proposed it as a distinct species in a section of its own. A small to medium-sized riparian poplar, *P. mexicana* has linear, willow-like juvenile leaves, deltoid to round mature leaves, and dry, bright yellow, blunt buds. It consists of two subspecies (Table 2), one native to the east coast of Mexico (subspecies *mexicana*) and one native to the west coast (subspecies *dimorpha*). Both forms

Table 3. Some natural and artificial hybrids among species in the genus *Populus*.[a]

Hybrid parents	Hybrid designation	Common name and notes
P. alba × *P. adenopoda*	*P.* ×*tomentosa* Carr.	May also be a tri-hybrid containing genes from *P. tremula*
P. alba × *P. grandidentata*	*P.* ×*rouleauiana* Boivin	
P. alba × *P. tremula*	*P.* ×*canescens* Sm.	Gray poplar
P. alba × *P. tremuloides*	*P.* ×*heimburgeri* Boivin	
P. angustifolia × *P. balsamifera*	*P.* ×*brayshawii* Boivin	Brayshaw's poplar
P. angustifolia × *P. deltoides*	*P.* ×*acuminata* Rydb.	Lanceleaf cottonwood; synonym *P.* ×*andrewsii* Sarg.
P. angustifolia × *P. fremontii*	*P.* ×*hinkleyana* Corr.	
P. angustifolia × *P. balsamifera* × *P. deltoides*	None	Trihybrid
P. balsamifera × *P. deltoides*	*P.* ×*jackii* Sarg.	Jack's hybrid poplar or heartleaf balsam poplar; also known as *P. balsamifera* var. *subcordata* or *P. candicans*; Balm-of-Gilead is a clone of this hybrid
P. deltoides × *P. nigra*	*P.* ×*canadensis* Moench	Euramerican poplar; synonym *P.* ×*euramericana* Guin.
(*P. deltoides* × *P. nigra*) × *P. balsamifera*	*P.* ×*rollandii*	Trihybrid; very similar to *P.* ×*jackii*
P. fremontii × *P. deltoides*	?	
P. fremontii × *P. nigra*	P. ×*inopina* Ecken.	
P. grandidentata × *P. tremuloides*	*P.* ×*smithii* Boivin	Synonym *P.* ×*barnesii* Wag.
P. laurifolia × *P. nigra*	*P.* ×*berolinensis* Dippel	Berlin or Russian poplar; synonyms *P.* ×*rasumowskyana* Schneid. or *P.* ×*petrowskyana* Schneid.
(*P. laurifolia* × *P. nigra*) × *P. balsamifera*	None	Trihybrid
(*P. laurifolia* × *P. nigra*) × *P. deltoides*	None	Trihybrid
P. tremula × *P. tremuloides*	*P.* ×*wettsteinii*	Often a triploid

Table 3 (*concluded*).

Hybrid parents	Hybrid designation	Common name and notes
P. trichocarpa × *P. deltoides*	*P.* ×*generosa* Henry	Interamerican poplar; synonym *P.* ×*interamericana* Brockh.
P. trichocarpa × *P. fremontii*	*P.* ×*parryi* Sarg.	Parry cottonwood

[a]These documented hybrids have spontaneously formed in areas where the natural range of species overlap or where exotic taxa have been planted near naturally growing poplar trees. In general, wherever two poplar taxa are grown together, hybrids will form (Eckenwalder 1996), especially among species within a section or between species in sections *Aigeiros* and *Tacamahaca*.

are found along the rivers of the coastal plain. Subspecies *mexicana* also occurs in the mountains of Oaxaca and Chiapas, making it the most southerly distributed New World poplar.

Section Turanga

Like section *Abaso*, Section *Turanga* at one time was monotypic, but it now comprises three riparian species (Table 1), among them the principal poplars native to the continent of Africa. There was a proposal to place the poplars in section *Turanga* in a separate genus, *Euphratodendron*, because of certain distinctive characteristics of their wood anatomy and flower structure, but these trees still remain poplars. The primitive floral characteristics of members of this section are archetypical. *Populus euphratica*, the most well-known member of the section, is an extraordinary species. A small- to medium-sized tree native to North and Central Africa, as well as western and central Asia, its distinctive features include slender, sympodial branches (i.e., the shoot apex aborts); small, downy buds; and leathery leaves, which vary in shape from linear on juvenile plants to toothed and rounded on mature plants. It is usually a tree with a short stocky bole and branchy crown. The species often grows in shrubby thickets along watercourses, but in favorable locations in central Asia it can attain commercial quality. Euphrates poplar can tolerate hot, arid, saline conditions or waterlogging and may offer some potential in hybridization programs whose objective is to create tolerance to such conditions. In fact, recent artificial hybrids with *P. deltoides*, *P. nigra*, and *P. simonii* have been successful (Zsuffa et al. 1996). A few clonal selections are planted in Morocco.

A recent addition to *Turanga* is *P. ilicifolia*, a native of equatorial East Africa, making its natural range the southernmost among the world's poplars. This species heretofore had been considered a form of *P. euphratica*. It roots well from stem and branch cuttings and grows into a large tree up to 30 m in height and 1.5 m in diameter, making it a candidate for tropical poplar culture. *Populus pruinosa* is another recent addition to section *Turanga*, but it also was once considered only a subspecies of *P. euphratica*. Although also a riparian poplar, this remarkable poplar has a wide ecological amplitude. In the eastern part of its range in the Sinkiang region of China, *P. pruinosa* grows in desert conditions, where mean annual rainfall is less than 50 mm and the salt content of the soil can range from 2 to 3%!

Section Leucoides

Section *Leucoides* contains three species, none of which are of great economic significance, although they are important ecologically. *Populus heterophylla*, the swamp cottonwood, is a medium to large tree distributed along the eastern U.S. coastal plain and the Mississippi Valley north to Michigan.[2] Leaves are ovoid to cordate and are borne on a round petiole. Twigs are dull brown or gray with a distinctive orange pith and stout, resinous, reddish-brown buds. Never a common tree, swamp cottonwood inhabits mixed-species forests in swamps, sloughs, and along river borders, usually on heavy clay soils with high water tables. Swamp cottonwood is among the most flood-tolerant of poplars; it will thrive on sites that are too wet for eastern cottonwood. Experience with propagation of swamp cottonwood is minimal, although cuttings appear difficult to root. It does, however, produce root suckers.

The two other species in section *Leucoides* are Asian. *Populus lasiocarpa*, native to the humid and semi-humid subtropics of China, is a striking tree with huge (up to 35 cm long and 25 cm wide) heart-shaped, bright green leaves with red midribs and petioles. It grows rapidly, but does not propagate easily from cuttings. It is one of the better poplars for ornamental uses. *Populus glauca* (formerly *P. wilsonii*), also distributed in central and western China, is similar to the previous species.

Section Aigeiros

For many years, *Aigeiros* has been the most important section of *Populus* for plantation culture in North America or the world as a whole. Ecologically, this section includes some of the major riparian poplars in the Northern Hemisphere. *Aigeiros*, which includes cottonwoods and black poplar, contains three species, two of which are native to North America (Table 1). Both the pure species and their hybrids are of great commercial significance.

The most familiar and important member of this section is *P. deltoides*, the eastern cottonwood. Distributed over most of the eastern and midwestern U.S. and southern Canada to the foothills of the Rocky Mountains, eastern cottonwood is a medium to large tree that develops a long, straight bole and small, round crown under forest conditions. But when growing in the open, the stem typically

[2]For a detailed portrayal of the natural ranges of most poplar species indigenous to North America, see the *Atlas of United States Trees*, Vol. 1 (1971) and Vol. 3 (1976) by former U.S. Forest Service dendrologist Elbert L. Little, Jr.

becomes many-forked with a massive crown (Fig. 2A). Cottonwood leaves are deltoid to cordate in shape, with roundly toothed margins, and they hang from long, flattened petioles. The twigs are stout, angular to ribbed in cross section, and produce slightly resinous, outcurved buds. Bark on young trees is smooth and greenish-yellow to gray, becoming ashy-gray and deeply furrowed on older trees.

Eastern cottonwood is a tree usually associated with bottomlands, alluvia, and riparian corridors, where optimum growth occurs on silty or sandy loam soils. In the lower Mississippi Valley, the best stands are found on the land between the river and the levees that is seasonally flooded. After flood waters recede, the raw sediment that is left behind can be colonized by many thousands of cottonwood seedlings per hectare. In the northern part of its range, cottonwood will grow almost anywhere, and it can withstand droughty conditions. In the southern Lake States, for example, cottonwood is a common invader of old fields, gravel pits, and other disturbed upland sites, as well as the sand dunes along the shore of Lake Michigan (Fig. 9).

Like other poplars, eastern cottonwood is highly intolerant of shade and competition; therefore, it must establish dominance early. It grows in pure stands or in association with other bottomland or pioneer species. The southern form of eastern cottonwood (subsp. *deltoides*; Table 2) is often regarded as the fastest-growing native tree in North America. On the best sites in the Mississippi River Delta, height growth can average over 4 m/year for the first 5 years (Knowe et al. 1998), and over 145 m^3/ha (40 cords/acre) of pulpwood can be produced in a 12-year rotation (McKnight 1970). At maturity, eastern cottonwood can attain large size, reaching heights of over 50 m and diameters of more than 1.5 m.

Eastern cottonwood usually produces good seed crops yearly when mature, and it will rapidly invade newly exposed, moist soils. The cottony seed can be carried long distances by the wind or on the surface of moving water. During seed shedding in areas with a high density of cottonwood, the air may be filled with cotton, and lawns may appear as though snow had just fallen. For this reason, cottonwood is regarded as a nuisance by many homeowners and municipalities, and it is generally not recommended for ornamental use. Vegetative reproduction via stump sprouts by this species is vigorous if trees are young when cut. Cottonwood also can be propagated with high survival rates by hardwood stem cuttings, although some genotypes root poorly. Therefore, rooting ability is an important selection criterion in cottonwood genetic improvement programs. In the Mississippi Delta of the southern U.S., trees can grow to over 5 m in height during the first year after a cutting is planted (McKnight 1970). The northern form of cottonwood (subsp. *monilifera*) generally does not root as easily from hardwood stem cuttings nor does it grow as fast as the southern subspecies.

Eastern cottonwood is an important commercial timber species, particularly in the southern U.S. where it is used for match stock, excelsior, sawtimber, veneer, and pulpwood. It has been widely planted on industrial land in the Mississippi

Fig. 9. Eastern cottonwood primarily is associated with riparian areas or wetlands throughout the eastern two-thirds of North America, but in areas with adequate rainfall it will grow almost anywhere. These cottonwoods have naturally established on a sand dune along the Lake Michigan shore.

River Delta, where the rich alluvial soils combined with the inherently fast growth rate of improved clones produce high wood yields. In addition to forest plantings, cottonwood has been extensively used in shelterbelts and for reclamation of strip-mine spoils. Indigenous people used this cottonwood for various pharmaceutical aids (Moerman 1998). Decoctions of twigs, leaves, and bark were used to treat wounds, snakebites, venereal disease, worms, and bronchial infections; poultices of leaves were applied to bruises, sores, and boils.

Five natural varieties of eastern cottonwood were proposed at one time, but Eckenwalder (1977) recognized just three subspecies (Table 2). Eastern cottonwood also has shown a substantial predisposition to hybridize, both naturally and under controlled conditions (Table 3). The most important group of hybrids worldwide is *P. deltoides* × *P. nigra*, known collectively as *P.* ×*canadensis* (synonym *P.* ×*euramericana*)[3] or Euramerican hybrids. This hybrid usually has *P. deltoides* as the maternal parent, because the reciprocal cross rarely is successful.

---

[3]According to the rules that govern plant nomenclature, *P.* ×*euramericana*, the binomial by which this hybrid is widely known and which at one time was adopted by the International Poplar Commission, does *not* take precedence over that proposed earlier by Moench (or Mönch) — *P.* ×*canadensis* (Boom 1957; Rehder 1940). Moench's binomial was first and should be the one used, and we will follow that convention in this book.

Many of the hybrid poplar clones now under cultivation worldwide belong to this group; e.g., cultivar 'I-214' in Europe and cultivar 'Eugenei' in North America. Intersectional hybrids of *P. trichocarpa* with *P. deltoides* (*P.* ×*generosa* or Interamerican hybrids) also have become prevalent in poplar culture, especially in the states of Washington and Oregon, the Canadian province of British Columbia, and western Europe. Both Euramerican and Interamerican hybrids show hybrid vigor (heterosis) in which certain hybrid offspring outgrow or outperform both of their parents in some way. Worldwide *P. deltoides* remains the most important poplar taxon in genetic improvement programs and plantation forestry.

The plains cottonwood (*P. deltoides* subsp. *monilifera*) — the state tree of Wyoming — deserves special mention, if only because of its enigmatic taxonomy. Charles Sprague Sargent, pioneer dendrologist and director of the Arnold Arboretum of Harvard University, originally included this cottonwood as a variety of *P. deltoides*. But that designation did not stand. Early in the 20th century, full-fledged species status (*P. sargentii*) was awarded to this western tree. In the 1927 and 1953 editions of the *Checklist of United States Trees*, the species status of plains cottonwood was upheld. In the latest checklist (Little 1979), however, Sargent's view once again held sway; plains cottonwood was listed as *P. deltoides* var. *occidentalis*. Most authorities have since followed Little's lead. Nonetheless, Eckenwalder (1977) claimed that plains cottonwood so closely resembles the cottonwood of the Great Lakes Region and other northern areas that they all should be included as subsp. *monilifera*. Oddly, the only unambiguity seems to be the common name, just the reverse of the usual situation in plant nomenclature where Latin names eliminate ambiguity.

The habitat of plains cottonwood is the high steppes of the U.S. and southern Canada. In that habitat, plains cottonwood is often the only tree found, so it has quite naturally become part of the legend and romance of the old West. It was a sure sign of water in an arid land, and its shade provided relief from the blazing sun. Plains cottonwood was virtually the only firewood or building material — save sod — readily available. In winter, the sweet and nutritious inner bark and branches were fed to horses by indigenous people and settlers. Indians also ate the buds, seed capsules, and inner bark. Buds were boiled to make a yellow dye, leaves could be fashioned into a flute-like instrument, and young trees were used as "sacred poles" during ceremonies (Moerman 1998). In a region where trees are not abundant, plains cottonwoods often served as trail markers and gathering sites. Some of them achieved lasting notoriety. The Lone Sentinel Cottonwood that stood on the banks of the Arkansas River in Kansas, for example, marked the place where the town of Dodge City sprang up.

Plains cottonwood is lesser in stature than the southern cottonwood (subsp. *deltoides*), with a spreading crown consisting of small leaves and hairier twigs. It occurs in the region below 2100 m elevation that can be classed as semiarid, growing in open stands or as solitary individuals along stream banks and in other moist places. Plains cottonwood has been tested for planting in windbreaks, as wildlife cover, and for ornamental purposes in the Great Plains Region, and growth and survival have been good despite severe cold, wind, and drought conditions. Because of the stressful condition under which it grows, plains cottonwood is considered by some to be more susceptible to insect and disease pests than other eastern cottonwoods, but research has not borne this out.

Another arid-zone poplar in section *Aigeiros* is *P. fremontii*, the common, low-elevation cottonwood of northern Mexico and the southwestern U.S. from western Texas to the San Joaquin Valley of California. It is a large tree (up to 30 m in height and 1.2 m in diameter) with small deltoid leaves that turn bright yellow in the fall. Besides providing welcome shade, the tree has a fibrous inner bark from which Mohave women made skirts. Indians also ate or chewed green catkins and capsules; used decoctions of the inner bark and leaves to prevent scurvy and treat sores, cuts, and bruises; and wove baskets from the young twigs (Moerman 1998). The Spanish word for poplar is *alamo*, and the Franciscan mission where the heroic but doomed Texans held off the Mexican army in 1836 was named for the adjacent grove of Fremont cottonwoods (Rupp 1990). This poplar is widely planted as an ornamental and for fuel around ranches and in towns throughout its native region (Fig. 9).

Taxonomic confusion also typifies the other southwestern cottonwoods. The Rio Grande cottonwood is native to western Texas and New Mexico. Once regarded as a separate species (*P. wislizeni*) or a variety of *P. fremontii*, it now is placed as a subspecies under *P. deltoides* (Eckenwalder 1977). Among the uses for this poplar by indigenous people, the catkins, capsules, and buds were collected for chewing gum or food (Moerman 1998). Another cottonwood found in New Mexico, Texas, and northern Mexico, formerly designated *P. arizonica*, is now included under *P. fremontii* as var. *mesetae*. Yet a fourth southwestern cottonwood found in central and southwestern Texas and northern Mexico has been proposed — *P. palmeri*. This rather obscure, medium-sized tree with ovate leaves now is placed in the synonymy of *P. deltoides*.

The last species in section *Aigeiros* is native to Eurasia, although its widespread and ancient cultivation has obscured the exact limits of its original natural range. *Populus nigra*, the black poplar, is the "type" species for the section, but it actually differs in many traits from the two cottonwoods in the section. In fact, it more closely resembles section *Tacamahaca* poplars in some traits, and its chloroplast DNA has ties to *P. alba* of section *Populus* (Smith and Sytsma 1990). Eventually, the North American cottonwoods may have to be placed in a section of their own, leaving *P. nigra* as the sole member of section *Aigeiros*.

Black poplar is a large tree that often produces an irregular, branchy crown. The typically crooked, buttressed bole can be massive, frequently producing large burs or epicormic branches. Many stands, however, produce straight, well-formed trees; e.g., along the Danube River in Eastern Europe. The leaves of black poplar are rhombic–ovoid, dark green in color, with finely toothed margins and long, flattened petioles. The twigs are round in cross section, reddish, and produce gummy, outcurved buds. The juvenile bark is yellow–white, becoming dark and deeply fissured with age.

Black poplar is an aggressive species that seeds into wastelands, riverbanks, and other exposed sites. It sprouts vigorously from stumps and, to some extent, suckers from roots. Propagation from stem cuttings also is very easy. At maturity, black poplar may reach heights of 40 m and diameters of nearly 2 m. The high human population density in most of its natural range, combined with a long history of exotic poplar introductions and disruption of riparian ecosystems, threatens the genetic integrity of this species. Therefore, conservation and restoration of natural riparian ecosystems dominated by black poplar are being given high priority in many European countries.

The large geographic distribution of *P. nigra* and its widespread cultivation have given rise to a plethora of named varieties and cultivars. Prominent among the cultivated clones of black poplar have been those exhibiting a columnar (fastigiate) growth habit; variety *italica* (Lombardy poplar) and variety *thevestina* (Fig. 10). Lombardy is the oldest and best-known columnar variety, having been introduced from central Asia into cultivation in the Po Valley of Italy in the 18th century (Zsuffa 1974). Lombardy poplar is a male tree that may represent propagules from a single original tree or a family of similar trees. As a consequence, some recognize it as a cultivar, others as a true variety. The dramatic silhouette produced by this distinctively columnar tree has become almost a synonym for the fastigiate growth habit. Lombardy poplar has been used extensively throughout the world as a landscape ornamental, and it may be the most widely planted of all poplars (see Part B of this book). Currently, a breeding program in Turkey is aimed at producing columnar types with a broader genetic diversity.

In North America, Lombardy poplar and other *P. nigra* varieties are more susceptible than native poplars to the canker caused by *Dothichiza populea*. This canker is largely responsible for the pitiably short life span of black poplars in certain regions of the eastern United States and Canada. In addition, massive infection of the stem by a wetwood bacterium frequently occurs, contributing to their rapid decline in plantings. Nonetheless, large, magnificent trees over 40 years old can be found in the northern Great Lakes and in the western U.S. and Canada.

Growth rates of *P. nigra* in North American tests have been variable, depending upon site and clone used, but they compare very favorably with other poplar species or hybrids. The major role black poplar has played in North American poplar culture, however, is as a parent in hybrid crosses produced naturally or by poplar breeders (Table 3). The Euramerican clones (*P.* ×*canadensis*) have been especially significant. The original North American Euramerican clones were 19th-century emigrants from Europe, but during the 20th century many new hybrids were created by indigenous breeding programs. The Euramerican hybrids remain fundamental to North American poplar culture. Recently, intersectional hybrids

Fig. 10. The striking columnar (fastigiate) form of the *P. nigra* varieties *italica* (Lombardy) and *thevestina* have long made them popular in North America for windbreaks and line plantings.

between *P. nigra* and the Asian *P. suaveolens* (formerly *P. maximowiczii*) have shown great promise for plantation wood production, and a few of these clones are now planted widely.

Section Tacamahaca

The fifth section in *Populus*, *Tacamahaca*, collectively known as the balsam poplars, contains nine species (Table 1). These largely riparian poplars are distributed in the northern latitudes, and several species are of commercial importance. Three species are native to North America (*P. balsamifera*, *P. trichocarpa*, and *P. angustifolia*), with the remaining six found in Asia. No *Tacamahaca* poplars are indigenous to Europe.

Balsam poplar (*P. balsamifera*) is the most widely distributed *Tacamahaca* poplar in North America, occurring transcontinentally across the northern U.S. and Canada to Alaska. It grows farther north than any North American poplar, occurring in shrubby form on Alaska's North Slope. According to Little (1979), the Latin binominal *P. balsamifera* (literally "bearing resin"), by which this species has long been known, was used synonymously for eastern cottonwood for many years. Apparently, the early description by Swedish taxonomist Karl von Linne was too vague to distinguish between the two. Thus, *P. tacamahaca* was proposed for balsam poplar, but this name did not gain wide acceptance.

Balsam poplar has ovate leaves, dark lustrous green above and pale metallic green below, often with rusty brown blotches. The leaf margins are finely toothed and the petioles round. The twigs also are round in cross section, with out-curved buds that are saturated with a fragrant, amber-colored resin that is one of this species most distinctive characteristics. In fact, the word balsam is said to be a derivation of the ancient Hebrew *bot min*, meaning "the chief of oils" (Rupp 1990). The bark of balsam poplar is smooth, greenish to reddish brown, turning gray with scaly ridges as it ages. The wood is soft, light brown in color with a grayish sapwood. It is considered inferior to aspen for bleached pulp by the forest industry because of the high resin content and dark color of the heartwood.

Balsam poplar is a medium-sized tree that reaches its largest proportions in the Northwest. Its habitat usually is moist or wet soils, such as the borders of streams, lakes, swamps, and depressions, although it will also grow on dry sites. Balsam poplar is an intolerant pioneer species that will rapidly invade disturbed wet sites by suckering or seeding. Although relatively short-lived, it will outlive quaking aspen and may eventually dominate on sites where the two coexist. Like aspen, balsam poplar commonly acts as a pioneer species, eventually succeeded by more tolerant hardwoods and conifers. Owing to its adaptability to dry conditions, resistance to cold, and ease of propagation by stem or root cuttings, balsam poplar has been planted for shelterbelts, although forest plantations of this species are virtually unknown. The hybrid with *P. deltoides* (*P.* ×*jackii*; Table 3), however, has been planted for wood production in eastern Canada.

Although the wood of balsam poplar is not of great commercial importance, the tree has other useful attributes. In his compilation of Native American ethnobotany, Moerman (1998) lists balsam poplar among the 10 plants with the greatest number of uses by indigenous people. The buds were used to make a salve or a decoction for coughs, lung infections, and sprains; a poultice of root scrapings was applied as a disinfectant and for headaches or other pains; a decoction of the inner bark was used as an eyewash, a tonic, and a treatment for tuberculosis and venereal disease; and branches and leaves were used in sweat baths for rheumatism and other pains. Young twigs were fed to horses when other food was not available, and the inner bark and sap were consumed by humans. Smoke from the buds and bark repelled flying insects; the inner bark also could be smoked in pipes when mixed with tobacco. Bees apparently use the resin of balsam poplar to seal cracks in their hives (Rupp 1990).

A note on the Balm-of-Gilead poplar is necessary here. Both the name and identity of this vegetatively propagated female tree are another source of confusion. Originally, the commercial cultivar Balm-of-Gilead was considered a clone of heartleaf poplar, a variety of balsam poplar given the name *P. tacamahaca* var. *candicans* (later becoming *P. balsamifera* var. *subcordata*). Other authorities maintained heartleaf poplar's species or hybrid status with the binomial *P. candicans* or *P.* ×*candicans*. Eckenwalder (1996) considered Balm-of-Gilead to be a *P.* ×*jackii* hybrid, which fits with the heart-shaped leaves. To add to the confusion, Balm-of-Gilead (or just "Balm") also is a colloquial name generally applied to balsam poplar in parts of its natural range. The leaves of Balm-of-Gilead and heartleaf poplars are larger and more cordate than normal balsam poplars, and they are whitish and hairy below, particularly on the veins and petioles. Balm-of-Gilead has been planted extensively as an ornamental tree and for windbreaks.

The western counterpart of balsam poplar is black cottonwood (*P. trichocarpa*), which occurs largely in the Pacific coastal states and western Canadian provinces from southern California to Alaska, and east through the Inland Empire. Morphologically, it is similar in most respects to balsam poplar, except that the capsules of black cottonwood split into three parts when mature rather than the two parts of balsam poplar, and the bark on older trees of black cottonwood is furrowed rather than scaly. In fact, where the ranges of the two species overlap in Alberta and the northern Rocky Mountains of the U.S., distinguishing one from the other may be impossible unless capsules can be examined. In this zone of overlap, hybrids between the two inevitably occur, complicating identification to the species level. Hybrid trees can have capsules in each catkin that split into either two or three parts.

Black cottonwood is a fast-growing tree component of moist bottomland, riverine, or alluvial ecosystems where it grows in pure stands or in association with other low-elevation hardwoods and conifers (Fig. 11). This species reaches its best development in climatic regions dominated by moist Pacific Ocean air. Black cottonwood grows to the largest size of any native poplar or any western hardwood; in the Puget Sound area, heights of 50 m and diameters over 1.5 m can be achieved. Mature forest-grown trees in coastal habitats develop long clear boles; from a commercial standpoint, this species has no equal among poplar taxa in stem form. The crowns of such trees are typically narrow, cylindrical, and round-topped. Trees growing in more arid locations east of the coastal mountain ranges are smaller in stature with broader, deliquescent crowns.

Like balsam poplar, black cottonwood can be easily propagated from stem cuttings and it readily sprouts from cut stumps, although root suckers are rare. Plantation culture of black cottonwood, however, has not been widely practiced in its native range, although growth rates can be quite impressive. Rather, during the last decades of the 20th century, clones of Interamerican hybrids (*P. trichocarpa* × *P. deltoides* = *P.* ×*generosa*; Table 3)[4] developed at the University of Washington began to be planted on a commercial scale in the Pacific Northwest. These

[4]In 1972, van Broekhuizen proposed the Latin binomial *P.* ×*interamericana* for hybrids of *P. trichocarpa* × *P. deltoides*, and this appellation has been widely used, including by the International Poplar Commission. Henry in 1914, however, assigned *P.* ×*generosa* to this hybrid and, although less descriptive, by the rules of botanical nomenclature this name takes precedence and will be the one used in this book (cf. Eckenwalder 1984; Rehder 1940).

Fig. 11. The northwestern black cottonwood in a typical riparian habitat, here in the eastern Cascade Mountains of Washington.

Interamerican hybrids are more productive than pure black cottonwood (Heilman and Stettler 1985), and they became the basis for a new hardwood pulpwood and sawtimber industry in a region where hardwoods had hardly been utilized. Plantations of Interamerican hybrids have shown spectacular growth rates on alluvial soils along the lower Columbia River and in irrigated plantations in the high desert east of the Cascade Mountains (Fig. 12). After 4 years growth, biomass yields of 50–140 tons/ha can be achieved in intensive culture systems; 15-year sawtimber volumes can be 146 m^3/ha (25 000 board ft/acre) or more (Heilman et al. 1990; Scarascia-Mugnozza et al. 1997). Hybrids of black cottonwood with *P. nigra* and *P. suaveolens* are now entering commercial production to complement the Interamericans. Poplar growers in the Low Countries of Europe also have recognized the virtues of black cottonwood, and clones of the pure species and its hybrids are now considered to be higher-yielding alternatives to traditional Euramerican clones.

Tribal groups in the Northwest used black cottonwood in many of the same ways as balsam poplar (Moerman 1998). But certain of the pharmaceutical uses were unique. For example, buds were mixed with chewed and warmed mountain goat kidney fat by the Bella Coola and applied as a face cream. They also mixed an infusion of buds with sockeye salmon oil and rubbed this tonic on the scalp for baldness. Because of this tree's large size, straightness, and relatively soft wood,

Fig. 12. An 8-year-old plantation of a *P. trichocarpa* × *P. deltoides* (Interamerican or TD) hybrid clone on industrial land in the lower Columbia River Valley of Oregon. Professor Reinhard Stettler of the University of Washington (pictured) led the pioneering breeding project that produced this extraordinary hybrid.

it was used by many tribes for dugout canoes. The distinctive bud resins were used as a glue, a base for paints, a yellow dye, and to conceal human scent while stealing enemy horses. The bark could be used for food storage containers and for sheathing lodges. Finally, black cottonwood was an omen; when its leaves shimmered when no wind was perceptible, bad weather surely was on the way.

The third *Tacamahaca* species indigenous to North America is *P. angustifolia*, the narrowleaf cottonwood. Widely distributed in the Rocky Mountains and Plains from southern Canada to northern Mexico, narrowleaf cottonwood is of little commercial importance at the present time. This medium-sized tree is characteristically found growing along streams and on moist upland flats in the foothill-mesa and montane life zones of the Rocky Mountain region (Fig. 13). It is distinguished by its narrow, willow-like leaves and the prominently ridged bark of older trees. Narrowleaf cottonwood is found in pure stands or growing in association with other intolerant, riparian species. Cuttings of narrow-leaf cottonwood root readily, and it has been used sparingly as a plantation tree or as an

Fig. 13. Narrowleaf cottonwood is one of the poplars so characteristic of riparian areas throughout the arid West.

ornamental in the Rocky Mountain region. Hybrids of this species with balsam poplar, plains cottonwood, and Fremont cottonwood are common, in some areas outnumbering the pure species (Table 3). The indigenous people of the Rocky Mountains used the buds of narrowleaf cottonwood as chewing gum or food, fed young twigs to horses, smoked the inner bark in pipes, and fashioned the wood into cradleboards (Moerman 1998).

Of the Asian *Tacamahaca* poplars, *P. suaveolens* (includes the former *P. cathayana*, *P. koreana*, and *P. maximowiczii*) has the widest distribution, occurring from Turkestan to eastern Siberia and Japan. In some parts of this vast area, it is frequently one of the few woody species to attain tree size. The Japanese poplar (formerly *P. maximowiczii*, but see p. 334) is the best known Asian poplar among North American poplar growers, if only because it has been a common parent in hybridization work. Some of the fastest-growing hybrid crosses made by the original North American poplar breeding project at the Oxford Paper Company (Stout and Schreiner 1933), including the Kingston, Oxford, and Androscoggin clones, used a Japanese poplar as the female parent. In its native range, Japanese poplar is a fast-growing tree that attains heights of up to 30 m. It grew so large in some areas that sea-going dugout canoes were fashioned from it. Leaves are leathery, shiny dark green above and whitish below, with glandular-toothed margins. When grown in North America, it is one of the first poplars to leaf out in the spring. The twigs of Japanese poplar are round, hairy, and reddish when young, with fragrant buds.

In eastern North America, Japanese poplar is highly susceptible to stem cankers caused by *Septoria musiva*, which severely deform or kill trees. Although exceptionally fast growing, most Japanese poplar hybrids — especially those with *P. trichocarpa* — cannot be recommended for planting in the eastern states and provinces because of canker problems. However, some recent hybrids with *P. nigra* introduced to North America by Canadian tree breeders — e.g., NM-6 — have shown good canker resistance. Damage by wind, snow, and ice has occurred to Japanese poplar hybrids in some localities.

Another Asian poplar known to North American poplar growers because of its inclusion in early hybridization work is *P. laurifolia*. The Strathglass poplar, for example, is a clone of a cross between *P. nigra* and *P. laurifolia* that showed early promise but is not widely planted now. The laurel poplar is indigenous to Siberia and northwest China, but it has been occasionally cultivated in Europe. It is a modest-sized tree with small, narrowly ovate leaves and slender, sharply angled twigs.

Populus yunnanensis, the most southerly of the balsam poplars, is a tall tree native to southwestern China. It is characterized by the bright red color of leaf midribs and petioles. Yunnan poplar is adaptable to many soil conditions and thrives in hot climates with long growing seasons. At low latitudes, the leaves are semipersistent throughout the year. This poplar has been planted in France and in New Zealand where it has proved to be very resistant to *Melampsora* rust. Another poplar native to China is *P. szechuanica*. This tree has large leaves and may be the tallest poplar in Asia, reaching 40 m in height.

The remaining Asian *Tacamahaca* poplars are currently of minor importance in North American forestry, although important in their native range. Simon poplar (*P. simonii*) is a tree of variable stature native to northwestern China and Korea. On moist, fertile sites, trees can reach 30 m in height and 1.5 m in diameter. In China, plantation culture of Simon poplar goes back over 2000 years, and today it is the most widely planted poplar in that country. It also has been planted in Europe and North America as an ornamental — especially the drooping cultivar 'Pendula' — and in Canada for shelterbelts, although it has suffered from winter injury and dieback there. Leaves are small and rhomboid in shape. Simon poplar has been used to some extent in breeding work. *Populus ciliata*, extensively distributed in the mixed forests of the lower slopes of the Himalayan Mountains, reaches large size and attains economic significance there. Resembling balsam poplar, it is a pioneer tree on disturbed sites and vegetatively reproduces by root suckers. The former *P. tristis* may be a hybrid of this species with *P. nigra* or *P. balsamifera*, but its origins are obscure and warrant further investigation.

Section Populus

The final section in the genus is *Populus* (formerly *Leuce*), an ecologically unique and complex grouping containing the aspens and white poplars (Table 1). The

10 members of this section are of great economic importance, and they are distributed over most of the Northern Hemisphere. As a group, they are distinguished by their predisposition to sucker and by their adaptation to upland habitats.

Populus tremuloides, the trembling or quaking aspen (also known colloquially as popple), is the most widely distributed tree in North America, occurring in the northern states, Rocky Mountains, and Cascade Mountains in the U.S., and transcontinentally across Canada to Alaska. It is a slender tree with straight to crooked form and a small rounded crown. Quaking aspen is a small to medium-sized tree in much of its range. The most impressive stands of quaking aspen occur in the central Rocky Mountains, where trees can attain heights of 30 m or more and diameters of nearly 1 m (Fig. 14A). At high elevations, high latitudes, on south- and west-facing slopes, or on the prairie fringe, however, aspen may only be a twisted shrub (Fig. 14B).

Aspen's distinctive bark is smooth, greenish-white to creamy, and frequently darkened by warty bands. On old trees, the very lowest part of the stem will break into dark, shallow ridges. The leaves of mature quaking aspen are small, round to oval, with a finely toothed margin, but on young suckers the leaves are much larger and more elongated. The petiole of quaking aspen is long and characteristically flattened, causing the leaves to flutter in the slightest breeze; hence the common name. The wood of aspen is light in color, soft, and straight-grained, and it is widely used for pulp, paper, oriented-strandboard, and other wood products.

Quaking aspen grows on a variety of habitats, ranging from wet clayey soils to coarse, droughty sands. But it reaches its best development on well-drained, loamy soils high in lime, with a water table within 1.5 m of the surface. This aspen is a common pioneer on sites disturbed by logging, fire, or other natural disruptions and is regarded as highly intolerant of shade and competition. It can form extensive pure, even-aged stands that commonly serve as a nurse crop for more tolerant hardwoods and conifers. In the Rocky Mountains, two-storied or uneven-aged aspen stands also occur. Mixed even-aged stands of aspen and other hardwoods or conifers are common, especially in the more southerly parts of its range. The shallow and widespreading root system produces abundant sucker regrowth if the stand is logged, killed by fire, or windthrown, and growth is very rapid during the first few years. Clones resulting from suckering can vary in size from several trees to many thousands. The area covered by individual aspen clones is especially large in the Rocky Mountains. In fact, the largest organism in the world may be a clone of interconnected aspen suckers in the Wasatch Mountains of south-central Utah that covers 43 ha, contains some 47 000 individual stems, and weighs an estimated 6 million kg (Mitton and Grant 1996). This extraordinary clone has been nicknamed *Pando*, a Latin word meaning "I spread."

Aspens cannot be propagated by hardwood stem cuttings, as is the case with poplars in other sections of the genus, but young suckers can be used as propagules if

Fig. 14. Quaking aspen has the largest natural range of any tree indigenous to North America. Within this range its form can vary considerably. (A) An imposing clone growing on a moist, rich site in west-central Colorado. (B) A shrubby clone eking out an existence in a thin, rocky soil on a west-facing slope in Wyoming. Photo by Kathleen McKevitt.

they are rooted under mist in a greenhouse. Sexual reproduction by quaking aspen is abundant, and the light cottony seed, which matures in early spring, is carried long distances by the wind. Normally, few seeds germinate and survive past the seedling stage. But if seeds find a favorable habitat on a moist, recently disturbed site, seedlings will establish abundantly. For example, substantial areas in and adjacent to Yellowstone National Park that burned intensely during the 1988 fires now support aspen stands of seed origin. Quaking aspen grows rapidly during the first 20 years, generally reaching maturity after 30–40 years. In the Lake States, longevity of clonal stands of quaking aspen decreases with increasing mean annual temperature and is lowest on dry sites or on sites low in exchangeable calcium (Shields and Bockheim 1981). In the Rocky Mountains, quaking aspen attains its maximum age, with individual trees reaching 200 years old. Clonal

Fig. 14 (*concluded*).

age, or the age of the vegetatively propagated genome, is another matter. Certain western quaking aspen clones may be 10 000 years old or more, having become established by seed after the Pleistocene glaciers receded. This subject can become quite metaphysical; some people actually consider western aspen clones to be immortal (Mitton and Grant 1996).

The area of disturbance-dependent aspen is nearly everywhere on the decline because of fire exclusion, severe browsing by deer and elk, and the inability of logging to keep up with stands that reach a decadent stage, die, and succeed to another vegetation type (Bartos and Campbell 1998). For example, U.S. Forest Service inventory data for the Lake States Region shows that during the period from the early 1980s to the mid 1990s, the area of aspen (includes bigtooth aspen) declined by 5% in Minnesota, 11% in Wisconsin, and 21% in Michigan. Many regard this decline as the greatest problem facing managers of aspen land, so restoration of aspen is an active area. In addition to the ungulate browsing that threatens its widespread existence, quaking aspen provides food and prime habitat in a more benign way for ruffed grouse, beaver, and a host of other animals and birds.

Because of its ubiquitous distribution in northern and Rocky Mountain regions, quaking aspen was important in the life of indigenous peoples. Moerman (1998) lists nearly 40 tribal groups that used this species. Poultices made from shredded roots, bark, or leaves eased the pain and stiffness of rheumatism or were applied to wounds and bee stings. Decoctions of roots or bark treated venereal disease, heartburn, stomachache, heart disease, worms, or colds and were used as a laxative or purgative. A powdery substance on the outer bark was scraped off and used as a styptic, an antiperspirant/deodorant, or to prevent growth of hair. As with several other poplars, the inner bark was an important food for people and horses or it was smoked in pipes — often mixed with tobacco — at ceremonial occasions. Aspen logs were used by the Cheyenne and Crow to construct Sundance lodges, and many tribes used young trees for teepee poles. Large trees sometimes were fashioned into dugout canoes.

Variation within a species with as wide a distribution as quaking aspen would be expected. In fact, this poplar may be the most genetically variable plant species ever studied (Mitton and Grant 1996). Clonal variation in growth rate and other traits is common. For example, time of leafing out among clones growing in the same area can vary by over 3 weeks. Five natural varieties of quaking aspen have been described (Table 2). In addition, natural hybrids between *P. tremuloides* and *P. grandidentata* (*P.* ×*smithii*) have been identified where the ranges of the two species overlap (Table 3). These hybrids are not common, however, because *P. tremuloides* generally flowers 1–2 weeks before *P. grandidentata*. Hybrids with other taxa in section *Populus* also have been identified; e.g., *P. tremula* × *P. tremuloides* (*P.* ×*wettsteinii*), a hybrid that has shown promise for forest plantations in the Great Lakes Region. Interspecific hybrids of *P. tremuloides* with taxa in other sections, however, are rare.

Bigtooth aspen (*P. grandidentata*) shares many of the same attributes as quaking aspen, but it is not nearly as widely distributed, occurring mainly in the Northeast, Lake States, and northern Midwest of the U.S. and in adjacent Canadian provinces. This aspen reaches fairly large size and is considered superior to quaking aspen by Lake States foresters because of its rapid growth rate and excellent form (Fig. 15). Bigtooth aspen can be distinguished from quaking aspen by its larger, coarsely toothed leaves, which are silvery on the underside (especially on the large leaves of young suckers), buds that diverge from the twigs, and the olive–green to brownish-orange tinge of the bark on older trees. Like most poplars, the inner bark of bigtooth aspen is edible; the Ojibwe Indians scraped and boiled it, producing a concoction something like eggs (Moerman 1998).

The site adaptability of bigtooth aspen is more restricted than quaking aspen, and it is most often found on well-drained, medium- to coarse-textured upland soils (Fig. 6B). It also is one of the most intolerant tree species and will only reproduce successfully in openings devoid of any overstory. Bigtooth aspen is a fast-growing tree that may reach heights of 20 m and diameters of 60 cm in 50 years. Reproduction by root suckers is common, and they may grow to over 2 m during the first growing season. Browsing whitetail deer seem to prefer bigtooth aspen suckers to those of quaking aspen, which can partially negate these impressive first-year spurts of growth. The hybrid of bigtooth aspen with white poplar (*P.* ×*rouleauiana*; Table 3), which often arises spontaneously where the two species grow together, is an especially impressive tree that has piqued the

Fig. 15. Bigtooth aspen is an important timber and wildlife tree in the Great Lakes Region. Its rapid growth rate and superior stem form endear it to foresters, while deer and ruffed grouse browse its nutritious foliage and buds.

interest of many poplar growers. If this hybrid could be vegetatively propagated in an efficient manner, it certainly would become widely deployed in plantations.

The European aspen (*P. tremula*) is a tree with an immense natural distribution, occurring over most of Europe, northern Africa, and northern Asia. Several geographic races have been recognized. For example, species status had been conferred on the Chinese *P. davidiana*, but this aspen now is considered a variety of *P. tremula*. European aspen is a medium-sized tree that is similar to quaking aspen in most respects; in fact, some taxonomists argue that quaking aspen and European aspen should be combined into one nearly circumpolar species. European aspen is a commercially important tree, especially in Finland and Scandinavia, although it does not rival quaking aspen in this respect.

Several aspens are strictly Asian. The Japanese aspen (*P. sieboldii*) is a medium-sized tree with twigs and buds covered by a semipersistent white down. Native to central and western China, *P. adenopoda* is a large tree with glandular leaves. *Populus gamblei* is a Himalayan species that was first described in 1906 from the vicinity of the city of Darjeeling. It is unrelated to any other Eurasian species, with the possible exception of *P. adenopoda*.

White poplar (*P. alba*) is a tree that varies in form from broad-crowned, crooked, and multi-stemmed to tall and straight. It is widely distributed over northern Africa, southern Europe, and central Asia. It was one of the first trees introduced to colonial North America as an ornamental and has remained popular for this purpose. White poplar appears to have become naturalized in many areas in North America, although — aside from suckers — much of this reproduction may actually be natural hybrids with native aspens. White poplar is a striking tree that grows to large size. The bark is metallic gray to chalky white on young trees, becoming black and deeply furrowed at the base in older trees. The leaves are among the most distinctive in the genus, varying in shape from coarsely toothed and deltoid to maple-like with three to five distinct lobes. Leaves are dark green above with a covering of thick white felt below.

White poplar is adapted to a wide range of sites, and on bottomland soils it can attain magnificent timber proportions. It is regarded as somewhat tolerant of drought, wind, salinity, and high temperatures, but it does suffer from low temperatures and frost. It produces abundant seed and, like the aspens, suckers vigorously. Rooting of white poplar from hardwood cuttings is possible, but cuttings from some trees root well whereas those from others do not. White poplar has shown excellent growth rates in the milder portions of the northeastern United States and southern Canada. Because of this inherent vigor, white poplar and, especially, its hybrids with aspens can be considered promising for timber planting.

Nine natural varieties of white poplar are recognized (Table 2). The distinctive columnar form of var. *pyramidalis* — the Bolleana poplar — makes it a popular tree for ornamental and line plantings, and var. *nivea* has been recommended for windbreak plantings. The hybrid between *P. alba* and *P. tremula* originally was described as a distinct species under the name gray poplar, but it is now recognized as a hybrid (*P.* ×*canescens*) that has arisen in regions where the range of both parent species overlap (Table 3). Gray poplars are intermediate between the two parent species in morphological characteristics, and they thrive on dry or saline soils better than white poplar. Chinese white poplar has been proposed as another species in *Populus*, but it is now considered a hybrid (*P.* ×*tomentosa*) between *P. alba* and *P. adenopoda* or a tri-hybrid of the previous two taxa with *P. tremula*. It has long been cultivated in several Chinese provinces.

The final three white poplars in section *Populus*, although North American, are little known outside their native localities in Mexico. All need to gain wider acquaintance. *Populus monticola* is endemic to montane regions of southern Baja California, although there has been a suggestion that it really is *P. alba* var. *subintegerrima* introduced by early Spanish settlers and gone natural. *Populus guzmanantlensis* is a tropical poplar from the Sierra de Manantlán in the southwestern state of Jalisco. More widely distributed than Manantlán white poplar, *P. simaroa* occurs in scattered locations in the mountains surrounding the Rio Balsas drainage in the states of Mexico and Guerrero. *Populus simaroa* is unusual because it is deciduous during the summer wet season and leafs out during the winter dry season. The two latter species are closely related and could be considered varieties or subspecies of a single species. Because they were first described in 1989 and 1975, respectively, neither is yet well-enough studied for a firm decision to be made about their taxonomic status.

Conclusions

This chapter has provided a general introduction to the genus *Populus*. In addition, anecdotal and background information about poplars was provided that should be useful to growers and others with interests in this fascinating tree genus. In general, poplars are exceptionally tractable subjects for tree culture in natural stands and for various kinds of plantings because of their favorable inherent traits:

- Across its six sections and 29 species, the genus is inherently variable in the extreme and predisposed to hybridization, so trees of virtually any combination of traits may be found in the wild or created through breeding and biotechnology;
- Clonal trees can be easily propagated, either by taking advantage of natural root suckering or by using hardwood stem cuttings;
- Species and, especially, hybrids in the genus are exceptionally fast growing and can have excellent stem form;
- A pure-species or hybrid clone of poplar can be found — or created — that is adapted to virtually any habitat capable of supporting trees, including those needing remediation or rehabilitation;
- Poplar wood is a very useful raw material for a wide range of end products, from paper to OSB to lumber;
- The high photosynthetic capacity and biomass production of poplars makes them ideal trees for carbon sequestration;
- Many of the forms taken by poplar trees lend themselves well to amenity or windbreak plantings, and people love to listen to their "leaf talk;"
- *Populus* is a well-studied tree genus and poplars are "model" research organisms, so almost any question asked about them can be answered.

Acknowledgements

I am indebted to Jim Richardson and Reinhard Stettler for their critical reviews of an early draft of this manuscript.

References

Altman, N. 1994. Sacred trees. Sierra Club Books, San Francisco, CA.

Bartos, D.L., and Campbell, R.B., Jr. 1998. Decline of quaking aspen in the Interior West — examples from Utah. Rangelands **20**(1): 17–24.

Boom, B.K. 1957. *Populus canadensis* Moench versus *P. euramericana* Guinier. Acta Bot. Neerland. **6**: 54–59.

Bradshaw, H.D., Jr. 2000. Why study the genetics of hybrid poplar? Poplar Molecular Genetics Cooperative. <http://poplar2.cfr.washington.edu/pmgc/whypopgen.htm> (accessed 01 June 2000).

Burns, R.M., and Honkala, B.H. (*Technical coordinators*). 1990. Silvics of North America. Vol. 2, hardwoods. U.S. For. Serv. Agric. Handbk. No. 654.

DeByle, N.V., and R.P. Winokur (*Editors*). 1985. Aspen: ecology and management in the western United States. U.S. For. Serv. Gen. Tech. Rep. RM-119.

Dickmann, D.I., and Stuart, K.W. 1983. The culture of poplars in eastern North America. Department of Forestry, Michigan State University, East Lansing, MI.

Eckenwalder, J.E. 1977. North American cottonwoods (*Populus*, Salicaceae) of sections *Abaso* and *Aigeiros*. J. Arnold Arbor. **58**: 194–208.

Eckenwalder, J.E. 1984. Natural intersectional hybridization between North American species of *Populus* (Salicaceae) in sections *Aigeiros* and *Tacamahaca*. II. Taxonomy. Can. J. Bot. **62**: 325–335.

Eckenwalder, J.E. 1996. Systematics and evolution of *Populus*. *In* Biology of *Populus* and its implications for management and conservation. *Edited by* R.F. Stettler, H.D. Bradshaw, Jr., P.E. Heilman, and T.M. Hinckley. NRC Research Press, Ottawa, ON. pp. 7–32.

Edlin, H.L. 1963. A modern sylva or a discourse of forest trees. 6. Poplars — *Populus* species. Quart. J. For. **57**: 200–210.

Elias, T.S. 1987. The complete trees of North America. Gramercy Publishing Co., New York, NY.

FAO. 1958. Poplars in forestry and land use. FAO United Nations, Forestry and Forest Prod. Stud. No. 12, Rome, Italy.

FAO. 1980. Poplars and willows in wood production and land use. FAO United Nations, Forestry Series No. 10, Rome, Italy.

Farrar, J.L. 1995. Trees of the northeastern United States and Canada. Iowa State University Press, Ames, IA.

Graham, S.A., Harrison, R.P., Jr., and Westell, C.E., Jr. 1963. Aspens: Phoenix trees of the great Lakes Region. University of Michigan Press, Ann Arbor, MI.

Harlow, W.H. 1957. Trees of the eastern and central United States and Canada. Dover Publications, Inc., New York, NY.

Harlow, W.H., Harrar, E.S., Hardin, J.W., and White, F.M. 1996. Textbook of dendrology, 8th edition. McGraw–Hill, Inc., New York, NY.

Heaney, S. 1996. The spirit level. Farrar Straus Giroux, New York, NY.

Heilman, P.E., and Stettler, R.F. 1985. Genetic variation and productivity of *Populus trichocarpa* T. & G. and its hybrids. II. Biomass productivity in a four-year plantation. Can. J. For. Res. **15**: 384–388.

Heilman, P.E., Stettler, R.F., Hanley, D.P., and Carkner, R.W. 1990. High yield hybrid poplar plantations in the Pacific Northwest. Pacific Northwest Exten. Pub. PNW-356.

Knowe, S.A., Foster, G.S., Rousseau, R.J., and Nance, W.L. 1998. Height-age and height-diameter relationships for monocultures and mixtures of eastern cottonwood clones. For. Ecol. Manage. **106**: 115–123.

Little, E.L., Jr. 1971. Atlas of United States Trees. Vol. 1, Conifers and important hardwoods. U.S. For. Serv. Misc. Publ. No. 1146.

Little, E.L., Jr. 1976. Atlas of United States Trees. Vol. 3, Minor western hardwoods. U.S. For. Serv. Misc. Publ. No. 1314.

Little, E.L., Jr. 1979. Checklist of United States trees. U.S. For. Serv. Agric. Handbk. No. 541.

McKnight, J.S. 1970. Planting cottonwood cuttings for timber production in the South. U.S. For. Serv. Res. Pap. SO-60.

Mitton, J.B., and Grant, M.C. 1996. Genetic variation and natural history of quaking aspen. BioScience, **46**: 25–31.
Moerman, D.E. 1998. Native American Ethnobotany. Timber Press, Portland, OR.
Rehder, A. 1940. Manual of cultivated trees and shrubs hardy in North America. The MacMillan Company, New York, NY.
Rogers, J.E. 1917. Trees worth knowing. Doubleday, Page, & Co., New York, NY.
Rupp, R. 1990. Red oaks and black birches: The science and lore of trees. Garden Way Publishing, Pownal, VT.
Scarascia-Mugnozza, G.E., Ceulemans, R., Heilman, P.E., Isebrands, J.G., Stettler, R.F., and Hinckley, T.M. 1997. Production physiology and morphology of *Populus* species and their hybrids grown under short rotation. II. Biomass components and harvest index of hybrid and parental species clones. Can. J. For. Res. **27**: 285–294.
Shields, W.J., Jr., and. Bockheim, J.G. 1981. Deterioration of trembling aspen clones in the Great Lakes Region. Can. J. For. Res. **11**: 530–537.
Smith, R.L., and Sytsma, K.J. 1990. Evolution of *Populus nigra* (Sect. *Aigeiros*): Introgressive hybridization and the chloroplast contribution of *Populus alba* (Sect. *Populus*). Am. J. Bot. **77**: 1176–1187.
Stettler, R.F., Zsuffa, L., and Wu, R. 1996. The role of hybridization in the genetic manipulation of *Populus*. *In* Biology of *Populus* and its implications for management and conservation. *Edited by* R.F. Stettler, H.D. Bradshaw, Jr., P.E. Heilman, and T.M. Hinckley. NRC Research Press, Ottawa, ON. pp. 87–112.
Stout, A.B., and Schreiner, E.J. 1933. Results of a project in hybridizing poplars. J. Hered. **24**: 216–229.
Uva, R.H., Neal, J.C., and DiTomaso, J.M. 1997. Weeds of the Northeast. Cornell University Press, Ithaca, NY.
Zsuffa, L. 1974. The genetics of *Populus nigra* L. Ann. Forest. **6/2**: 29–53.
Zsuffa, L., Giordano, E., Pryor, L.D., and Stettler, R.F. 1996. Trends in poplar culture: some global and regional perspectives. *In* Biology of *Populus* and its implications for management and conservation. *Edited by* R.F. Stettler, H.D. Bradshaw, Jr., P.E. Heilman, and T.M. Hinckley. NRC Research Press, Ottawa, ON. pp. 515–539.

CHAPTER 2
Poplar breeding strategies

Don E. Riemenschneider, Brian J. Stanton, Gilles Vallée, and Pierre Périnet

Introduction

Plant breeding is an artful enterprise that is undertaken with the objective of producing varieties that are more useful to people than wild plants. Plant breeding strategies exist in bewildering array, each matched to the biological characteristics of the species and populations to which they are applied. Breeding strategies differ, depending on whether a species is self-pollinated or cross-pollinated in nature. Strategies differ, depending on whether the species can be asexually propagated or whether varietal reproduction, which is a prerequisite to commercial deployment, depends on seed and is thus subject to genetic recombination. Strategies also depend on the ecological setting within which the production system must be sustained. The presence of significant damaging agents such as insects and disease can often dictate the kind of artificial selection pressure that must be applied to a population to achieve efficient and, more importantly, sustainable production. Overall, breeding strategies have to consider the various constraints imposed by the environment within which the trees will be planted. Each environment possesses different attributes including biological, climatic, physiographic, socio-economic, market, and other factors.

Regardless of diversity in method, all breeding strategies have elements in common. The first step is to identify available parents and to evaluate those parents in an environment coincident with, or similar to, the commercial zone of deployment. This first step is often practiced without rigorous selection in poplar breeding programs because rapid production of hybrid progeny may be the sole initial

D.E. Riemenschneider. USDA Forest Service, North Central Forest Experiment Station, Forestry Sciences Laboratory, 5985 Highway K, Rhinelander, WI 54501, U.S.A.
B.J. Stanton. Research Associate, Fort James Corporation, 349 Northwest 7th Avenue, Camas WA 98607, U.S.A.
G. Vallée. Principal Leader (retired), Ministère des Ressources Naturelles du Québec (MRN), 124 des Frênes Ouest, QC G1L 1G5, Canada.
P. Périnet. Project Leader, Direction de la Recherche Forestière, MRN, 2700, rue Einstein, Sainte-Foy, QC G1P 3W8, Canada.
Correct citation: Riemenschneider, D.E., Stanton, B.J., Vallée, G., and Périnet, P. 2001. Poplar breeding strategies. Part A, Chapter 2. *In* Poplar Culture in North America. *Edited by* D.I. Dickmann, J.G. Isebrands, J.E. Eckenwalder, and J. Richardson. NRC Research Press, National Research Council of Canada, Ottawa, ON K1A 0R6, Canada. pp. 43–76.

focus. Second, genetic variability is generated within a breeding population through the controlled mating of parents that are presumed to be sufficiently different from one another to produce a variable population of offspring. Third, progeny must be subjected to testing and selection. Testing is commonly practiced on an empirical basis by observation and measurement of the progeny population grown within the zone of proposed commercial deployment under conditions that at least approximate commercial culture. Selection may be based on a single trait or on multiple traits, depending on commercial needs and the ecological complexity of the proposed production system. Selection may be based on observation of individual plants or on joint evaluation of the individual and various relatives. Selection may also be based on direct knowledge of crop genotype using molecular markers (Bradshaw 1996), although this method presupposes that a relationship has already been established between a marker and some interesting trait, a relationship that can only be established through previous empirical testing. Fourth, a system must be developed that places sufficient propagules in the hands of growers to support the commercial production system. Propagule expansion is highly dependent on the biological characteristics of the species. Seed-based production systems, such as for hybrid corn, require vast seed production systems where paired inbred lines are crossed to produce F_1 (first generation hybrid) seed. Vegetatively-based production systems, such as for potatoes and poplars, require techniques that sequentially increase the number of vegetative propagules (i.e., tubers and cuttings). Poplar breeding utilizes planting systems called stool-beds to increase the commercial production of selected varieties — clones, in this case (see Chap. 5 in this volume).

Overlain on the art of plant breeding is the science of quantitative genetics (Riemenschneider et al. 1996). Quantitative genetics aids in choosing and optimizing breeding strategies by providing a mathematical means to predict the outcome of various alternative approaches. Thus, quantitative genetics simplifies breeding by reducing the need for empirical testing of alternative strategies. The ability to predict the outcome of different strategies depends on knowledge of how much genetic variation is attributable to different kinds of gene action, how much variation is attributable to the interaction of crop genotype and environment and, in perennial crops, the predictability of phenotype over different plant ages.

Poplar breeding and its strategies exist within this matrix of biological complexity. Poplar breeding strategies generate genetically variable breeding populations by the mating of diverse parents. Poplar breeding populations are subjected to empirical testing and selection for characteristics that depend on the biology of the population, the needs of the commercial producer, and the complexity of the ecology within which the production system exists. Importantly, there exists sufficient variation within the genus *Populus* and among the environments in which poplars are grown, that no single breeding strategy can conceivably apply to all situations. In some cases, poplars have been commercially deployed on marginal agricultural land. In other cases, agricultural land is required for food crops and is therefore not widely available for poplar culture, as in Quebec. Thus, an important selection criterion in the latter case has been tolerance of medium fertility and acidic non-agricultural soils (pH 5.0–5.5). Large variation in local environments can lead to equally large variation in the pedigrees within which adapted genotypes are sought (Table 1). Other breeding programs are substantially more focused, relying on a less diverse array of pedigrees. Breeding programs in Minnesota (cooperative public and corporate, Table 2) and Washington (corporate, Table 3) are good examples.

This chapter describes variation within the genus *Populus* as it applies to choice of breeding strategy, several alternate breeding strategies from which a method suited to any particular case might be selected and optimized, and some peculiarities of testing and selection that are especially applicable to poplar breeding strategies. We conform to the binomial *Populus maximowiczii* in the text and tables that follow even though the species may now be included in *P. suaveolens*. This is to avoid confusion to the reader.

Implications of clonal propagation

All poplars may be propagated asexually, but not with equal ease or by a common method. Dormant unrooted hardwood cuttings of species from the section *Tacamahaca* produce roots within days after planting in warm, moist soils. Hardwood cuttings of species from the section *Aigeiros* will also root under field conditions, but rooting may be slower and less reliable than from the *Tacamahaca* poplars. Species of the section *Populus* generally root poorly or not at all from hardwood cuttings but may be readily propagated using cuttings from actively growing shoots maintained under constant mist, especially after the base of the cutting has been treated with an auxin-containing root promoter. Sprouts from pieces of large roots may also be rooted under mist.

Asexual propagation confers substantial advantage to poplar breeding compared to crops that can only be propagated by seed. The advantage derives from the different kinds of genetic effects that control important traits and the ease with which they can be captured by selection. Genetic effects are divided into several categories. Additive effects are the value conferred to the genotype by having superior alleles occur at different loci. Additive effects can be captured in populations and are the basis for the improvement of seed-propagated crops such as pines and the old open-pollinated corn varieties. Non-additive effects result from specific combinations of specific alleles throughout the genome of the plant. Non-additive effects are difficult to capture in seed-propagated crops because sexual recombination results in genetic shuffling that changes favorable arrangements. Single-cross hybrid corn breeding using inbred parental lines was developed to overcome this problem. It is much easier, however, to capture non-additive effects in clonally propagated crops because randomly-occurring gene combinations can be reproduced by cloning the selected genotype. Thus, clonal propagation allows all genetic effects to be captured, regardless of their complexity. If

Table 1. In 1968, a project was initiated by the Quebec government for the selection of clones and genetic improvement of poplar under the direction of Dr. Gilles Vallée (Vallée 1970, 1971). The objective was to develop and select clones well-adapted to the different ecological regions of Quebec for wood production in poplar plantations. The table shows the number of families and seedlings obtained per successful artificial cross of hybrid poplar since 1971.

No. of families	Pedigree[a]	No. of seedlings
9	*P. alba* × *P. alba*	1045
29	*P. alba* × *P. grandidentata*	1459
1	*P. alba* × *P. nigra*	3
8	*P. alba* × *P. tremula*	163
19	*P. alba* × *P. tremuloides*	482
4	*P. alba* × *P.* ×*canescens*	34
1	*P. alba* × *P.* ×*jackii*	7
4	*P. alba* × (*P. alba* × *P. grandidentata*)	108
7	*P. alba* × (*P. tremuloides* ×*tremula*)	286
47	*P. balsamifera* × *P. balsamifera*	3736
1	*P. balsamifera* × *P. deltoides*	2
7	*P. balsamifera* × *P. maximowiczii*	613
26	*P. balsamifera* × *P. nigra*	3810
9	*P. balsamifera* × *P. trichocarpa*	2103
5	*P. balsamifera* × *P.* ×*canadensis*	689
4	*P. balsamifera* × *P.* ×*canadensis*	296
10	*P. balsamifera* × *P.* ×*jackii*	938
1	*P. balsamifera* × *P.* ×*rollandii*	90
2	*P. balsamifera* × (*P. balsamifera* × ?)	150
5	*P.* ×*canescens* × *P. grandidentata*	113
4	*P.* ×*canescens* × *P. tremuloides*	746
1	*P.* ×*canescens* × *P.* ×*canescens*	2
1	*P.* ×*canescens* × (*P. alba* × *P. grandidentata*)	9
1	*P.* ×*canescens* × (*P. grandidentata* × *P. tremuloides*)	13
4	*P. cathayana* × *P. balsamifera*	1040
7	*P. cathayana* × *P. deltoides*	595
1	*P. cathayana* × *P. nigra*	416
1	*P. cathayana* × *P. simonii*	245
1	*P. cathayana* × *P.* ×*canadensis*	323
1	*P. cathayana* × *P.* ×*jackii*	320
1	*P. cathayana* × (*P. balsamifera* × ?)	65
35	*P. deltoides* × *P. balsamifera*	2043

Table 1 (*continued*).

No. of families	Pedigree[a]	No. of seedlings
72	*P. deltoides* × *P. deltoides*	19351
2	*P. deltoides* × *P. grandidentata*	5
20	*P. deltoides* × *P. maximowiczii*	1362
26	*P. deltoides* × *P. nigra*	1840
5	*P. deltoides* × *P. simonii*	602
15	*P. deltoides* × *P. trichocarpa*	1287
6	*P. deltoides* × *P.* ×*canadensis*	661
13	*P. deltoides* × *P.* ×*generosa*	780
22	*P. deltoides* × *P.* ×*jackii*	980
4	*P. deltoides* × *P.* ×*rollandii*	235
4	*P. deltoides* × (*P. balsamifera* × ?)	22
15	*P. grandidentata* × *P. alba*	1332
13	*P. grandidentata* × *P. grandidentata*	721
10	*P. grandidentata* × *P. tremula*	895
8	*P. grandidentata* × *P. tremuloides*	341
3	*P. grandidentata* × *P.* ×*canescens*	128
5	*P. grandidentata* × (*P. alba* × *P. grandidentata*)	152
1	*P. grandidentata* × (*P. alba* × *P.* ×*jackii*)	21
1	*P. grandidentata* × (*P. tremuloides* × *P. tremula*)	588
1	*P. maximowiczii* × *P. alba*	2
52	*P. maximowiczii* × *P. balsamifera*	13073
1	*P. maximowiczii* × *P. deltoides*	4
17	*P. maximowiczii* × *P. maximowiczii*	1673
57	*P. maximowiczii* × *P. nigra*	6152
20	*P. maximowiczii* × *P. trichocarpa*	6009
6	*P. maximowiczii* × *P.* ×*canadensis*	669
21	*P. maximowiczii* × *P.* ×*generosa*	5791
38	*P. maximowiczii* × *P.* ×*jackii*	4071
8	*P. maximowiczii* × *P.* ×*rollandii*	1447
2	*P. maximowiczii* × (*P. balsamifera* × *P. nigra*)	145
2	*P. maximowiczii* × (*P.* ×*jackii* + *P.* ×*rollandii*)	4
13	*P. nigra* × *P. balsamifera*	885
1	*P. nigra* × *P. maximowiczii*	58
2	*P. nigra* × *P. nigra*	138
2	*P. nigra* × *P. trichocarpa*	54
1	*P. nigra* × *P.* ×*generosa*	126

Table 1 (*continued*).

No. of families	Pedigree[a]	No. of seedlings
2	*P. nigra* × *P.* ×*jackii*	168
3	*P. tremula* × *P. alba*	626
2	*P. tremula* × *P. grandidentata*	565
8	*P. tremula* × *P. tremuloides*	304
23	*P. tremuloides* × *P. alba*	1445
12	*P. tremuloides* × *P. grandidentata*	415
1	*P. tremuloides* × *P. simonii*	1
35	*P. tremuloides* × *P. tremula*	4357
44	*P. tremuloides* × *P. tremuloides*	6840
20	*P. tremuloides* × *P.* ×*canescens*	1280
1	*P. tremuloides* × *P.* ×*canadensis*	8
2	*P. tremuloides* × *P.* ×*generosa*	2
1	*P. tremuloides* × *P.* ×*jackii*	2
3	*P. tremuloides* × (*P. alba* × *P. grandidentata*)	109
2	*P. tremuloides* × (*P. alba* × *P.* ×*jackii*)	144
2	*P. tremuloides* × (*P. tremuloides* × *P. tremula*)	660
4	*P. trichocarpa* × *P. balsamifera*	182
1	*P. trichocarpa* × *P. deltoides*	1
9	*P. trichocarpa* × *P. maximowiczii*	476
1	*P. trichocarpa* × *P. trichocarpa*	817
2	*P. trichocarpa* × *P.* ×*jackii*	46
2	*P. trichocarpa* × *P.* ×*rollandii*	16
26	*P.* ×*canadensis* × *P. balsamifera*	744
1	*P.* ×*canadensis* × *P. deltoides*	1
13	*P.* ×*canadensis* × *P. maximowiczii*	804
16	*P.* ×*canadensis* × *P. nigra*	1530
8	*P.* ×*canadensis* × *P. trichocarpa*	449
1	*P.* ×*canadensis* × *P.* ×*canadensis*	3
10	*P.* ×*canadensis* × *P.* ×*generosa*	197
22	*P.* ×*canadensis* × *P.* ×*jackii*	3156
1	*P.* ×*canadensis* × (*P. balsamifera* × ?)	19
3	*P.* ×*canadensis* × (*P. balsamifera* × *P. nigra*)	176
1	*P.* ×*canadensis* × (*P.* ×*euramericana* × *P. deltoides*)	1
1	*P.* ×*canadensis* × (*P.* ×*jackii* + *P.* ×*rollandii*)	2
2	*P.* ×*generosa* × *P. alba*	298
16	*P.* ×*generosa* × *P. balsamifera*	3342

Table 1 (*continued*).

No. of families	Pedigree[a]	No. of seedlings
6	*P.* ×*generosa* × *P. deltoides*	605
20	*P.* ×*generosa* × *P. maximowiczii*	1515
12	*P.* ×*generosa* × *P. nigra*	679
3	*P.* ×*generosa* × *P. trichocarpa*	270
8	*P.* ×*generosa* × *P.* ×*canadensis*	216
2	*P.* ×*generosa* × *P.* ×*generosa*	643
9	*P.* ×*generosa* × *P.* ×*jackii*	2488
2	*P.* ×*generosa* × *P.* ×*rollandii*	34
1	*P.* ×*generosa* × (*P. balsamifera* × *P. nigra*)	117
1	*P.* ×*generosa* × (*P. maximowiczii* × *P. balsamifera*)	9
7	*P.* ×*jackii* × *P. balsamifera*	566
4	*P.* ×*jackii* × *P. deltoides*	115
1	*P.* ×*jackii* × *P. grandidentata*	10
15	*P.* ×*jackii* × *P. maximowiczii*	1292
20	*P.* ×*jackii* × *P. nigra*	1322
1	*P.* ×*jackii* × *P. simonii*	6
16	*P.* ×*jackii* × *P. trichocarpa*	1986
4	*P.* ×*jackii* × *P.* ×*canadensis*	284
3	*P.* ×*jackii* × *P.* ×*generosa*	147
7	*P.* ×*jackii* × *P.* ×*jackii*	419
1	*P.* ×*jackii* × *P.* ×*rollandii*	31
1	*P.* ×*jackii* × (*P. alba* × *P. grandidentata*)	3
1	*P.* ×*jackii* × (*P. balsamifera* × ?)	1
2	*P.* ×*jackii* × (*P. balsamifera* × *P. nigra*)	1000
1	*P.* ×*jackii* × (*P. maximowiczii* × *P. balsamifera*)	120
1	*P.* ×*rollandii* × ?	300
5	*P.* ×*rollandii* × *P. balsamifera*	252
15	*P.* ×*rollandii* × *P. maximowiczii*	2628
36	*P.* ×*rollandii* × *P. nigra*	1884
5	*P.* ×*rollandii* × *P. trichocarpa*	898
2	*P.* ×*rollandii* × *P.* ×*jackii*	310
1	*P.* ×*rollandii* × (*P. balsamifera* × *P. nigra*)	100
1	*P.* ×*rollandii* × (*P. maximowiczii* × *P. balsamifera*)	211
1	(*P. alba* × *P. grandidentata*) × *P. tremula*	150
4	(*P. alba* × *P. tremula*?) × *P. grandidentata*	228
1	(*P. alba* × *P. tremula*?) × *P. tremula*	5

Table 1 (*concluded*).

No. of families	Pedigree[a]	No. of seedlings
1	(*P. alba* × *P. tremula*?) × (*P. alba* × *P. grandidentata*)	72
2	(*P. balsamifera* × *P. nigra*) × *P.* ×*canadensis*	1000
2	(*P. nigra* × *P. maximowiczii*) × *P. trichocarpa*	313
1	(*P. nigra* × *P. maximowiczii*) × *P.* ×*rollandii*	12
4	(*P. nigra* × *P. trichocarpa*) × *P. balsamifera*	805
1	(*P. nigra* × *P. trichocarpa*) × *P. trichocarpa*	8
1	(*P. nigra* × *P. trichocarpa*) × *P.* ×*canadensis*	1
1	(*P. nigra* × *P. trichocarpa*) × *P.* ×*generosa*	2
3	(*P. nigra* × *P. trichocarpa*) × *P.* ×*jackii*	100
1	(*P. nigra* × *P. trichocarpa*) × (*P. balsamifera* × ?)	55
1	(*P. nigra* × *P. trichocarpa*) × (*P. balsamifera* × *P. nigra*)	166
3	(*P.* ×*canadensis* × *P. maximowiczii*) × *P. balsamifera*	357
3	(*P.* ×*canadensis* × *P. maximowiczii*) × *P.* ×*jackii*	40
1	(*P.* ×*maximowiczii* × *P. balsamifera*) × *P. balsamifera*	78
2	(*P.* ×*maximowiczii* × *P. balsamifera*) × *P. nigra*	81
1	(*P.* ×*maximowiczii* × *P. balsamifera*) × *P.* ×*jackii*	60

[a]Several Latin names in this table do not reflect current taxonomic priority. See Chap. 1, especially Tables 1 and 3, for correct synonyms.

Table 2. Numbers of families and seedlings produced from different pedigrees by the Minnesota Hybrid Poplar Research Cooperative. Large numbers of new open pollinated families were collected from northwestern Minnesota, an area previously unsampled but of current commercial production interest. Clones are selected for deployment from northwest to east central Minnesota.

No. of families	Pedigree	No. of seedlings (approximate)
126	Open pollinated *P. deltoides* (natural stands)	7500
88	*P. deltoides* × *P. deltoides*	7000
70	*P. deltoides* × *P. maximowiczii*	5600
64	*P. deltoides* × *P. nigra*	5100
57	(*P. deltoides* × *P. maximowiczii*) × *P. deltoides*	4600
8	Various F_2	600
29	Other	2300

Table 3. Numbers of families and seedlings produced from different pedigrees by the breeding and selection program at Fort James Corporation, Camas, WA. Selected genotypes are commercially deployed east of the Cascade Range in northwestern Oregon.

No. of families	Pedigree	No. of seedlings
111	*P. deltoides* × *P. deltoides*	6810
228	*P. deltoides* × *P. trichocarpa*	9829
233	*P. trichocarpa* × *P. deltoides*	3233
38	*P. deltoides* × *P. maximowiczii*	1918
49	*P. trichocarpa* × *P. maximowiczii*	4195
14	(*P. trichocarpa* × *P. deltoides*) × (*P. trichocarpa* × *P. deltoides*)	562
40	(*P. trichocarpa* × *P. deltoides*) × *P. deltoides*	1164
6	*P. deltoides* × (*P. trichocarpa* × *P. deltoides*)	345
6	(*P. trichocarpa* × *P. deltoides*) × *P. maximowiczii*	161
2	*P. deltoides* × (*P. trichocarpa* × *P. maximowiczii*)	58

non-additive effects are high compared to additive effects, the advantage conferred by clonal propagation can be substantial. The facility with which poplars, especially *Tacamahaca*, *Aigeiros*, and their hybrids, can be clonally propagated, in combination with the large non-additive genetic effects that are probably common in most hybrid poplar progeny populations, forms the basis for the success of poplar breeding worldwide.

The concept of genetic gain in poplar breeding

Breeding and selection result in new varieties, populations, clones, etc. that are superior, according to some criterion, when compared to some standard. Recurrent breeding and selection programs are often compared and contrasted using the concept of genetic gain (see Shelbourne [1969] for several good examples). Programs are designed so that each generation is superior to the previous one, and methods are optimized so that superiority per unit cost is maximized. The difference between generations in the value of one or more selection criteria is referred to as genetic gain. Poplar breeding is inherently difficult to describe using this method because the standard of comparison is rarely the previous generation of the pedigree under selection. For example, the most common commercial clones in the North Central U.S. are DN-34, DN-5 (both *P. deltoides* × *P. nigra*), NM-6 (*P. nigra* × *P. maximowiczii*), and a few others. Yet, no or few offspring of these clones are to be found among the tens of thousands of new hybrid progenies developed within the region. Many pedigrees are currently under development in Quebec (Table 1), and it is unlikely that all of them have traceable ancestry to commercial plantings or large yield plots. Overall, there often seems little generational connection between new populations and past standards. As a result, gain relative to standards is not always predictable according to quantitative genetic theory and can only be obtained by empirical experiment.

Selection criteria

Tree growth is a uniformly applied selection criterion because it forms the biological basis for economic return. Melampsora leaf rust (*Melampsora medusae*) and Marssonina leaf spot (*Marssonina brunnea*) are two major leaf diseases considered in most breeding programs throughout North America. Rooting ability is also a major selection criterion because, in the *Aigeiros* and *Tacamahaca* sections at least, a dormant unrooted hardwood cutting is the commercial propagule of choice. Selection for rooting ability may be overt, or may be incidental as experimental clones are moved through multiple-stage selection strategies (Riemenschneider et al. 1996). Other selection criteria, like many aspects of poplar breeding, can differ among environments. Selection for resistance to Septoria canker (*Septoria musiva*) is absolutely necessary in the Central and Eastern United States and in southern Quebec and Ontario. Yet, even though the causal organism exists in the Pacific Northwest, canker incidence is low or non-existent, and selection for resistance is generally relaxed or absent. Cold winter temperatures and short growing seasons require selection for cold hardiness and resistance to frost crack and sunscald in Eastern Canada and the North Central United States. Aspens, especially *P. tremuloides*, can be highly susceptible to Hypoxylon canker (*Hypoxylon mammatum*) in all regions, and selection for resistance to this disease is needed.

Selection for wood characteristics is often problematic because objectives can vary, depending on proposed industrial use and the percent of total furnish supplied by cultivated poplars. For example, wood color is important when poplars are used as a raw material in mechanical pulping processes. Wood specific gravity is important when poplar chips are to be substituted for aspen in the manufacture of oriented strand board products without modification of press equipment and process. In general, the most important need is to select enough clones based on growth and cultural characteristics so that sufficient variation remains in the population to select for wood characteristics.

Breeding strategies

In the following, we discuss several common breeding strategies that represent generally accepted methods (Bisoffi and Gullberg 1996). The discussion necessarily begins with a brief enumeration of the possible species with which a breeding program might start. Not all species are equally inter-fertile, so a discussion of patterns of incompatibility and methods for overcoming incompatibility is included. We then discuss different mating and breeding strategies that can be used to develop populations of pure species and produce improved inter-specific hybrids. We further discuss common strategies for the testing and selection of superior genotypes and the kind of knowledge needed to do so efficiently. We make the point, using a numerical example, that selection during early stages of testing makes it very difficult to obtain the kind of quantitative genetic estimates required to optimize testing and selection strategies.

Species selection

The genus *Populus* consists of 22–85 species divided unequally among six taxonomic sections (Eckenwalder 1996). Disagreement regarding species number has been traced to two factors: the taxonomic classification of hybrid populations and philosophical differences between "lumpers" and "splitters" (Eckenwalder 1996). Regardless of classification, variability within the genus, combined with a multitude of alternative breeding strategies, makes the discussion of all possible breeding strategies difficult.

For example, at least five major *Populus* species are indigenous to North America: *P. deltoides* in section *Aigeiros*; *P. balsamifera* and *P. trichocarpa* in section *Tacamahaca*; and *P. grandidentata* and *P. tremuloides* in section *Populus*. Exotic species such as *P. maximowiczii* (now included in *P. suaveolens*), *P. nigra*, *P. tremula*, etc., further increase the number of breeding options. Three exotic poplar species have been introduced into or across North America in sufficient numbers to assess their potential as parental germplasm. These are *P. maximowiczii*, *P. nigra*, and *P. trichocarpa*, the last being an intra-continental introduction. In Quebec, *P. maximowiczii* shows good resistance to insects and diseases, except for Septoria canker, and a very good adaptation to medium fertility and acid soils (pH 4.8) with shoot growth of 1.5 m per year. Out of three provenances of *P. maximowiczii* introduced from Japan in 1988, representing 28 progenies coming from high elevation (475, 580, and 920 m) on Hokkaido island, the provenance Kamikawa (475 m) is the best one for growth in southern Quebec (Périnet 1999). Hybrids between *P. deltoides* and *P. maximowiczii* have also demonstrated good growth in the North Central U.S., although susceptibility to Septoria canker is a common concern.

P. nigra has shown good resistance to foliar disease and insect damage, and medium resistance to *Septoria musiva* in Quebec. Twenty-eight progenies representing 21 provenances from Belgium, the Netherlands, Hungary, Yugoslavia, and Bulgaria have been tested in southern Quebec. *P. nigra* is being used as a parent in inter-specific breeding with *P. deltoides* in Quebec and in the North Central U.S.

Fifty-seven provenances of *P. trichocarpa* were tested in southern Quebec. Seeds were collected by Dr. Koster for the IUFRO working group on poplar provenances in Oregon, Washington, British Columbia, and Alaska. Provenance assessment from a test on acid soil with a low fertility at the Villeroy populetum located 75 km west of Quebec City gives the following results: provenances east of the Cascade Range (inland provenances) are more susceptible to *Melampsora medusae* at the beginning of September than the western provenances (coastal). Northern provenances are more susceptible to Melampsora rust than southern provenances within both coastal and inland collections. Inland provenances are more resistant to frost damage, sunscald, and frost cracks, and have a lower incidence of epicormic branches compared to coastal origins. In general, *P. trichocarpa* is very susceptible to Septoria canker and to the poplar-and-willow borer (*Cryptorhynchus lapathi*). A general trend of growth reduction was also observed from south to north. Several hundred *P. trichocarpa* clones from Idaho and British Columbia have been tested in the North Central U.S. (Riemenschneider et al. 1994), and about 30 of those clones have been selected for canker resistance and inclusion in inter-specific breeding.

Species incompatibility

Inter-specific hybridization has been, in nearly all cases, the main improvement strategy pursued by poplar breeders worldwide. Hybridization among the species that comprise the three main commercial sections of the *Populus* genus is oftentimes, however, encumbered by a lack of reproductive compatibility that isolates different species to different extents. Generally, greater effort is required to cross species belonging to different sections than those more closely related. As an example, members of the *Populus* section are relatively highly inter-fertile, but nearly completely isolated from those of the *Tacamahaca* and *Aigeiros* sections. Moreover, while crossing between *Aigeiros* and *Tacamahaca* is entirely feasible, the effort is typically less productive than when crossing species within either section.

Inter-specific reproductive incompatibility is manifested in several ways in *Populus*. Pre-zygotic isolation by ineffectual pollination and fertilization separates members of the *Populus* section from those of *Tacamahaca* and *Aigeiros*. Post-zygotic blockages, especially embryo abortion are the common obstructions to hybridization between *Tacamahaca* and *Aigeiros* sections. The latter obstacle is most pronounced when the *Tacamahaca* representative is used as the seed parent. For example, in crosses between *P. deltoides* and *P. trichocarpa*, embryonic development is aborted prior to dehiscence when *P. trichocarpa* serves as the seed parent, while female *P. deltoides* selections are capable of producing viable seeds, but with more variable germination compared to intra-specific seeds. Similar reciprocal crossing effects have been observed in the *P. deltoides* × *maximowiczii* hybrid pedigree, with crosses in the direction of a *P. deltoides* female being the more productive (Zsuffa et al. 1999). Post-zygotic obstacles continue to be expressed at the germinative, cotyledonal, and seedling stages, although these are not as great or as troublesome as the block exhibited during embryonic developmental.

Accommodation of pre-zygotic incompatibility in the design of production breeding programs is difficult and impractical. One technique employs mentor or recognition pollen, a pollen mixture of both species involved in hybridization in which the pollen belonging to the species matching the intended female has been killed by radiation, repeated freeze–thawing cycles, or with methanol or ether treatments (Knox et al. 1972; Stettler and Guries 1976). Proteins released from the dead pollen activate stigmata that become receptive to the functional pollen of the unrelated species. Double fertilization is effected, resulting in inter-specific seed. Although mentor and recognition pollen mixes have been used to a limited extent in experimental crossing of the *Populus* and *Aigeiros* sections, it is not used regularly in applied poplar breeding. Circumvention of this pre-zygotic barrier has also been completed using complex *Populus* hybrids (e.g., *P. alba* × *tremula* crossed with *P. alba* × *grandidentata*) as female parents in crosses with *Tacamahaca* and *Aigeiros* male selections (Ronald 1982). The impetus behind this technique has been to improve backcross recovery rates of adventitious rooting ability.

More effort has gone into the development of techniques to overcome post-zygotic blocks targeting embryo inviability. A variety of techniques to recover aborted embryos are in use today, the individual choice dependent upon the maturation stage at which embryo development ceases. These vary from microculture (in vitro culture) of individual embryos to culture of whole carpels (Kouider et al. 1984; Noh et al. 1986; Raquin et al. 1993; Savka et al. 1987). These techniques are used extensively in hybridization programs in the Pacific Northwest, principally with female *P. trichocarpa* selections, where premature capsule abscission and splitting in response to aborted embryo development is quite frequent. Likewise, in vitro germination of nearly mature whole ovules can also be applied to the more fecund *P. deltoides* × *trichocarpa* families to enhance normal seed germination rates (T. Chen, personal communication).

Beyond the use of in vitro culture techniques, controlled hybridization success in the face of post-zygotic embryo abortion can be improved somewhat by provenance and individual breeder selection and by the direction in which the cross is designed. In hybridizing *P. trichocarpa* and *P. deltoides*, the most important crossing determinant may be selection of individual female *P. trichocarpa* genotype. Stettler et al. (1996) suggested that embryo abortion may occur in response to the asynchronous development of the maternal carpel tissue and that of maturing hybrid embryos. The coincidence of these two processes is quite different for northerly *P. trichocarpa* in which the normal carpel maturation period lasts 4–8 weeks, and *P. deltoides* from southerly latitudes in which carpels mature over a 12–20 week period. Abortions regularly occur in these combinations of species and provenances, with most of the variation in the rate of abortions due to the female P. *trichocarpa* genotype. When pollination involves male *P. deltoides* from more northerly provenances in which the normal schedule of reproductive development is more closely matched to *P. trichocarpa*, the abortion phenomenon may not be expressed to the same extent, although such crosses are still difficult to make (B. Michiels, personal communication). In the reverse cross direction with *P. deltoides* female parents, embryo abortion is much reduced or non-existent, and embryo rescue techniques may not be required or may be limited to in vitro germination of nearly mature ovules.

Not all hybrid combinations between *Tacamahaca* and *Aigeiros* poplars exhibit the same intensity of abortion. For example, in the *P. trichocarpa* × *P. nigra* and the *P. nigra* × *P. trichocarpa* pedigrees, embryo abortion does not appear to limit artificial breeding success.

Advanced generation techniques such as backcross breeding and three-way hybridization are also affected to varying degrees by the same reproductive incompatibilities that constrain F_1 hybridization. Yet, poplar breeders pursuing either of these strategies must also contend with the negative effects arising from the reformation of linkage groups arising from meiotic recombinations during F_1 parental gametogenesis. This effect can be managed in three-way hybridizations to a certain extent by the way in which the cross is designed. For example, hybridization of *P. trichocarpa* (T), *P. deltoides* (D), and *P. maximowiczii* (M) may meet with better success when *P. deltoides* enters the cross as the pure species parent in combination with an F_1 *P. trichocarpa* × *P. maximowiczii* parent. Designing the cross in this manner frequently yields a hybrid of favorable form and growth rate as compared to a cross design in which the pure species representative is from *Tacamahaca* (e.g., *P. maximowiczii* crossed to an F_1 *P. trichocarpa* × *P. deltoides*). The genomic constitution of the former hybrid (50% *P. deltoides*, 25% *P. trichocarpa*, and 25% *P. maximowiczii*) differs from the latter hybrid (25% *P. deltoides*, 25% *P. trichocarpa*, and 50% *P. maximowiczii*). More importantly perhaps, the two resulting pedigrees may also differ in the construction of their genomes. Reduction division during gametogenesis of *P. trichocarpa* × *P. maximowiczii* F_1 parents involves chromosomal exchanges between genomes of species from the same section. These meiotic recombinations may be more harmonious than those that occur during the gametogenesis of an F_1 *P. trichocarpa* × *P. deltoides* parent that brings chromosomal segments from the *Tacamahaca* and *Aigeiros* sections together into the same linkage groups. Such discordant combinations may further contribute to the variation in performance between TM × D hybrids and their TD × M and DM × T counterparts.

Mating designs

Breeding always begins with the controlled mating of individuals, which may come from the same or different species. Artificial crosses are commonly made by using the method of "*plançons-en-pots*" (Fig. 1) where floral cuttings are stuck in pots for rooting and have access to an aerated water reservoir at the bottom (Joennoz and Vallée 1972). Female flowers can be pollinated by hand (Fig. 1), and if all goes well, the female catkins develop and produce large amounts of seed (Fig. 2). All mating systems serve two goals: (1) to produce

Fig. 1. Branches from female poplar trees can be rooted in pots with various support media or in aerated water. Female flowers are pollinated by hand, using a small brush or an air-driven device. (Photo by Natural Resources Research Institute, Duluth, MN.)

Fig. 2. Female catkins develop normally on well-rooted female branches after pollination. Time to capsule maturation and seed shed depends on the species of the female parent and can take from 4 to 8 weeks. (Photo by Natural Resources Research Institute, Duluth, MN.)

progeny for testing, selection, and breeding and (2) to provide a family structure that can be used to attribute observed sources of variation to various kinds of gene action, for example, whether different alleles at a locus act additively or interact to produce dominance effects. Single-pair mating (Table 4) crosses one male with one female only, each parent being used in only one cross. The advantage of single-pair mating is that it utilizes the most parents for a fixed number of families, compared to other strategies. The disadvantage is that additive genetic effects are confounded with some non-additive effects. Thus, it is impossible to obtain "clean" estimates of variation due to any one kind of gene action. A nested mating scheme (Table 4) crosses one male with several different females (or the converse), each male being crossed with a different set of females. This scheme, also know as Design I (Comstock and Robinson 1948), permits an estimate of additive genetic variance (among males) and dominance variance. Crossing all males with all females is known as a factorial mating scheme, also known as Design II (Comstock and Robinson 1948). The factorial mating scheme allows two estimates of additive genetic variance (among males, among females) and an estimate of dominance variance (male × female interaction). A variation on the factorial scheme is to cross subsets of males and females, then replicate the sets across the entire population of parents (Table 4). The complete factorial mating scheme utilizes few parents given the number of families produced, a disadvantage that can be countered by the use of subsets. The most cumbersome mating scheme encountered in crop breeding literature is the diallele. However, the diallele scheme requires monoecious plants where the same parents are used as both male and female, and is thus not applicable to poplars.

There are two important points to remember regarding selection of a mating scheme. First, there is no one optimum mating design; choice of design should be

Table 4. Consolidated diagram of the most common mating systems used in controlled crossing. We assume a population of six female parents and six male parents; the strategies can be extended to any number of parents, limited only by the logistical capabilities of the breeder. "s" represents those crosses made under single-pair mating, "n" represents crosses made in a nested mating scheme where each male parent is crossed to two different female parents (only three males are used in this hypothetical scheme; reality would require an additional six females to complete the strategy). "p" represents crosses made in three sets of 2 × 2 factorials. A complete factorial mating scheme would fill all 36 cells.

	Male 1	Male 2	Male 3	Male 4	Male 5	Male 6
Female 1	s, p, n	p				
Female 2	p, n	s, p				
Female 3		n	s, p	p		
Female 4		n	p	s, p		
Female 5			n		s, p	p
Female 6			n		p	s, p

dictated by the relative importance of number of parents versus the need to estimate different kinds of gene action. Second, estimates of gene action are derived by equating observed sources of variation to expectations of gene action that assume, among other things, that alleles are in linkage equilibrium in the breeding population. The assumption may be valid when parents from a previously random mating population of a single species are crossed, or when estimates are made from advanced generation inter-specific populations. However, the assumption is inarguably untrue when individuals from different species are bred in the first (F_1) generation. Thus, strength of inference can only be judged cautiously, in light of the history and nature of the breeding program and with full consideration of the assumptions involved in the estimation.

Non-recurrent F_1 breeding

Most programs today have come to rely on F_1 hybridization, mostly because F_1 progeny populations are highly variable, and it is likely that at least some genotypes with high commercial utility can be selected from within those populations. The ability of some progeny to exceed the performance of both parents is termed heterosis, and may be attributable to heterozygosity in the F_1 generation and the random occurrence of favorable combinations of alleles from the two parents. Further, non-recurrent F_1 breeding can be practiced using tested parents from closed breeding populations or using parents randomly obtained from the wild. Inter-specific hybridizations among parent trees of the best phenotypes from local or exotic species have been made by virtually every breeding program in North America, representing the Pacific Northwest, the North Central U.S., Ontario, and Quebec. Selected trees of *P. deltoides* are commonly used as one parent in eastern North America because it is a native species, resistant to Septoria canker, and frequently available as select clones derived from long-term testing. Natural populations of *P. deltoides* are found growing on acid soils in southern Quebec, which contributes to regional objectives. The method is simple, straightforward technically, and will not be discussed further.

Recurrent intra-specific breeding

Simple intra-specific recurrent breeding and selection (Fig. 3) is a long-standing strategy. Parents from a single species are crossed using various mating designs, as described above. Individuals are selected using among- and within-family selection (see testing and selection strategies below) based on one or more characteristics (selection criteria). Selected individuals are then inter-mated using one of the previously described mating designs to produce the next generation. The procedure continues indefinitely, hence the term "recurrent." The population under recurrent selection is improved in regard to selection criteria based on narrow-sense (additive) genetic effects. Non-additive effects can be exploited in

Fig. 3. Simple intra-specific recurrent selection. Individuals from the same species are crossed to produce progeny populations. Progeny are selected according to one or more selection criteria, then inter-mated to produce the next generation.

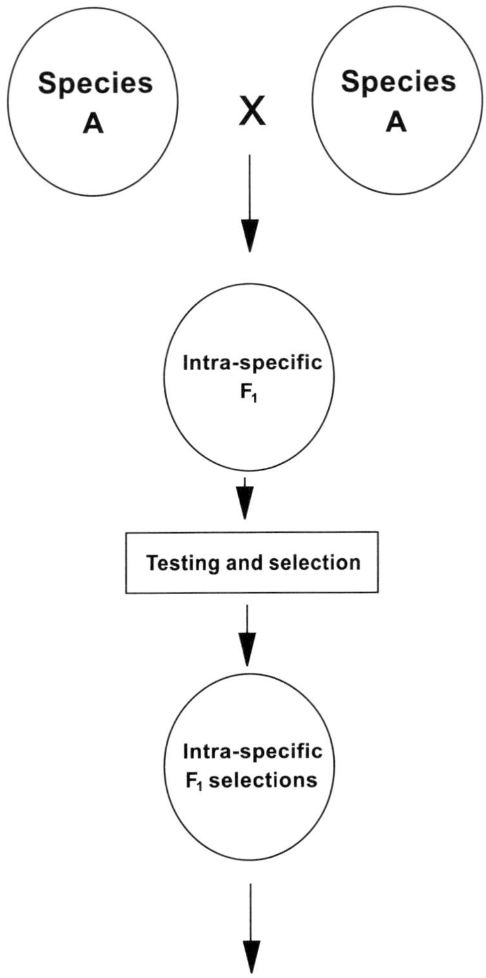

each generation by clonal testing of progeny, vegetative increase, and clonal deployment as has been discussed in a previous section. Intra-specific recurrent breeding coupled with clonal selection is most often practiced where a pure species is targeted for commercial deployment. Development of pure *P. deltoides* clones in the southern or north central U.S. is a good example.

Recurrent inter-specific breeding

Recurrent inter-specific breeding (Fig. 4) has not been widely applied as a poplar improvement strategy. The method involves initial crosses among individuals

Fig. 4. Inter-specific (meaning between species) recurrent breeding begins as does intra-specific breeding except that parents represent two species. The frequency of alleles from different species varies in advanced generations, depending on the kind of gene action affecting selection criteria.

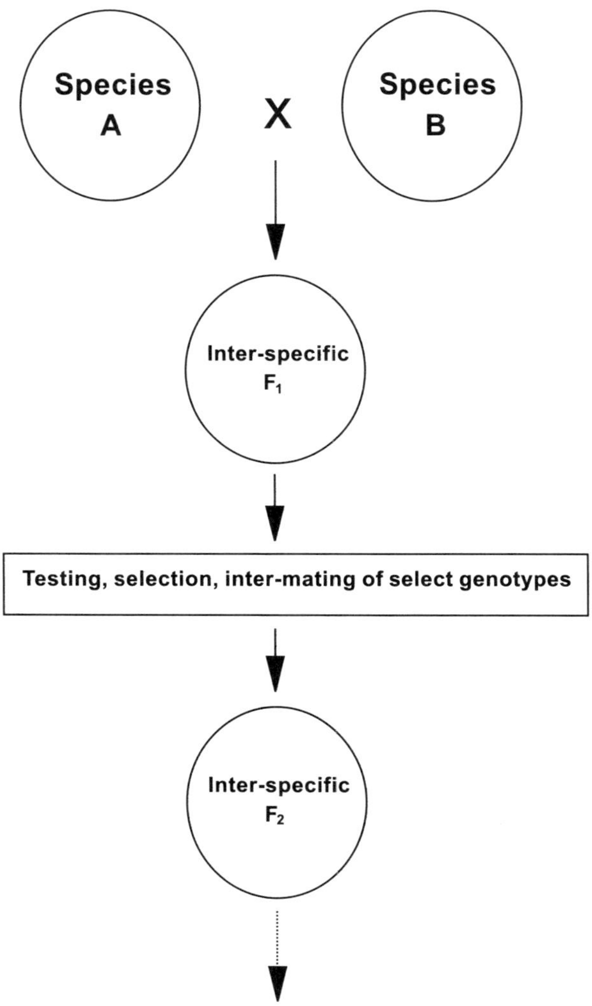

from two or more species. Selected F_1 individuals would then be inter-mated at random to produce an F_2 generation, etc. Each generation would contain segregating parental genes that could, in the long term, exist at any frequency. Positive alleles at additive loci would be driven towards fixation, regardless of the parental species from which they originated. Alleles at loci that act according to over-dominance should be maintained at intermediate frequency. Theoretically, valuable combinations of alleles at different loci (the non-additive effects we discussed in combination with the implications of clonal propagation) should occur at random in each generation. Also, the benefit of those valuable combinations could be captured by vegetative propagation. The strategy may have some merit as a poplar breeding method, but it has not been widely applied except in support of molecular analysis (Bradshaw 1996). This breeding method bears strong resemblance to the development of "synthetic varieties" in the agronomic crop breeding literature.

Intra–inter-specific breeding

Recurrent intra-specific breeding can be combined with non-recurrent F_1 breeding to take advantage of additive effects and F_1 hybrid vigor (Fig 5). Here, recurrent breeding of two or more parental species is conducted based on performance of intra-specific progeny per se (as contrasted with reciprocal recurrent selection which is discussed later). Select individuals are inter-mated con-specifically to produce subsequent generations of each parental species. Select individuals are also mated between species to produce new F_1 progeny. Additive (GCA) effects contribute to inter-specific performance to the extent that intra-specific GCA is a sound predictor of inter-specific breeding values (Stettler et al. 1996). The validity of that contention depends on the relative degree of additive and over-dominance effects on the respective performances of intra-specific and inter-specific progenies. As a long-term breeding strategy, the method would result in improvement of each of the constituent parental species (certain and predictable) and hybrids (less certain and predictable).

Reciprocal recurrent selection

Reciprocal recurrent selection (RRS) was proposed in 1949 as a breeding method for open-pollinated crops that would take advantage of both additive and dominance genetic effects (Comstock et al. 1949). In RRS, mating is done between individuals of one species or population and those of another (Fig. 6). Selection is based on the performance of progeny in hybrid combination. Then, the next intra-specific generation is produced by mating among individuals within species that have been selected based on their performance as inter-specific parents. Thus, each cycle of breeding requires two generations: one upon which to base selection, a second for production of the next generation. The effect of the breeding strategy on gene frequency is to drive positive alleles at additive loci to fixation in both parental populations, and to drive different alleles to fixation in different populations at over-dominant loci. Thus, one population will have one allele of an over-dominant locus, the other population will have another allele. Then, in hybridization, all progeny will be heterozygous for all over-dominant loci. It has been proposed that the most appropriate method for hybrid tree parental improvement may be RRS, although recurrent selection for general combining ability merits consideration (Shelbourne 1969; Stettler et al. 1996; Li and Wyckoff 1991).

Application of RRS to poplars raises some questions, however. First, the method is cumbersome and time-consuming, and it is doubtful that much progress could

Fig. 5. Intra–inter-specific breeding is probably the most common approach to the breeding of hybrid genotypes. Recurrent intra-specific breeding is practiced for both parental species, and the select genotypes from each are used to produce new F_1 inter-specific hybrids. Parents are selected on their performance per se.

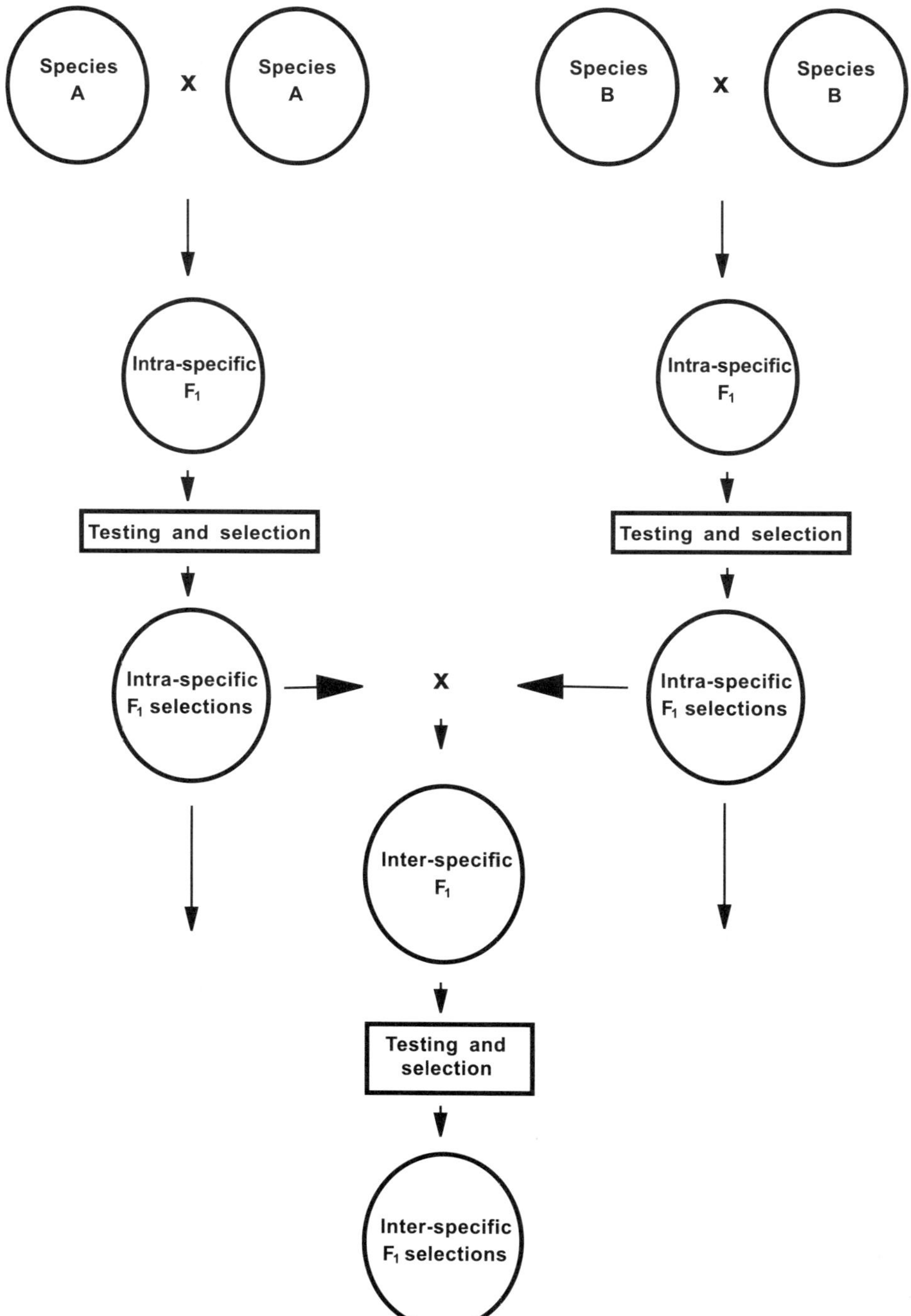

Fig. 6. Reciprocal recurrent selection and breeding has been suggested for use in poplar breeding. Trees within each species are selected based on their breeding value in hybrid combination with another species. Then, trees within each species are crossed to produce the next generation. Each cycle of breeding requires two generations: one to produce the inter-specific hybrids for testing, one to produce the intra-specific progeny which constitute the next generation.

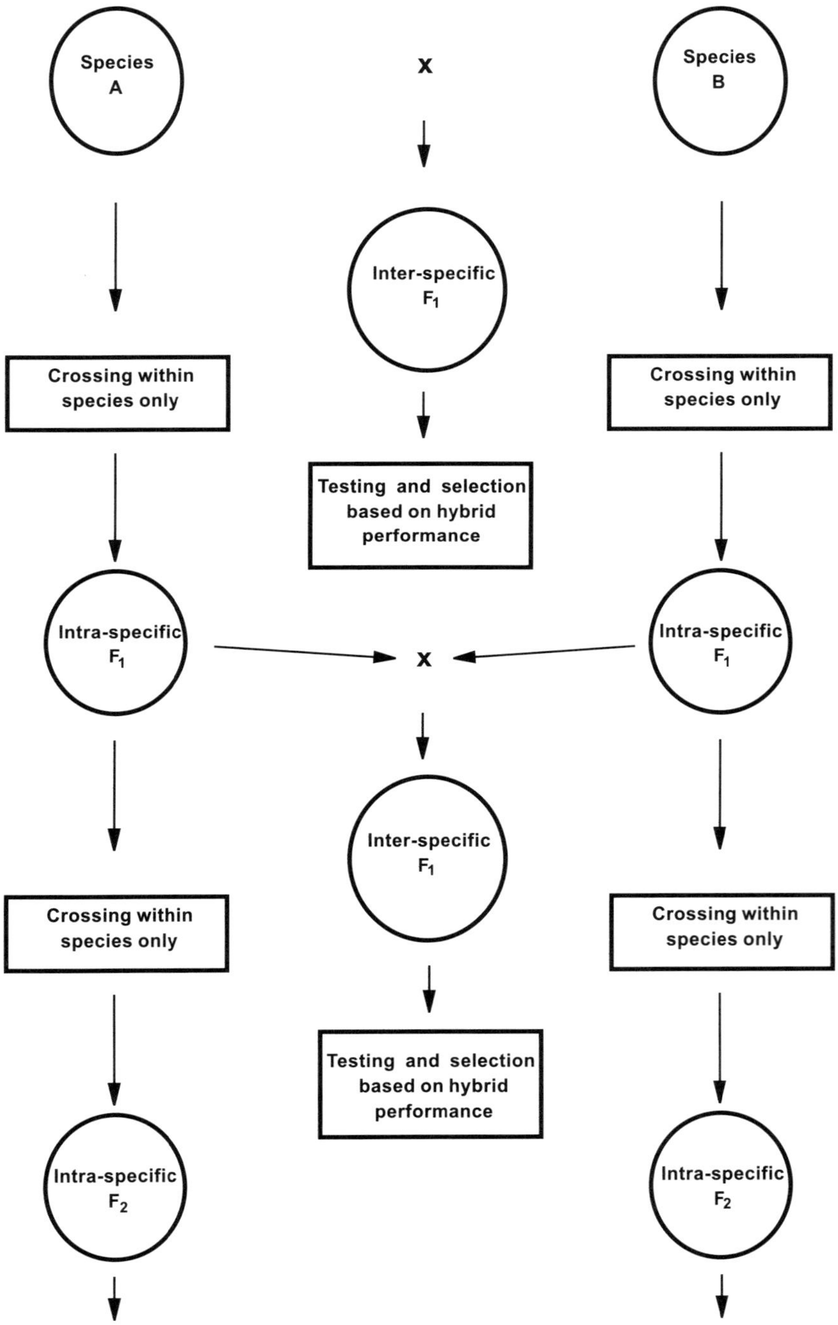

be made given the logistical realities of tree improvement. Second, the severity of reproductive incompatibilities can range from situations in which a RRS program is actually unworkable (*P. nigra* × *P. deltoides*) to instances where its implementation is feasible but at a much higher cost than other strategies (*P. trichocarpa* × *P. deltoides*). Third, RRS was invented for the improvement of a seed-based production system (i.e., hybrid corn). Varietal increase of a seed-based crop whose improvement is based on heterozygous over-dominant loci requires fixation of different alleles in different populations. The random occurrence of favorable genotypes in segregating populations has no value because genotypes are not preserved. Poplars are a vegetatively propagated crop, and the occurrence, by random chance, of heterozygous, over-dominant loci is to be expected at some frequency. Those genotypes can be directly exploited by cloning, as we have discussed previously. These are much different propositions and illustrate the hazards of transferring breeding systems between crops with different biologies. Last, early empirical studies of RRS in corn demonstrated that RRS resulted in greater inter-population heterosis compared to intra-population breeding, but also that RRS resulted in less intra-population improvement in the two parental lines compared to intra-population breeding (Moll and Stuber 1971). This could be of serious concern in poplars, where breeding programs may have the goal of developing commercial selections of both parental species and hybrids.

A modified RRS approach has been proposed for the Italian *P. deltoides* × *P. nigra* program. Here, the female component of the *P. deltoides* recurrent population and the male component of the *P. nigra* recurrent population are evaluated using *P. deltoides* × *P. nigra* test crosses (Bisoffi and Gullberg 1996). Because of the extremely low reproductive output of *P. nigra* × *P. deltoides* crosses, evaluations of *P. deltoides* males and *P. nigra* females are made using intra-specific test crosses.

Backcross breeding

Backcross breeding is a strategy used to introduce one, or a few, missing characteristics into a species or population that is otherwise of high utility. The strategy starts with hybridization between two species that, collectively, express all needed characteristics. Further breeding follows with selections in each generation being crossed with one of the species (the recurrent parent), while selection is practiced for the trait being introduced from the other species (the non-recurrent parent) (Fig. 7). The backcross breeding strategy may have strong merit in certain regions of North America. For example, *P. deltoides* grows well in eastern North America, sometimes equaling the growth of hybrid clones in field tests in Minnesota, Iowa, Wisconsin, and Michigan. *Populus deltoides* is also highly resistant to Septoria canker, which is not true of many hybrids. Deployment of *P. deltoides* is mostly hindered because cuttings root erratically or not at all under field conditions.

We have applied backcross breeding to correct poor rooting in the North Central U.S. and in Quebec. Crosses have been made between *P. deltoides* and

Fig. 7. Backcross breeding begins with F_1 inter-specific hybridization followed by the crossing of select F_1's to one of the parental species (the recurrent parent). Backcross breeding is designed to introduce one or more traits from the non-recurrent parent into a population that is otherwise identical to the recurrent species.

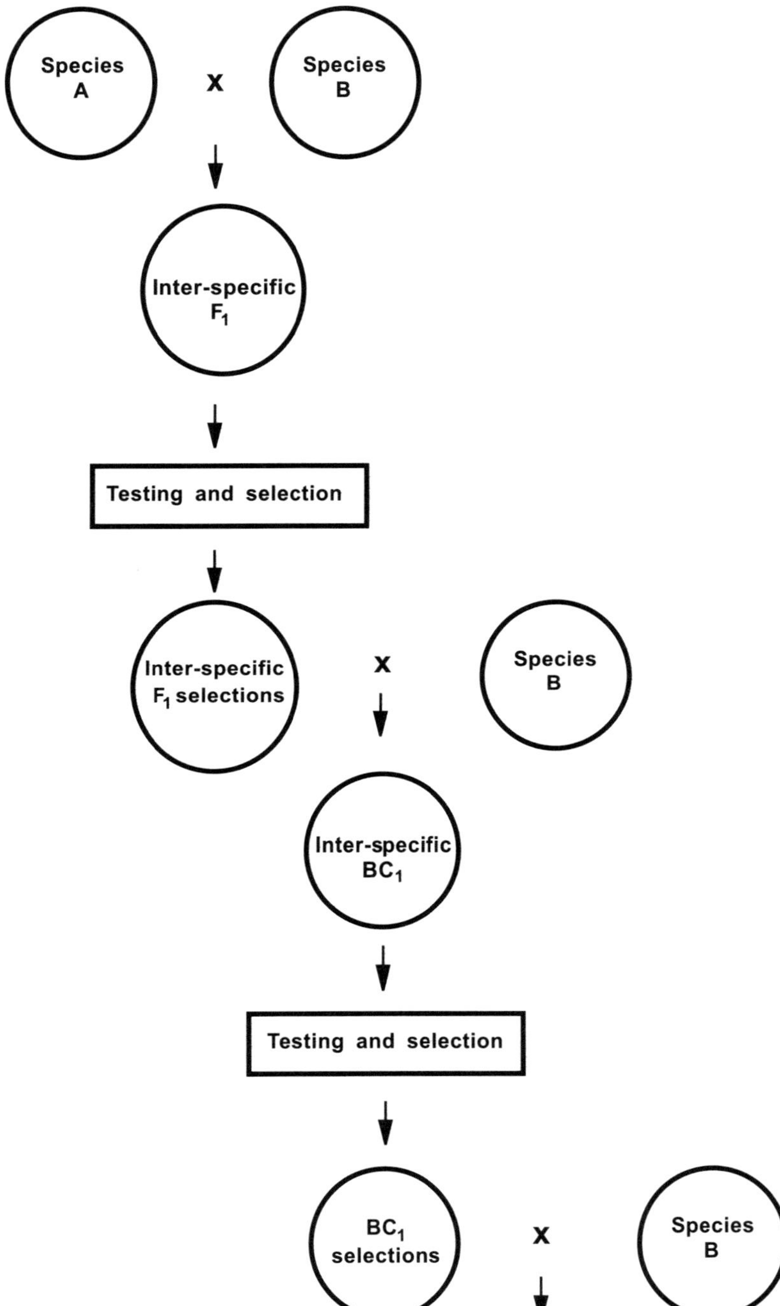

P. trichocarpa or *P. maximowiczii*. Hybrids are backcrossed to *P. deltoides* to recover canker resistance, while selection is imposed for good adventitious rooting ability. Likewise, Septoria canker resistance and improved productivity can also be attained with a *P. nigra* × *P. maximowiczii* backcross program with *P. nigra* as the recurrent.

More complex hybridization methods

In the Pacific Northwest and Quebec (Table 1), three-way hybridization of *P. trichocarpa* × *P. deltoides* F_1's and *P. maximowiczii* is being promoted as a way of improving the rust resistance of otherwise superior *P. trichocarpa* ×*deltoides* hybrid selections. Another potential three-way hybridization scheme would use *P. alba* × *P. tremula* F_1's crossed with *P. trichocarpa* as the initial step in introducing adventitious rooting ability into the *Populus* section. This three-way hybridization is also a good example of using complex breeding strategies to overcome inter-specific incompatibility.

Testing strategies

Testing and selection strategies suited to the identification of commercial intraspecific and hybrid poplar clones differ from strategies applied to most tree species and many crop plants because of the large numbers of progeny that can be produced quickly and because the progeny can be propagated asexually. As an example of the rate at which population sizes can grow, consider breeding progress in Quebec since 1971. Eighty provenance and progeny trials have been established in different regions of Quebec, representing provenances and progenies of *P. deltoides*, *P. balsamifera*, *P.* ×*jackii*, *P.* ×*rollandii*, *P. grandidentata*, and *P. tremuloides*. In addition, 7808 crosses have been made of which 1381 families have produced 129 125 seedlings (Table 1). So far, a total of 4094 clones have been screened in the nursery, representing 575 trees selected in natural stands, 2760 trees selected from progeny and provenance trials, and 759 clones introduced principally from Europe and Ontario. Since the beginning of the project, 108 clonal trials have been established.

Asexual propagation permits the utilization of total genetic variation, including non-additive effects, as long as the commercial propagule is vegetative, such as a cutting. However, the effective use of total genetic variation requires clonal testing, which can be laborious given large population sizes as exemplified above. The problem is substantial because, while it is very easy to produce large numbers of control cross seedlings, often numbering in the tens of thousands, it is more difficult to test the resultant clones. For example, consider the hypothetical need to test the yield potential of the clones derived from 25 000 seedlings. We might require two replications of 100 tree plots at each of three locations within some commercial zone of deployment. At a test spacing of, say, 3 m, the resultant test would occupy over 13 000 ha, a test equal to a reasonable commercial operation.

As a result, the most common selection strategies involve the application of some kind of multi-stage testing across vegetative generations. First, large seedling populations are produced through controlled mating. Then, individual seedlings are selected, propagated vegetatively, and evaluated in small plot nursery or field tests. Further propagation and selection is characterized by an increasingly small population of clones with increasingly large replication (Fig. 8).

The testing strategy followed by the Ministry of Natural Resources, Quebec, is a good example of the above. First, seedling progeny are planted in two or three different sites for a given region. Then, trees are selected from the progeny test based on growth and resistance to cankers, leaf diseases and insects, and are cloned. Second, clones are tested in stoolbeds for rapid juvenile growth, other desirable phenotypic characteristics, tolerance or resistance to leaf diseases and stem cankers, and rooting ability (stem cuttings for *Aigeiros* and *Tacamahaca* hybrids and root cuttings for *Populus* hybrids). Artificial inoculation techniques with *Septoria musiva* and *Hypoxylon mammatum* have been developed and are

Fig. 8. The relative ease with which large populations of hybrid seedlings can be produced, combined with the need for clonal testing, results in the application of multi-stage testing strategies. Different stages of testing vary in time and cost, and the overall strategy can only be optimized using knowledge of the variance/covariance structures that describe genotypic relations among stages.

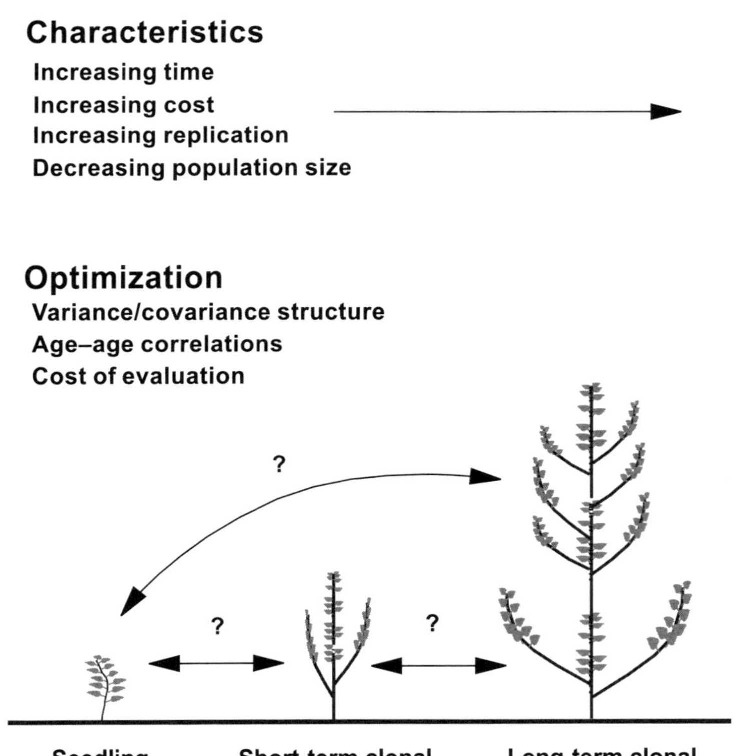

used for clonal selection (Bussières and Vallée 1987; Mottet et al. 1991; Mottet 1992). Finally, remaining clones are tested in trials with 8–10 replicates of 2 or 4 trees per plot for each clone. Trials are established on 2–3 representative sites for a given breeding region. Trees from each clonal test are measured and assessed for resistance to diseases, insects, and climatic damage, and for their phenotypic development.

Multi-stage testing strategies require some means of allocating selection intensity across stages. Thus, only some seedlings are entered into short-term clone tests, and only some of those clones are entered into longer-term large-plot clone tests. Deciding how to allocate selection intensity is crucial. For example, if correlations between traits that can be observed in different stages is high, then selection for those traits should be applied early. To defer selection to a later stage only increases the cost of the breeding and testing program. Conversely, if correlations between traits at an early stage (say, seedlings) and a later stage (say, large clone plots) are low, then the application of selection at an early stage results in the mistaken identification of superior clones and failure to improve anything. So, it would seem that the first need is to understand how traits are correlated across testing stages. Given knowledge of such correlations — or covariance — it is relatively easy to construct a mathematical expression of selection intensities and population correlations and variances that would predict the outcome of various testing strategies and enable the determination of optimum practice. For example, to optimize allocation of selection intensity between the two stages of selection we can create an equation to describe response to selection in each stage in terms of the selection intensities (i_1 and i_2) and the appropriate variances and covariance. We would like to find values of i_1 and i_2 that yield the best possible clones, while testing as few clones as possible in stage 2. Clearly, we need estimates of phenotypic variance among seedlings and among clone means, the genotypic covariance between the seedling and a clone mean, and the broad sense genotypic variance among clones. It is this need that gives rise to critical problems in optimizing even our simplified selection system. We use the following experimental and numerical example to illustrate how the act of being efficient, that is, discarding some genotypes at early stages regardless of knowledge, precludes the ability to optimize in the future.

A factorial mating design was completed with three female F_1 hybrids (*P. trichocarpa* × *P. deltoides*) and four males (*P. deltoides*) in a backcrossing strategy. The female F_1 hybrid parents derived from crosses between *P. trichocarpa*, originating from the coastal Pacific Northwest (west of the Cascade Mountains), and *P. deltoides*, originating from natural populations in Minnesota. The recurrent male *P. deltoides* were selected from Minnesota breeding populations. Seedlings were grown in a nursery in Rhinelander, WI. The heights and stem calipers of each tree were determined at the end of the first two growing seasons. Stems were harvested and resultant cuttings used to establish a clonal test in a randomized complete block design with five cuttings per ortet per replication. Height and stem caliper of the resultant trees were measured at the end of the first growing season. Data were analyzed by analyses of variance and covariance. This population and the data obtained therefrom are used to illustrate some issues surrounding multi-stage testing.

First, consider the estimation of $\sigma_{Gs,c}$ the genotypic covariance between a seedling and clone mean. Importantly, this statistic is not directly estimable using analysis of covariance because the seedling is an unreplicated experimental unit. There can be, therefore, no way to biometrically separate variation among seedlings within families that is attributable to genetic effects from variation attributable to environmental effects. The only possible way to estimate the statistic is to establish the entire population in both stages and then, by applying different hypothetical selection intensities, to identify the covariance value that yields predicted response equal to observed response. However, the entire untruncated population must be established in stage 2 to accomplish the estimation.

Alternatively, we could elect to plant only selected clones in stage 2 and content ourselves with the estimation of the correlation between seedlings and their respective clone means as a guide to optimization. However, truncation of the population in stage 1 has a considerable effect on the inter-stage correlation (Table 5). Note that the correlation estimated from the entire population (r_p = 0.479) differs greatly from correlations estimated from the population after stage 1 selection, where the proportion retained is 0.5 or less (Table 5). There is, overall, no way to obtain estimates describing untruncated populations using populations that have been subjected to selection.

So, why are the above discussion and numerical example worth considering? The several-hundred-year history of poplar breeding has been characterized by two results. First, success has been achieved in the absence of much genetic knowledge. Given two biological characteristics that define poplars — large non-additive genetic effects in populations derived from diverse parents and the ability to vegetatively propagate select genotypes — it is very difficult to breed a few randomly chosen parents and not identify a few clones with high commercial utility. Second, such readily obtained success has been principally responsible for the propagation of intuition, as opposed to knowledge, and the lack of experimental work designed to yield the data needed to optimize future work. Breeding undertaken strictly for the purpose of commercial advantage almost always practices selection at each stage of a multi-stage testing system, whatever that system may be. Yet, the act of selection makes the estimation of population parameters impossible; the knowledge we need to optimize our breeding strategies is rendered inaccessible by our practices.

Multiple trait issues

The application of multiple trait selection strategies is probably unavoidable in poplar breeding. Breeding objectives almost always include development of

Table 5. The effect of truncation selection for seedling stem diameter (test stage 1) on the variances among clones in subsequent tests (test stage 2) and on the correlation between seedlings and clones.

Selection intensity	Number of clones in second stage	Broad sense genotypic variance among clones	Phenotypic covariance between seedlings and clone mean	Phenotypic correlation between seedlings and clone means
1.00	547	2955	3.510	0.479
0.95	519	2876	2.589	0.381
0.90	492	2763	2.216	0.349
0.80	437	2742	2.032	0.356
0.70	382	2645	1.693	0.329
0.60	328	2617	1.443	0.303
0.50	273	2644	1.206	0.273
0.40	218	2439	1.006	0.260
0.30	164	2523	0.819	0.223
0.20	109	2767	0.902	0.252
0.10	54	2605	0.943	0.300
0.05	27	2185	0.583	0.217

fast-growing trees that are resistant to important diseases such as Melampsora rust and Septoria canker or to insect attack. As previously mentioned, commercial deployment is most often based on the planting of unrooted dormant hardwood cuttings, adding the additional selection criterion of rooting ability. Furthermore, some attention to wood quality is needed although the exact attributes and their optimum values are so process-dependent as to render generalizations difficult if not impossible.

Several quantitatively rigorous methods have been developed to address multiple-trait selection. These include index selection, tandem selection, independent culling levels, and their variants (Riemenschneider et al. 1996). Index selection is based on the incorporation of all selection criteria into an index of the form:

$$I = b_1X_1 + b_xX_2 + ... + b_nX_n$$

where I is the index value of a genotype and b_i is the weight given to trait X_i. The weights, b_i, are estimated so as to maximize the economic worth of the selected clones by mathematical manipulation of variances and covariances that need not be discussed here. Independent culling levels are easier to apply because minimum values for each criterion (truncation points) are set and only those individuals exceeding truncation points for all traits are selected. Tandem selection is the application of selection for different traits in different generations of the breeding program. Theoretical evaluation has demonstrated that index selection is always at least as efficient as other strategies and usually more efficient under most conditions.

In practice, poplar breeding commonly utilizes a mixture of the above strategies that is dictated by a combination of biological and logistical considerations. For example, during the first stage of evaluation (observation of the original seedling populations) it is possible to observe tree growth and the incidence of certain diseases. But it is not possible to observe rooting ability because seedling root systems develop from a bipolar embryo, whereas rooting of commercial significance relies on adventitious rooting from a dormant hardwood cutting — completely different ontogenies. Thus, selection for rooting ability is always deferred to a later stage of testing. Also, tests of wood quality are generally expensive and thus mostly deferred until the latest stages of testing when the number of clonal entries is much reduced and trees can be allowed to reach commercial size.

Multiple trait index selection can be rendered complex because of non-linear relations between phenotype and economic value (Namkoong 1979 and references therein). Several possible examples come readily to mind. First, the value of adventitious root formation is probably near zero until rooting is sufficient to result in survival of the cutting. Then, rooting may be linearly related to value because, over some range of rooting, survival and growth increase directly with root development. Past some point, however, increased rooting may confer little marginal return and, if allometrically competitive with shoot growth, actually cause a reduction in the aggregate value of the genotype. Disease resistance is another good example where the relationship between expression and value may have the same shape as the relationship between root development and value, albeit for different reasons. The relationship between value and various measures of wood quality may also be non-linear, depending on industrial end-use. For example, consider the relationship between specific gravity and value in the manufacture of oriented strandboard. Below the specific gravity needed to fill a press with a sufficient mass of wood, the value of specific gravity is zero at all points. Above some threshold, however, value increases in the form of a discontinuous function. Above the threshold, value may again be constant, or increase somewhat, depending on the nature of the specifics of the manufacturing process.

Understanding non-linearities between phenotype and value is important in the design of a testing and selection program. The value of wood quality is probably most easily understood because the shape of the curve is defined by mechanics and engineering, not biology. Understanding the value of rooting and pest resistances is different and would require some significant investment in physiological genetics. For example, the effect of differing incidences of Melampsora rust on tree growth could be quantified by evaluating leaf photosynthetic rate within a population of clones that differed in rust incidence. The effect of different adventitious rooting abilities could be determined by appropriate studies of upward and downward translocation of minerals and photosynthate, respectively. It will remain difficult to associate value with phenotype until such knowledge is forthcoming. Dickmann and Keathley (1996), in questioning the role of physiology in breeding, stated that "... physiologists working with tree improvement programs for *Populus* and other taxa need to re-evaluate their modus operandi." We strongly suggest that a new modus operandi might be found for physiological investigation in defining the relationship between trait expression and the economic worth thereof.

Conclusions and future work

The breeding of poplars and their hybrids to produce fast-growing, pest-resistant select clones has undoubtedly been a successful exercise in applied plant breeding. The Minnesota Hybrid Poplar Research Cooperative has produced a total population of approximately 30 000 seedlings of pure *P. deltoides*, F_1 hybrids between *P. deltoides* and *P. nigra* or *P. maximowiczii*, and backcross populations, using *P. deltoides* as the recurrent parent and *P. trichocarpa* as the non-recurrent parent. Several selections from breeding done in the early 1980s are outproducing current commercial standards by 20% to over 100%, depending on test site. Poplar breeding conducted in the Pacific Northwest, which began with crosses between *P. trichocarpa* and *P. deltoides* (Stettler et al. 1996) has given rise to an entire commercial production system. Breeding among select aspen genotypes (*P. tremuloides*, *P. grandidentata*, *P. tremula*, etc.) in Minnesota also has promise for increasing productivity (Li and Wyckoff 1991). Clonal trials in Quebec have allowed the recommendation of 50 selected clones for poplar plantations, representing approximately 10 clones for each of five poplar breeding regions of Quebec (Vallée 1995; Vallée et al. 1997; Périnet 1998). The expected production of these clones in well-managed plantations will vary from 10 to 15 $m^3\ ha^{-1}\ year^{-1}$ and could reach as much as 20 $m^3\ ha^{-1}\ year^{-1}$, according to site fertility and the length of the growing season.

The introduction of *P. maximowiczii* in Quebec has been a breakthrough in the production of hybrids well-adapted to acid soil with medium fertility for which clones have already been selected. For future breeding work in Quebec, an intensive sampling of native populations of *P. maximowiczii* should be done and other Asian species should also be introduced. The fast growth that is characteristic of hybrids between *P. deltoides* and *P. maximowiczii* throughout the North Central United States suggests that collaboration among institutions in the development of expanded parental foundation populations could contribute to increased efficiency of poplar breeding throughout eastern North America. This would be especially true where poplar breeding is conducted in a non-proprietary setting.

Given the genetic complexity introduced by the number of potential parental species, environmental settings, and different logistical constraints throughout North America, it is difficult to discern how any optimum breeding strategy might be selected from among all possible alternatives. It is certain that F_1 hybrid breeding has a future because of the long-proven success of the method. Additionally, F_1 hybrid breeding can be readily practiced in an opportunistic, non-recurrent fashion where periodic increases in logistical capacity can be used to greatly increase available populations that can be tested over time. Various strategies might be used to improve the performance of parental selections through intraspecific recurrent selection. The value of the most complex breeding strategy, reciprocal recurrent selection, remains unproven. Molecular studies suggest that many traits are subject to large non-additive effects (Bradshaw 1996), and so superficial evaluation suggests that RRS may be useful. However, the ability to vegetatively propagate many poplars also suggests that favorable inter- and intra-allelic interactions might be captured without the need for RRS via, for example, recurrent selection within hybrid populations. The choice of mating scheme is highly dependent on the reason for doing breeding, whether to simply advance generations or to develop an understanding of different kinds of gene action. Thus, no one mating scheme could be globally defined as optimum. Balanced representation of parents is probably universally important, as is the ability to define pedigrees so as to avoid inbreeding in the short term. Beyond these considerations, local circumstances are probably the predominant factors in operational decision-making.

The design of testing protocols is probably one area that could benefit from rigorous attention to method. It is clear from our preceding arguments that some kind of multi-stage testing strategy is usually needed. However, the kinds of variance and covariance estimates needed to optimize such strategies are mostly unavailable because unselected populations are rarely carried through all testing stages. Such experiments are clearly needed. Unselected populations could be established by random sampling, and the resultant "reference" populations used to guide future testing strategies.

Additional attention needs to be given to the non-linear relation between the phenotypic expression of some traits, such as disease resistance and rooting ability, and economic value. We suggest that such experimentation would provide an opportunity to connect breeding, genetics, and physiology, a connection that has been difficult to define in the past (Dickmann and Keathley 1996).

References

Bisoffi, S., and Gullberg, U. 1996. Poplar breeding and selection strategies. *In* Biology of *Populus* and its implications for management and conservation. *Edited by* R.F. Stettler, H.D. Bradshaw, Jr., P.E. Heilman, and T.M. Hinckley. NRC Research Press, Ottawa, ON. pp. 139–158.

Bradshaw, H.D., Jr. 1996. Molecular genetics of *Populus*. *In* Biology of *Populus* and its implications for management and conservation. *Edited by* R.F. Stettler, H.D. Bradshaw, Jr., P.E. Heilman, and T.M. Hinckley. NRC Research Press, Ottawa, ON. pp. 183–200.

Bussières, G., and Vallée, G. 1987. Technique d'inoculation artificielle de *Septoria musiva* Pk. pour la sélection de clones de peuplier. Proceedings of the Annual meeting of the Poplar

Council of Canada and the Poplar Council of United States, June 21–24, 1987, Cornwall, Ontario.

Comstock, R.E., and Robinson, H.F. 1948. The components of genetic variance in populations of biparental progenies and their use in estimating average degree of dominance. Biometrics, **4**: 254–266.

Comstock, R.E., Robinson, H.F., and Harvey, P.H. 1949. A breeding procedure designed to make maximum use of both general and specific combining ability. J. Am. Soc. Agron. **41**: 360–367.

Dickmann, D.I., and Keathley, D.E. 1996. Linking physiology, molecular genetics, and the *Populus* ideotype. *In* Biology of *Populus* and its implications for management and conservation. *Edited by* R.F. Stettler, H.D. Bradshaw, Jr., P.E. Heilman, and T.M. Hinckley. NRC Research Press, Ottawa, ON. pp. 491–514.

Eckenwalder, J.E. 1996. Systematics and evolution of *Populus*. *In* Biology of *Populus* and its implications for management and conservation. *Edited by* R.F. Stettler, H.D. Bradshaw, Jr., P.E. Heilman, and T.M. Hinckley. NRC Research Press, Ottawa, ON. pp. 7–32.

Joennoz, R., and Vallée, G. 1972. Recherche et développement sur le peuplier dans la région de l'Est du Québec. II Résultats d'hybridations artificielles chez les peupliers. Ministère des Terres et Forêts, Direction générale des forêts, Service de la recherche, Mémoire 13.

Knox, R.B., Willing, R.R., and Pryor, L.D. 1972. Interspecific hybridization in poplars using recognition pollen. Silvae Genet. **21**: 65–69.

Kouider, M., Skirvin, R.M., Saladin, K.P., Dawson, J.O., and Jokela, J.J. 1984. A method to culture immature embryos of *Populus deltoides* in vitro. Can. J. For. Res. **14**: 956–958.

Li, B., and Wyckoff, G.W. 1991. A breeding strategy to improve aspen hybrids for the University of Minnesota aspen/larch genetics cooperative. Proc. IEA Task V Activity Groups. Iowa State University, Ames, IA. 9 pp.

Moll, R.H., and Stuber, C.W. 1971. Comparisons of response to alternative selection procedures initiated with two populations of maize (*Zea mays* L.). Crop Sci. **11**: 706–711.

Mottet, M.-J. 1992. Méthodes pour sélectionner des peupliers résistants au chancre hypoxylinien. Ministère des Forêts, Direction de la recherche forestière, Mémoire de recherche 104.

Mottet, M.-J., Bussières, G., and Vallée, G. 1991. Test précoce pour l'évaluation de la sensibilité de peupliers hybrides au chancre septorien. For. Chron. **67**: 411–416.

Namkoong, G. 1979. Introduction to quantitative genetics in forestry. Tech. Bull. No. 1588. USDA Forest Service, Washington, DC. 342 pp.

Noh, E.R., Koo, Y.B., and Lee, S.K. 1986. Hydridization between incompatible poplar species through ovary and embryo culture. Res. Rep. Inst. For. Genetics (Korea), **22**: 9–14.

Périnet, P. 1998. Poplar genetic improvement program in Quebec. *In* Proceedings of the 1998 Annual Meeting of the Poplar Council of Canada, September 21–24, 1998, Quebec City, Quebec pp. 56–58.

Périnet, P. 1999. *Populus maximowiczii*: Results of a progeny test planted in 1988 and breeding potential of this species for Quebec, poster abstract, International Poplar Symposium II, September 13–17, 1999, Orléans, France.

Raquin, C., Troussard, L., and Villar, M. 1993. In-ovary embryo culture as a tool for poplar hybridization. Can. J. Bot. **71**: 1271–1275.

Riemenschneider, D.E., McMahon, B.E., and Ostry, M.E. 1994. Population-dependent selection strategies needed for 2-year-old black cottonwood clones. Can J. For. Res. **24**: 1704–1710.

Riemenschneider, D.E., Stelzer, H.E., and. Foster, G.S. 1996. Quantitative genetics of poplars and poplar hybrids. *In* Biology of *Populus* and its implications for management and conservation. *Edited by* R.F. Stettler, H.D. Bradshaw, Jr., P.E. Heilman, and T.M. Hinckley. NRC Research Press, Ottawa, ON. pp 159–182.

Ronald, W.G. 1982. Intersectional hybridization of *Populus* sections Leuce-Aigeiros and Leuce-Tacamahaca. Silvae Genet. **31**: 94–99.

Savka, M.A., Dawson, J.O., Jokela, J.J., and Skirvin, R.M. 1987. A liquid culture method for rescuing immature embryos of eastern cottonwood. Plant Cell Tissue Organ Cult. **10**: 221–226.

Shelbourne, C.J.A. 1969. Tree breeding methods. New Zealand Forest Service, Wellington. 44 pp.

Stettler, R.F., and Guries, R.P. 1976. The mentor pollen phenomenon in black cottonwood. Can. J. Bot. **54**: 820–830.

Stettler, R.F., Zsuffa L., and Wu, R. 1996. The role of hybridization in the genetic manipulation of *Populus*. *In* Biology of *Populus* and its implications for management and conservation. *Edited by* R.F. Stettler, H.D. Bradshaw, Jr., P.E. Heilman, and T.M. Hinckley. NRC Research Press, Ottawa, ON. pp. 87–112.

Vallée, G. 1970. Rapport d'activités du Service de la recherche du ministère des Terres et Forêts du Québec au Comité canadien pour l'amélioration des arbres forestiers. Proceedings of the Twelfth Meeting, Committee on Forest Tree Breeding in Canada, Canadian Forest Service, Ottawa, ON. Part 2. pp. 25–32.

Vallée, G. 1971. Rapport sur l'état des recherches sur le peuplier au Service de la recherche du ministère des Terres et Forêts du Québec pour le Comité de recherche en génétique forestière du CRDF. La populiculture au Québec. Ministère des Terres et Forêts du Québec, Conseil de la recherche et du développement forestiers. Rapport n° 1. pp. 14–23.

Vallée, G. 1995. Projet du MRNQ sur l'amélioration génétique des peupliers dans la région du Saguenay–Lac-Saint-Jean. Proceedings of the 1995 Annual Meeting of the Poplar Council of Canada, September 26–29, 1995, Chicoutimi, Quebec. pp. 89–105.

Vallée, G., Gagnon, H., and Morin, S. 1997. Listes des clones recommandés selon les régions écologiques forestières du Québec. Ministère des Ressources naturelles du Québec, Direction de la Recherche forestière. Notice d'information. 5 pp.

Zsuffa, L., Lin, D., and Payne, P. 1999. One-way crossing barriers in some interspecific crosses of Aigeiros and Tacamahaca poplars. For. Chron. **75**: 833–836.

CHAPTER 3
Physiological ecology of poplars

Donald I. Dickmann, J.G. Isebrands, Terence J. Blake, Kevin Kosola, and John Kort

...foresters must understand how trees grow, and this requires some understanding of tree physiology.

Paul J. Kramer (1986)

Introduction

The late Paul Kramer, one of the 20th century's most eminent plant physiologists, summed up well the underlying theme of this chapter. As poplar physiologists with a combined experience of over a century, we believe that the success of poplar culture requires an intimate knowledge of the subject. To understand how trees grow, a knowledge of both structure and function are equally important; physiological processes (functions) take place in the context of a specific form, so anatomy and morphology also need to be considered.

This chapter is titled "physiological ecology" of poplars for a good reason. Ecology is the study of organisms in relation to the natural environment. The physiological processes that produce growth — photosynthesis, respiration, cell division and expansion, activity of growth-regulating hormones, absorption and use of water and nutrients, movement of substances in the plant, and so on — are

D.I. Dickmann. Department of Forestry, Michigan State University, East Lansing, MI 48824-1222, U.S.A.
J.G. Isebrands. USDA Forest Service, North Central Research Station, Rhinelander, WI 54501, U.S.A.
T.J. Blake. Faculty of Forestry, University of Toronto, 33 Willcocks Street, Toronto, ON M5S 3B3, Canada.
K. Kosola. Department of Forestry, Michigan State University, East Lansing, MI 48824-1222, U.S.A.
J. Kort. PFRA Shelterbelt Centre, Agriculture & Agrifood Canada, Indian Head, SK S0G 2K0, Canada.
Correct citation: Dickmann, D.I., Isebrands, J.G., Blake, T.J., Kosola, K., and Kort, J. 2001. Physiological ecology of poplars. *In* Poplar Culture in North America. Part A, Chapter 3. *Edited by* D.I. Dickmann, J.G. Isebrands, J.E. Eckenwalder, and J. Richardson. NRC Research Press, National Research Council of Canada, Ottawa, ON K1A 0R6, Canada. pp. 77–118.

set by a tree's genes. But these processes function in a fluctuating environment and are modified by it. It is therefore crucial to understand how temperature, sunlight, day length, water and nutrient availability, relative humidity, atmospheric pollutants, and other environmental variables regulate physiological processes.

Our purpose is to summarize those aspects of poplar physiological ecology that will be most useful for growers. We believe this knowledge, combined with scientific work in genetics and breeding (see Chap. 2) and developments in equipment and technology, can lead to an "optimized" poplar culture that will take us firmly into the 21st century (Fig. 1). Since more is known about the physiological ecology and genetics of poplars — from root tip to shoot tip — than almost any other tree genus (see Chap. 10), it is possible to lay a firm scientific foundation for optimal poplar culture.

A complete review of the current knowledge of poplar physiological ecology is beyond the scope of this chapter. For those interested in a deeper understanding of physiological ecology, several books can be recommended. *Biology of Populus* edited by Stettler et al. (1996) and *Ecophysiology of Short Rotation Forest Crops* edited by Mitchell et al. (1992) both contain chapters on the physiology and ecology of poplars and other short-rotation forest crops. In-depth treatments of the physiology of woody plants are also offered in the two volumes by Kozlowski and Pallardy (1997*a*, *b*). The field of plant physiological ecology is covered by Kozlowski et al. (1991) and Lambers et al. (1998). The literature cited in this chapter and the more extensive bibliography cited in the above texts will allow the interested reader to delve deeper into the subject of physiological ecology.

Fig. 1. Knowledge of physiological ecology is the foundation for optimized poplar culture.

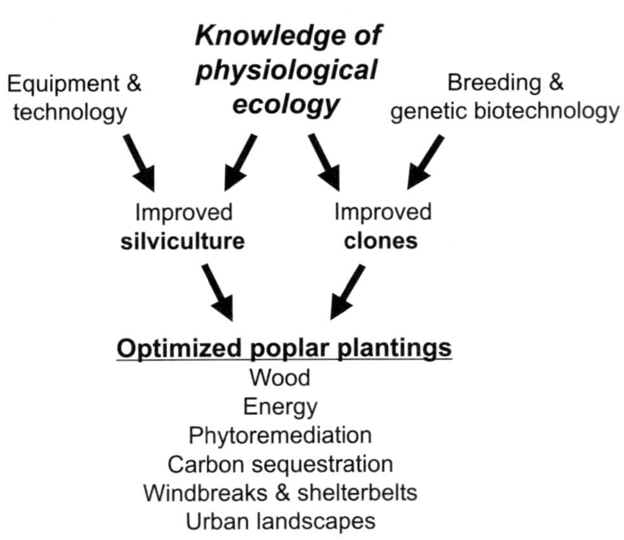

Structure and growth

Growth and morphology of the crown

Poplars reproduce both sexually and vegetatively (asexually). In nature, the life history traits of reproduction vary by species and ecosystem. In riparian ecosystems, poplars propagate by seed and asexually from abscised branchlets and sprouts from stumps or roots (Braatne et al. 1996). In upland ecosystems, seed reproduction and sprouting are equally important. Whether originating sexually or vegetatively, a young poplar propagule must quickly establish root contact in the soil, produce new leaves to produce carbohydrates through photosynthesis, and establish a shoot–root vascular connection. If any one of these fails, the propagule will die.

Poplar seeds range in size, germination time, and growth rates. The seed provides the protective coat for the embryo, which must be kept alive during seed dispersal and storage for a new plant to germinate, survive, and grow. The vascular development of the young poplar seedling begins with the embryonic cotyledons (first leaves) and proceeds to eventually form an intimate vascular connection between the shoot and the root that sets the stage for subsequent primary (extension) growth and secondary growth in girth by the cambium (Fig. 2). The vascularization of the cotyledons and the young leaves that follow them becomes the template for secondary xylem (wood) formation in the poplar seedling and,

Fig. 2. Primary–secondary vascular transition in a growing eastern cottonwood shoot. At this developmental stage, a fully functional pathway of transport between the shoot and roots is complete.

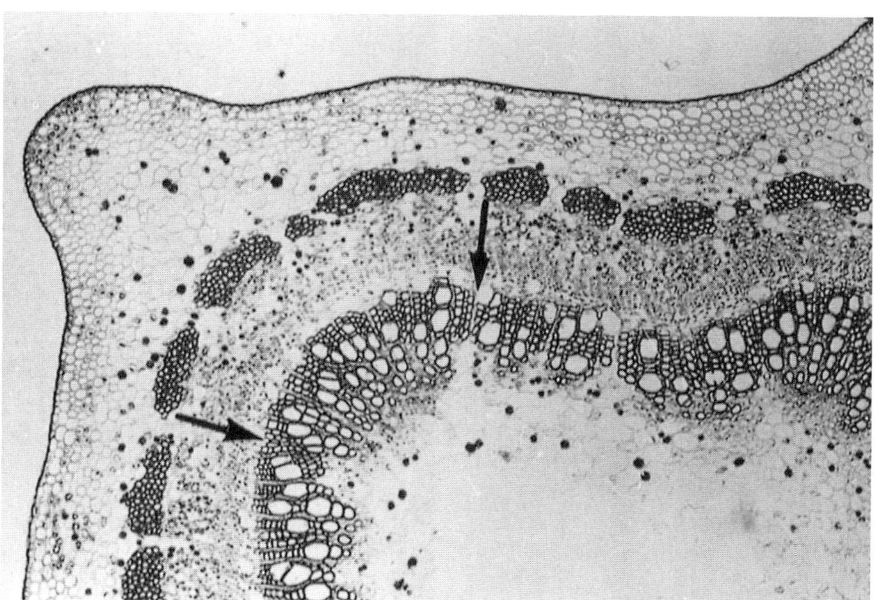

eventually, the entire tree. Thus, all of the vascular wood elements formed by the cambium of the stem (Fig. 3) and the roots are intimately connected and can be traced back to their origin in the developing leaves originating from the seed (Larson 1994).

The aboveground parts of poplar plants that propagate vegetatively from branchlets naturally, or artificially from hardwood cuttings, originate from dormant buds. Dormancy is a growth cessation period induced by prolonged cold temperatures and short day lengths (Kozlowski and Pallardy 1997*b*). In the spring, after the cold period ends and when day length increases and temperatures rise, vegetative buds on the cutting break dormancy, and the new shoot emerges from leaves preformed within the bud during summer and autumn of the previous year. Then, as in the case of the seed, the developing leaves of the new shoot initiate the vascular connection between the shoot and the stemwood of the branch or cutting, which in turn provides the connection to the adventitious roots that form. Thus, the stemwood of the branch or hardwood cutting provides the intimate connection between the developing sprout and the roots in the young vegetatively propagated poplar plant.

The morphology of leaves on a poplar plant is tied directly to their structural and functional development. The "leaf plastochron index (LPI)" is often used to quantify the morphological stage of leaf development of poplars. This index can then

Fig. 3. The secondary xylem (wood) that forms in poplar stems has its morphological origin in the vascular tissue of developing leaves.

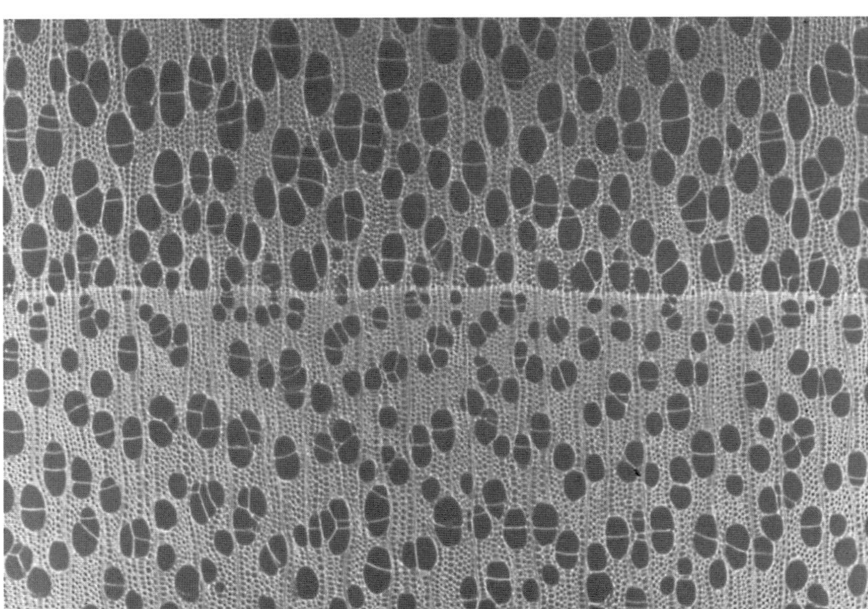

be related to structural development of the whole plant over time (Larson and Isebrands 1971). Briefly, LPI relates the morphological stage of leaves to anatomical, physiological, and biochemical processes taking place in the plant. LPI 0–7 are developing leaves, LPI 8–12 are recently mature, and leaves greater than LPI 12 are mature. These zones may comprise fewer leaves in a young plant or more leaves in a very vigorous plant. Young poplar plants show indeterminate growth and produce leaves continuously over the course of the growing season, from budbreak in the spring to bud set in the fall. This pattern gives rise to shoots made up of leaves of varying LPIs throughout the tree, i.e., different anatomical stages. This pattern distinguishes poplars from trees with determinate (sugar maple or hickory) or multiple flushing (oaks) growth patterns. The crown of a poplar tree is thereby a complex of leaves of many different ages, sizes, and stages of anatomical development, although within a clone the basic anatomical characteristics of leaves is similar. Leaf characters, however, can vary substantially among clones and species within the genus *Populus*.

The vertical profile of the crowns of hybrid poplar trees has been studied extensively (e.g., Heilman et al. 1996). Indeterminate long shoots, sometimes several meters in length, with distinct leaf internodes occur primarily in the upper crown. Determinate short shoots, at the extreme consisting of only a whorl of small leaves, occur in the lower crown. From the top to the bottom of the crown, the length and type of shoots intergrades between these two extremes, with many intermediate forms present. Tree age also affects shoot distribution; the crown of older trees comprises a larger proportion of determinate shoots than young trees. In fact, an old tree may consist of nearly 100% determinate shoots, although those in the upper crown may grow many centimeters in length in a season. These shoot types are physiologically important because leaves of different sizes, orientation, structures, and morphology occur on them. Leaves borne on long shoots in the upper crown are large, thick in cross-section, highly angled from the horizontal, and show high rates of photosynthesis per unit of leaf area. Short-shoot leaves are small, thin, nearly horizontal in orientation, and have relatively low rates of unit-area photosynthesis. Leaves in the upper crown also are the last to abscise in the autumn.

Leaves on long shoots contribute the most to growth of a poplar tree because of their long duration, abundance, large area, and high photosynthetic rate. Thus, geneticists and silviculturalists should aim to maximize the proportion of long shoots in the canopy of a poplar stand, whenever possible. Because hybrid poplars have a higher proportion of long shoots to short shoots than most native poplars and aspens, they often have a more rapid growth rate (Isebrands et al. 1983). Growth rate of poplars has been shown to be correlated with leaf size; clones with large leaves often grow faster. For example, a striking feature of highly productive *P. trichocarpa* × *P. deltoides* hybrids in the Pacific Northwest is their high number of long shoots and enormous leaves. Our emphasis on long shoots, however, is not meant to diminish the importance of other parts of the crown; the tree is an integrated whole and all parts of the crown are important.

Branch angle, size, and number also vary by clone and species, and they play an important role in determining crown architecture and canopy density of poplar stands. Many poplars produce both proleptic and sylleptic branches. Proleptic branches are formed from buds that have undergone a winter dormant period. In contrast, sylleptic branches emerge from buds during the same growing season in which they were formed, i.e., without an intervening dormant period (Fig. 4). There is significant variation in the number of sylleptic branches within the crowns of different poplar clones. Those clones with high numbers of sylleptic branches usually are more productive in the early years of the rotation (Scarascia-Mugnozza et al. 1999). Thus, branch composition of poplar trees has important implications for solar radiation interception and productivity. Branch characteristics, therefore, are important traits used for early selection of superior clones in breeding programs and for silvicultural manipulation of poplar stands.

Architecture of the root system

The root systems of poplars are especially complex in comparison with other tree genera. As with any poplar trait, a pronounced genotypic effect is evident; different species or clones of the same species or parentage differ in the number, size, depth, structure, and orientation of roots. An additional source of complexity is that poplar roots can originate in many different ways — from the radicle of a seed, from a cutting or abscised branch, or, in the case of suckers, from an existing root system. Species like *P. nigra* or *P. balsamifera* readily produce new roots in all three ways.

A germinating seed develops a root system that originates from the radicle of the seed embryo; in this respect, poplars are not different from other plants. What makes poplars unique is the variety of ways clonal plants can be produced. Vegetative propagation in poplars outside section *Populus* occurs by the formation of adventitious roots on the propagating plant part. Abscised twigs of riparian poplars form roots at their base if they are buried in riverbank silt or a sand bar, producing a new clonal plant by the process known as "cladoptosis." Humans make use of the predisposition of most poplars to form adventitious roots by taking hardwood stem cuttings or, less commonly, softwood (leafy) tip cuttings and rooting them. In contrast, clonally propagating aspens do not need to form a new root system at all; new sucker shoots spring from the fully functional root system of parent trees that have been cut or killed by a natural disturbance.

Populus is among the few tree genera that readily produce adventitious roots on hardwood cuttings (Fig. 5). In fact, this characteristic is one of the primary reasons why these trees are used extensively in plantations throughout the world (Zsuffa et al. 1996). Adventitious roots grow from preformed root primordia (also called latent root primordia) that develop in the inner bark of the stem; i.e., the beginnings of roots are created as stems develop. But these primordia will not grow unless they are placed in a cool, dark, moist environment such as the soil. Formation and growth of these primordia are thought to be stimulated by high

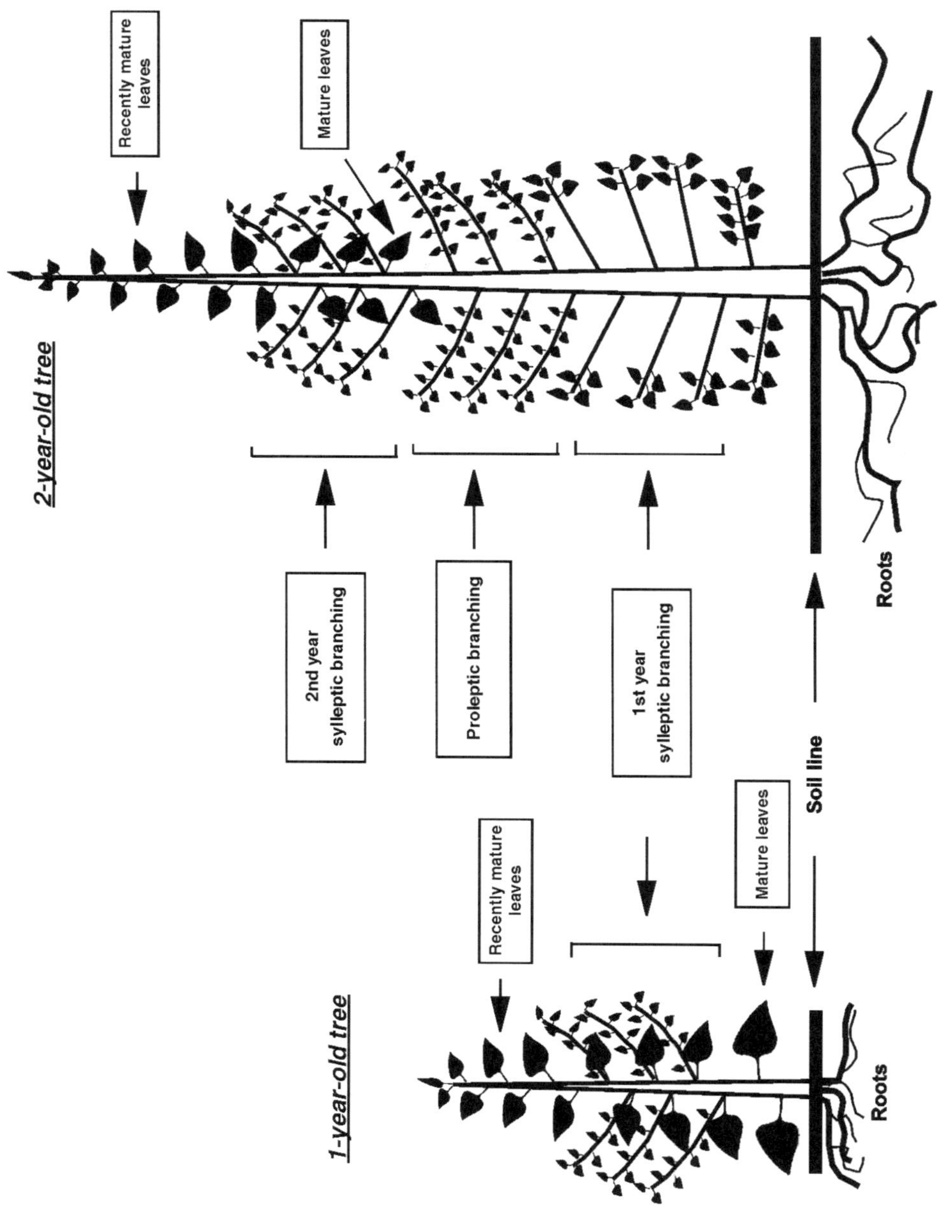

Fig. 4. Leaf maturity classes, proleptic branches, and sylleptic branches in 1- and 2-year-old hybrid poplar trees.

Fig. 5. The adventitious roots produced by a hardwood poplar cutting arise from preformed primordia initiated during stem growth and from basal callus tissue. Photo by Jon Johnson.

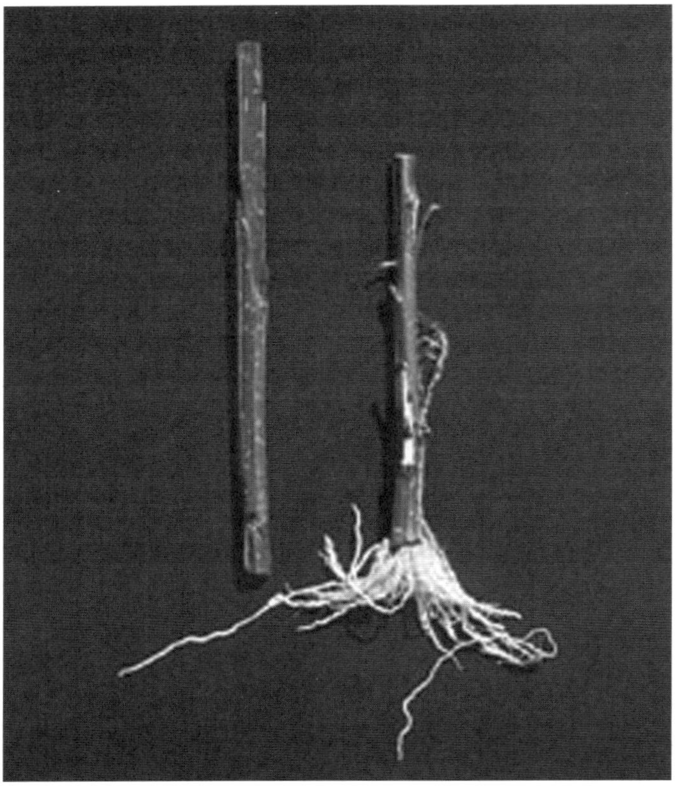

levels of the hormone auxin in the stem. Basal cuttings are the best to use for propagation, since the number of root primordia declines from the base to the tip of a 1-year-old shoot (Smith and Wareing 1974).

Adventitious roots also will form in the white, grainy callus tissue that grows at the base of a planted cutting. Adventitious rooting of poplar cuttings is under strong genetic control. Hardwood stem cuttings of species and hybrids from section *Populus* generally do not root (certain genotypes of *P. alba* are the exception). Even in sections *Tacamahaca* and *Aigeiros*, where stem cuttings generally produce many roots, considerable clone-to-clone variation in the extent and vigor of rooting occurs. Some clones of eastern cottonwood, for example, hardly produce any adventitious roots, while other clones root well.

The new roots produced adventitiously by a poplar cutting provide a large, succulent absorbing surface. These roots expand vigorously into the soil while the first leaves are flushing out of the terminal buds on the cutting. Although some of these initial roots may die, many continue to elongate to form the initial fine-root network. For example, in the Pacific Northwest U.S.A. Heilman et al. (1994)

found that planted cuttings of *P. deltoides*, *P. trichocarpa*, and their hybrid had formed between 30 and 50 first-order roots — i.e., roots attached directly to the cutting — by the end of the first growing season. These roots varied in length from 40 cm to over 1 m in length, and the majority of them was produced from the base of the cutting. Most of the roots that survive the first growing season become suberized (i.e., impregnated with suberin, a waxy substance) and undergo secondary thickening in subsequent years, forming the structural architecture of the root system (Fig. 6). The cambium of the cutting continues to divide after planting, producing growth rings and an increase in girth, while vertical sinker roots from basal callus elongate and thicken, forming one or several taproots (Fig. 7). The development of the root system is highly dependent upon the properties of the soil. For example, an impermeable hardpan layer or a water table close to the surface will impede vertical root penetration.

Fig. 6. Five-year development of the taproot of a Euramerican hybrid poplar tree established from a hardwood cutting in the nursery and then transplanted as a 2-year-old rooted plant into the field. From Faulkner (1976).

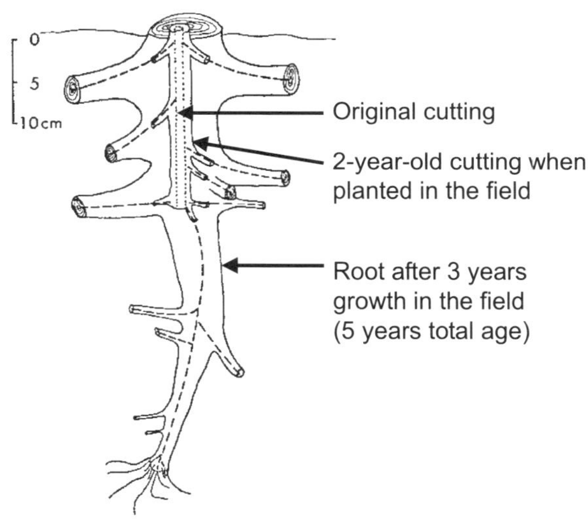

Fig. 7. Structure of the root system of a 5-year-old Euramerican hybrid poplar established in the field from a rooted hardwood cutting. From Faulkner (1976).

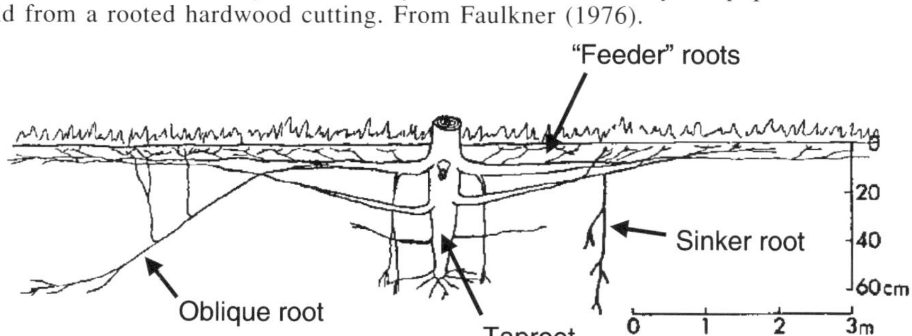

As the shoots produced by a cutting grow in height, there is a proportional extension of the horizontal root system. Growing strongly away from the taproot are coarse horizontal roots, many located between 5 and 20 cm from the soil surface (Fig. 6). Horizontal roots can be found several tree lengths away from the base of the stem in a planted stand, although the closer the spacing of trees, the more restricted lateral root growth becomes. Many of the roots from adjacent trees in a clonal stand become grafted together, forming an interconnected network. Vertical "sinker" roots branch from the horizontal roots and explore the soil to depths of 1–3 m or more. Poplar trees have at least four orders of root branching (i.e., branches upon branches upon branches upon branches), although this aspect of root morphology is not well understood (Pregitzer and Friend 1996). Different poplar clones produced from rooted cuttings have a similar root architecture, although clonal variation has been observed (Dickmann and Pregitzer 1992). These differences are mainly related to the relative allocation of carbon to the root system, morphological characteristics, and the longevity of individual roots.

The root system of a seedling poplar begins in an entirely different way than one produced from a cutting. Because a stem cutting has considerable starch, sugar, and protein reserves, and leaves develop from preformed buds very quickly, abundant energy and carbon compounds are available to fuel rapid root growth. On the other hand, a tiny poplar seed has virtually no reserves and it takes some time for leaves of any size to develop. So initial root growth is slow. Nonetheless, a young seedling invests much of its limited photosynthetic capital into root growth; without early deep penetration of roots into the soil the seedling is doomed. Mahoney and Rood (1991), for example, found that 46-day-old *P.* ×*jackii* seedlings with only 1 cm^2 of leaf surface already had roots 17 cm long. At the end of the first growing season, these roots may grow to 1 m or more in length (Braatne et al. 1996). Nonetheless, of the countless poplar seedlings that germinate, few survive past the first few months of life, unless the weather is especially cool and moist (Fig. 8). Mortality is especially high on droughty upland sites. After several years of growth, a seedling-derived root system will look similar to one derived from a cutting.

The root system of clonal aspen stands is different from other poplars, especially after several clonal generations. The individual stems in an aspen sucker stand, which at the extreme may encompass an area many hectares in size, exist on a root system that is highly interconnected, at least for the first few decades. For example, Barnes (1966) excavated the root system of a Michigan aspen clone and found that 70% of suckers were growing on seven different branched root systems; up to 15 suckers were located on the same root system. In Utah, as many as 43 aspen sucker stems were connected on a single root system (Tew et al. 1969). These root interconnections gradually disintegrate, however, so that the individual stems of a mature aspen clone largely exist on their own independent root system.

Fig. 8. Natural eastern cottonwood seedlings established along the bank of a creek in Burlington, IA.

Aspen root systems are unique in a second way. The horizontal root of the parent tree on which a sucker forms becomes enlarged on the distal side; i.e., the side facing away from the parent tree. The root connection on the proximal side (toward the parent tree) remains small, and it eventually withers and dies (Fig. 9). The resulting J-shaped root structure is characteristic of clonal aspens. Suckers do not produce distinct taproots, although sinker roots are formed from horizontal roots and at the base of the sucker.

The fine absorbing "feeder" roots of most plants are concentrated near the mineral soil surface or in areas of high availability of water and nutrients, and poplars are no exception. Most fine poplar roots are in the top 10 cm of the soil where they form an extensive network (Fig. 7). But soil drainage, texture, and profile characteristics markedly influence the exact distribution of tree roots. For example, roots will proliferate in moist or highly fertile patches or layers in the soil. Management practices also can have a profound effect on the root system. Fertilization, especially with nitrogen, causes a shift of carbon allocation to the tops of trees, decreasing the relative proportion of roots. Irrigation also markedly affects roots. For example, roots will concentrate near the emitters in a trickle irrigation system, especially if the water contains dissolved fertilizer. Roots also will be particularly dense near the soil surface if sprinkler or flood irrigation is

Fig. 9. Excavated root system of a bigtooth aspen. The side that is distinctly swelled faced away from the parent tree. The root that connects the sucker to the parent tree eventually will wither and die.

employed. Cultivation disrupts the surface root system and can negate some of the benefits of weed control; therefore, only shallow cultivation should be used. Finally, the heavy equipment used in the establishment and tending of poplar plantations can lead to soil compaction and restricted root distribution, especially on soils with a high clay content.

Some poplars allocate more of their carbon resources to roots than others. In addition, some clones are very "plastic" in root production from one environment to the next, whereas others produce about the same proportion of roots regardless of the environment (Nguyen et al. 1990; Pregitzer et al. 1990). Generally, the harsher the environment the greater the proportion of total tree biomass contained belowground. But a note of caution is necessary here. Methods vary in their effectiveness in recovering roots, and many studies report data on root weights based on only partial root excavation (e.g., Friend et al. 1991). Therefore, data on root biomass or root:shoot ratios should be viewed somewhat skeptically. Fine-root length densities (root length per unit of soil volume) or fine-root length per unit of leaf area may be much more meaningful expressions of the physiological status of poplar root systems (Pregitzer et al. 1990; Pregitzer and Friend 1996).

Dynamics of fine roots

Fine "feeder" roots are small diameter (ca. 0.1–1 mm), relatively short-lived roots that function primarily in the uptake of water and nutrients. Young succulent white roots (Fig. 10) are thought to be the most important class of roots for absorbing water and for nutrient uptake. Older, brown roots (often referred to as "suberized") are considered to be less permeable to water and nutrients, although they are more abundant, especially during the dormant season. The balance of total root length (or surface area) in these two stages of development can have significant effects on the ability of the tree to acquire water and nutrients.

Fine roots grow rapidly (>10 mm per day) when moisture and fertility are high. Although fine roots constitute only a small percentage of total tree or stand biomass at any one time, they can account for a significant proportion of the biomass produced each year — more than 50% on stressful sites. Trees invest much in maintaining their fine root system; substantial amounts of carbon and nutrients are used by fine roots for growth, respiration, maintaining mycorrhizal fungi, uptake and transport of nutrients, and production of growth regulators. The fine-root network of a poplar tree in the upper soil horizon is extensive; fine-root length densities of planted poplars have been measured in the range of 2.4–6.3 cm of roots/cm^3 of soil volume (Heilman et al. 1994). These values are unusually high for trees and partially explain why poplars grow so fast.

Fine roots of poplars usually are infected by fungi that form mycorrhizae. Both ectomycorrhizal and endomycorrhizal (vesicular-arbuscular or VAM) associations are found (Fig. 11). With advancing stand age and the buildup of soil organic matter, infection tends to shift from endo- to ectomycorrhizal (Heilman et al. 1996). Although the additional cost of maintaining mycorrhizal vs. nonmycorrhizal roots has yet to be fully established for most poplars, respiration of heavily infected herbaceous root systems may be as much as 75–100% higher than non-infected roots (Harley and Smith 1983). As a result, three to four times

Fig. 10. Life history options for fine roots produced by a poplar seedling or cutting.

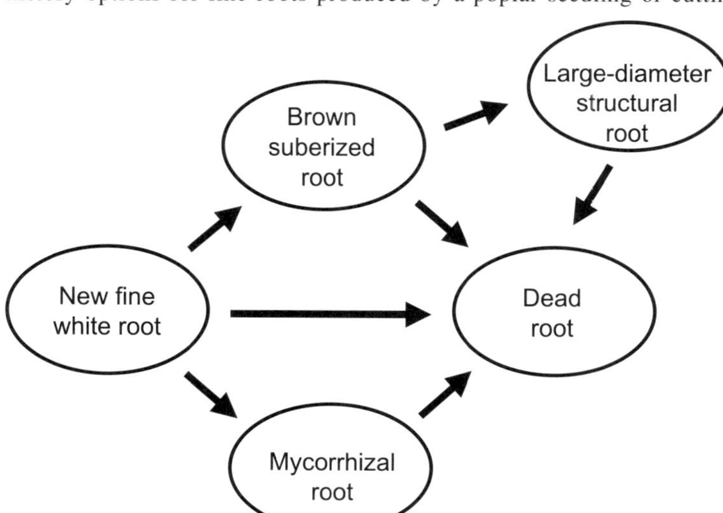

Fig. 11. Euramerican hybrid poplar fine roots showing both ectomycorrhizal (club-shaped lateral) and endomycorrhizal (main root segment on left) infections. Blue stain indicates fungal hyphae.

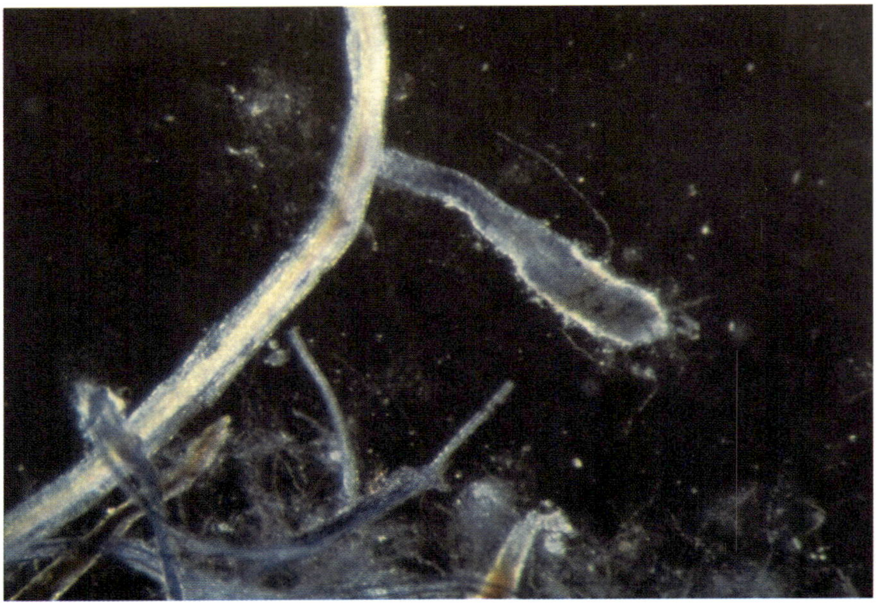

as much carbon can be allocated to root systems of mycorrhizal compared to non-mycorrhizal plants. These costs are not without benefit to the tree. The thread-like fungal hyphae of mycorrhizal roots effectively scavenge a greater volume of soil for water and mineral nutrients than non-mycorrhizal roots. Infection of fine roots by ectomycorrhizal fungi also can deter invasion by pathogenic fungi.

Production of new fine roots is a very dynamic process. New roots are rapidly produced as the root-absorbing area rapidly expands during the juvenile phase of growth, but production declines and equilibrates to some extent as trees mature. Seasonal variation in root production rates are pronounced and correlated with aboveground growth. A substantial amount of fine root production occurs before and soon after budbreak in hybrid poplars; this spring flush of new root growth is characteristic of many trees (Dickmann and Pregitzer 1992). During summer, root production rates fall off, especially during drought periods, but a minor peak of production may again occur in the fall. While new fine roots form, some existing roots die in a process known as root turnover (Fig. 10). The death rate pattern of fine roots tends to be the reverse of root production; low in the spring, increasing over the course of the growing season, and reaching a peak in early fall. Drought and other stresses tend to increase fine-root death. Little root production and death occur during the winter months. The very youngest roots die most frequently, probably because they are most susceptible to pathogenic fungi and grazing by soil arthropods and other root herbivores.

Fine-root production and growth are highly dependent on current photosynthesis in rooting cuttings or young trees. But roots become less dependent upon current photosynthesis and more dependent on local reserves of starch and sugar as a substantial root-storage mass develops as trees become older. Nonetheless, depleted root reserves eventually have to be restored by downward translocation of sugars produced in photosynthesis in order for the root system to continue normal functioning. The accumulation of root reserves has great significance to coppicing. For example, Dickmann et al. (1996) found that a strong spring flush of new fine roots occurred in hybrid poplars even though the stems had been cut back to the ground in early March. These new roots are important in sustaining rapidly growing coppice sprouts.

Carbon physiology

Light energy is intercepted by chlorophyll in poplar leaves and used to produce sugars from carbon dioxide (CO_2) and water by the process of photosynthesis. Net photosynthesis of a stand is the product of the photosynthetic rate — measured by the net uptake of CO_2 — of the entire surface area of individual leaves in the stand integrated over the growing season. Net primary production is the balance between net canopy photosynthesis and the quantity of carbon lost to respiration by all parts of the trees in the stand (Dickson 1989). Rapid growth and high production of poplar stands depends, therefore, upon maintaining a positive net primary production — i.e., an accumulation of carbon — in a changing environment.

Poplars have shown measured rates of photosynthesis that are among the highest in trees (Nelson 1984). Photosynthetic activity varies significantly among the leaves of a poplar tree, depending upon leaf position, orientation, size, and age, as well as time during the growing season. Whole leaf photosynthesis rates are greatest in the uppermost parts of the crown where light levels are high. Poplars attain maximum photosynthetic rates early in the growing season and early in the rotation, as they develop significant quantities of leaf area per unit land area rapidly. There are small differences among poplar clones in photosynthetic rates per unit of leaf area, but large differences in whole-leaf and whole-tree photosynthesis due to the size and duration of a clone's individual leaves and canopy. Poplars are known to photosynthesize late into the autumn, especially in the current terminal shoot and upper branches, where leaves are last to abscise. Clones with high whole-tree photosynthesis generally produce more wood or biomass. Thus, by manipulating photosynthesis variables, coupled with canopy characteristics, geneticists and physiologists can improve biomass production (Isebrands et al. 1983).

Carbon allocation is the process of carbon flow through the phloem from the leaves to the other components of the plant — branches, bark, wood, and roots (Ceulemans and Isebrands 1996). Carbon is transported in the phloem in the form of simple sugars, primarily sucrose. Carbon partitioning refers to the process of carbon transformation among the different chemical fractions within the tree — carbohydrates, amino acids, cellulose, lignin, fats, etc. Carbon allocation and partitioning patterns in poplars vary with the developmental stage of the leaves (Fig. 12). Developing leaves, i.e., those that have not reached full maturity, transport carbon primarily upward to younger developing leaves and the apex. Recently mature leaves transport carbon both upward and downward to the lower stem and roots, while older mature leaves transport primarily downward to the stem and roots. These patterns explain why premature defoliation of older leaves in poplars by insects, pathogens, or air pollutants causes a decrease in root growth and storage reserves in the autumn. Poplars are known to compensate for biotic and abiotic stresses with increased photosynthetic rates. However, if the stress condition continues too long, a poplar tree will not maintain the positive carbon balance necessary for substantial growth that season. Because roots are the biomass component that is farthest from the leaves, they are the most vulnerable to biotic and abiotic agents that disrupt canopy photosynthesis.

Carbon allocation and partitioning patterns are more complex in large trees because they have multi-branched canopies, and their interactions with environmental variables and silvicultural practices are more complicated. Carbon allocation from poplar branches is largely downward to the stem and roots (Fig. 12). Little carbon is transported to other branches, and almost no carbon is transported from branches upward to the current main-stem terminal and expanding upper shoots. Sylleptic branches are important contributors to early growth in poplars because they provide additional photosynthetic carbon for stem and root growth. There are substantial differences among poplar clones in the number of sylleptic branches, but those with many sylleptic branches usually are more productive (Scarascia-Mugnozza et al. 1999).

Poplars store reserves when excess carbon production occurs; i.e., when the seasonal carbon balance is highly positive. Carbohydrates and lipids are the major storage compounds of carbon in poplar. Reserves begin to build up in late summer, reach their highest in the autumn, and then are depleted in spring. Starch is the primary storage compound initially, but in late fall much of the starch is converted to sugars (Nguyen et al. 1990). These sugars are important in the development of winter hardiness. Winter hardiness also is presumed to be associated with buildup of growth inhibitors such as abscisic acid (ABA) or root to shoot feedbacks involving certain growth regulators. Poplars also store protein during the winter, primarily in twigs (Sauter et al. 1989). Much of this protein is derived from the nitrogen that is transported back into the stem from senescing leaves in the autumn. Reserves are normally used for the initial surge of growth that occurs in the spring — reserves in twigs for shoot growth from buds, reserves in the trunk for growth of the cambium, and reserves in the roots for fine-root growth. When the first leaves are fully expanded, they take over as the major carbon sources for growth. Reserves also serve as a hedge against catastrophic injuries such as defoliation by biotic agents, top kill by fire, wind, ice, or snow breakage, or felling by beavers or humans.

Fig. 12. Allocation of photosynthetically fixed carbon before and after bud set in two clones of *Populus trichocarpa* × *P. deltoides* during the first growing season. Recently mature, mature, and sylleptic leaves (large arrows) were treated with $^{14}CO_2$, and translocation of the radioactive carbon was followed. Note that lower mature leaves in clone 11-11 had abscissed prior to the "after bud set" treatment.

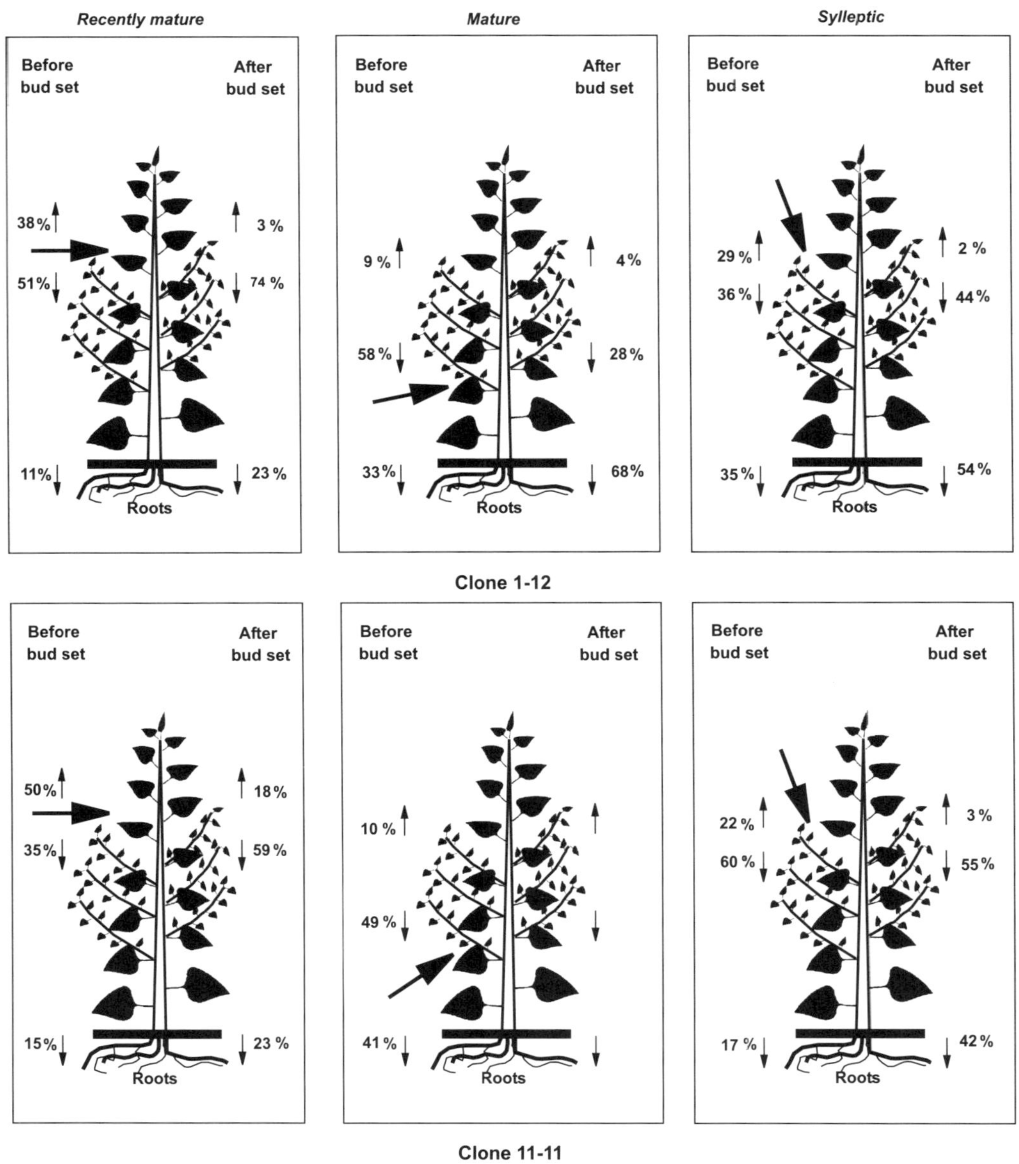

Effects of biotic and abiotic stresses

Poplars can withstand a single moderate defoliation by insects and pathogens without major growth losses, but repeated defoliations over several years can have devastating impacts. For example, after three successive years of early season defoliation by the gypsy moth (*Lymantria dispar*), clonal stands of *P.* ×*canadensis* 'Eugenei' showed top dieback, substantial reductions in diameter growth, and tree death. However, trees were able to recover from a single defoliation quickly. Carbohydrate levels in the twigs, trunk, and fine roots were depleted as trees adjusted to the defoliation, but they were back up to levels near those of undefoliated trees by the end of the season due to the high photosynthesis rate of the new leaves that formed after gypsy moth larvae completed their life cycle (Kosola et al. 2001). Defoliation by Melampsora leaf rust, which begins in midsummer, can be especially troublesome in high-yield plantations. Volume growth losses of 50–65% in rust-susceptible clones have been recorded (Widin and Schipper 1981), and tree death can occur after repeated defoliations.

Atmospheric pollutants also can substantially alter the carbon physiology of poplar trees. Elevated tropospheric ozone (O_3) causes a reduction in the photosynthesis rates of mature leaves of poplar clones and causes leaves in the lower canopy to prematurely senesce and die. Thus, overall whole-tree photosynthesis for the season is decreased. Moreover, this premature defoliation decreases normal carbon allocation from the lower leaves to the basal portions of the stem and to the roots so that seasonal diameter and root growth of trees decreases. Generally, elevated CO_2 increases photosynthesis rates and leaf area in poplars, thereby increasing whole-tree photosynthesis and subsequent growth of the trees. However, when poplars are exposed to elevated CO_2 and O_3 simultaneously, as they may be in future climates, the positive effect of CO_2 on growth is offset and often negated by O_3. In addition, the incidence and severity of stresses caused by poplar insects, diseases, and drought also may increase under elevated interacting CO_2 and O_3. Thus, in the future the overall effect of interacting stresses on tree growth may result in lower growth rates for many aspen and hybrid poplar clones (Isebrands et al. 2000). Certain poplar clones are more tolerant to air pollutants than others (Dickson et al. 1998), so selection for tolerance to atmospheric pollutants or interacting stresses may be very important as new generation clones are developed and tested by genetic improvement programs.

Coppice physiology

All poplars have the ability to re-sprout after sudden top dieback or harvest. Some poplars sprout from the stump (cottonwood poplars), some from roots (aspens), and others sprout from both the stem and roots (black and white poplars). Coppicing is a silvicultural method based on cutting mature trees near the base of the stem and relying on the new sprouts that will emerge to reproduce the stand. Coppicing also can occur when the tops of poplar trees are killed by natural biotic or abiotic factors. Pollarding is a similar method except that the stem is cut higher from the ground. Both methods have been used by farmers around the world for millennia for producing fuelwood, fence posts, and other small-diameter products. Furthermore, foresters throughout North America employ coppicing as a standard silvicultural practice for regenerating aspen (see Chap. 4). Although numerous physiologists have studied sprouting, the physiology of coppice still is not totally understood.

Coppicing affects the overall balance between the above- and below-ground parts of the poplar tree in more ways than just by altering the distribution of biomass. Coppicing eliminates the supply of photosynthetically produced carbohydrates, growth hormones, and inhibitors supplied by the crown to the root. Neither the growth rate of coppice sprouts nor the number of sprouts produced, however, is closely related to the starch reserve content in the roots. Sugar reserves in the roots, which are a small part of the total reserve content, are what fuel initial sprout growth. The number of sprouts is dependent on the quantity of suppressed buds that form at the root collar prior to coppicing, on the number of callus buds that form on the surface of a stump, or on the number of adventitious buds produced by horizontal roots. But the subsequent growth of the sprout and replenishment of depleted root reserves is highly dependent on an increase in the supply of carbohydrates from photosynthesizing leaves that form on coppice shoots. These leaves are larger, their photosynthetic rates are substantially higher, and they export a larger proportion of their photosynthetically fixed carbon than those on intact plants (Tschaplinski and Blake 1989). In addition, coppice shoots draw upon the abundant water, mineral nutrients, and cytokinin hormones supplied by the parental root system. If trees are cut back frequently or if they are cut during the summer, however, the increase in water stress that results can lead to the decline and eventual death of sprouts and their supporting root systems. Thus, a high carbon acquisition rate and abundant supplies of water and cytokinins from existing roots explain why coppice sprouts grow so rapidly, especially during their first year.

The major increase in root-to-shoot ratio that occurs following coppicing also explains the rejuvenation events that occur in the plant. Coppicing completely removes competing apices in the crown, but a fully functioning root system is left intact (Dickmann et al. 1996). Therefore, water stress is relieved, a large supply of essential nutrients is available, and growth-promoting hormones from the root are abundant, causing rejuvenated growth of sprouts. In certain cases, the growth of these sprouts approaches the theoretical maximum for a species (Blake 1983).

Water relations

The importance of water to a plant lies in its role in cell expansion, cooling of leaves, nutrient and carbon transport, photosynthesis, and as a solvent or reagent in key metabolic reactions. Water limits growth more than any other environmental variable. Whereas studies have established strong correlations between the rate of process variables and dry matter accumulation in irrigated and fertilized nurseries, it is much more difficult to relate physiological parameters and growth rate in plantation-grown trees, due to the confounding effects of water deficits (Blake et al. 1995). Water potential[1] and other water relations parameters usually correlate poorly with growth in the field where water stress limits growth on a daily basis (Blake et al. 1995; Bevilacqua and Blake 1997). There are a number of reasons why it is difficult to correlate growth in the field, which is cumulative, with physiological parameters, which are measured instantaneously. Therefore, physiological adjustments of plants to chronic water stress are complex and poorly documented.

Water economy

Plant productivity can be related to water use by considering "water use efficiency" (WUE), the increase in growth per unit of water use (sometimes the reciprocal also is used). WUE can be instantaneously measured in leaves as the rate of CO_2 uptake by photosynthesis per unit of water lost in transpiration. Whole-tree WUE based on weight change is usually much less than instantaneous foliar WUE. Depending on which method is used, factors such as diurnal variation in root respiration, relative carbon allocation to roots, and turnover of fine roots and leaves influence WUE. Nonetheless, WUE remained relatively constant in three hybrid poplar clones — 3.5–4.4 g of dry biomass per liter of water consumed — for a range of soil moisture contents, despite marked variation in root–shoot ratios (Souch and Stephens 1998). Because most of the carbohydrates transported to roots are consumed in other processes (Ericsson et al. 1996), they do not contribute to net biomass accumulation, and, therefore, WUE.

Although stomata influence both the rate of CO_2 uptake and water loss by leaves, the relationship between whole-tree WUE and leaf-level WUE is complex. When stomata close under soil drought, water loss declines more than CO_2 uptake, which increases WUE (Farquhar and Sharkey 1982). Since the allocation of carbon to roots increases under soil drought (Kramer 1983), an enhancement of whole-plant water economy is not necessarily reflected in aboveground WUE alone. An enhancement of fine-root growth eventually increases the water available aboveground, but the resulting increase in shoot growth may not be observed for some time.

The biomass (total or incremental) in a tree can be used to calculate its water use, as shown in eq. [1]. If whole-tree WUE remains constant in a particular clone,

---

[1]Water potential, often abbreviated with the Greek letter psi (Ψ), is a measure of the water status of a plant or plant part. Values of water potential are expressed in units of pressure (MPa) and are always negative. As a plant becomes water stressed, the solutes in its dehydrating cells become more concentrated and the cells become more flaccid. As a result, water potential decreases, i.e., it becomes more negative.

total and incremental biomass accumulation can be calculated from total and incremental water use. Assuming a WUE of 250 L/kg and applying this value to mature trees, water use can be estimated from biomass, as follows:

$$[1] \qquad W = 250(B)$$

where W is the amount of water used in L, and B is total-tree biomass in kg, including roots. The total (above- and below-ground) biomass (B) for a plantation-grown poplar tree can be calculated from diameter at breast height (DBH) in cm and height of the tree (H) in m, as follows:

$$[2] \qquad B = 0.0172(\text{DBH})^2 \times H$$

The coefficient 0.0172 in eq. [2] is derived from biomass data in *A Grower's Guide To Hybrid Poplar* (Boysen and Strobl 1991), and assumes that a constant 30% of tree biomass occurs in the roots, which may not always be the case. Drought-adapted clones are more effective in allocating carbon to roots under drought. Thus aboveground WUE would be higher for less well-adapted clones with a lower root-to-shoot ratio even though overall WUE is the same for both.

The value of high WUE to survival is clear, yet clones with high WUE are not necessarily the most productive. The ability to minimize water loss because of small stomatal size, number, and aperture, for example, may serve a plant well under drought conditions. These same traits will limit growth under well-watered conditions by reducing the availability of CO_2 for photosynthesis. Other physiological or morphological characteristics may also compensate for low WUE. For example, Dickmann et al. (1992) showed that the highly determinate poplar clone 'Tristis' was considerably less productive than the indeterminate clone 'Eugenei,' even though Tristis had a consistently higher instantaneous WUE. In other words, the longer shoot-growth period of Eugenei was more important than the higher WUE of Tristis.

Poplars are thought to consume large amounts of water to support their rapid growth rates, but close analysis of their water use disputes this contention. The stomata of poplars tend to open wide, so they conduct water vapor more efficiently than most hardwood and coniferous trees or agricultural crops. Poplar stomata, however, are poorly coupled to the atmosphere, i.e., even under conditions of intense solar radiation, warm temperatures, and brisk wind the unique characteristics of poplar canopies moderates the amount of water actually lost through transpiration. For example, Hinckley et al. (1994) calculated the maximum water loss from a stand of *Populus trichocarpa* × *P. deltoides* in late summer of their fourth growing season to be 4.8 mm/day, about the same as Douglas-fir. The actual weight of water lost by these trees varied from 20 to 51 kg/day, depending on tree size. The data of Hinckley et al. (1994) were recalculated by Braatne (1999) on a yearly unit-area basis (Table 1). These data clearly show that the use of water by poplars is no more profligate than annual or perennial agricultural crops. Clonal variation in water use, however, is significant. Bassman (2000)

Table 1. Water use by annual and perennial agricultural crops and hybrid poplars in eastern Washington (from Braatne 1999).

Crop	Estimated yearly water use (mm/ha/year)
Annual vegetable and grain crops	1505–2260
Alfalfa	1755–2825
Apples (with cover crop)	2135–3135
Hybrid poplars:	
1st year	625–880
2nd and 3rd year	1380–1630
4th year to harvest (7th or 8th year)	2010–2260

showed an almost two-fold difference in water use over an 85-day period among three hybrid poplar clones grown under irrigation in eastern Washington.

Root–shoot relations

When trees are exposed to drought they close their stomata, shed leaves, and increase root growth to postpone dehydration. Because shoot growth is inhibited as water potentials decline during a drought (i.e., become more negative), water deficits actually help to coordinate growth within a tree. The importance of internal water deficits as a correlating mechanism in trees was first proposed by 19th century botanists, who described a Darwinian "struggle for existence" among branches of a tree. Water potential in lateral shoots is always less than the main stem, particularly in the lower crown. Because branch tips in the lower crown have the lowest water potentials, drought-induced water stress commences there and progresses inwards towards the trunk, upper branches and, finally, to the terminal shoot. This progression can be explained by higher frictional resistance to water flow (i.e., lower hydraulic conductivity) in the xylem of branches compared to the main stem and hydraulic constrictions at branch nodes. Thus, during the early stages of a drought, water availability first starts to decline in the lateral shoots, which allows the main stem to continue to grow. This increases the dominance of the growing terminal apex over lateral branches (Zimmerman 1978, 1983). Foliage in the lower crown is the first to be shed during a drought partly because stomatal regulation of water loss is poor there; i.e., stomata of lower leaves do not close at low water potentials (Tschaplinski and Blake 1985). As trees become larger in size, water deficits cause fluctuations in shoot growth, which slows or ceases when water loss by transpiration exceeds the ability of roots to take up water.

A moderate decline in water potential to about –1.0 MPa may close stomata, but more severe moisture stress (–1.2 to –1.6 MPa) causes photosynthesis to cease (Kramer 1983). Under drought, phloem transport continues until photosynthesis reaches quite low levels (Ericsson et al. 1996), which tends to shift carbon allocation away from new leaves and branches and towards root initiation. Accumulation of ABA under drought also favors root initiation but inhibits root and shoot elongation. Finally, water potential in a plant comprises a gradient from least negative in the roots to most negative in the uppermost shoots. These reactions may explain why root growth is less sensitive to drought, compared to shoot growth, especially in drought-adapted clones.

The regular loss of branches and leaves in trees during a drought also occurs because of cavitation, i.e., the vessels in the wood fill with air and embolize. This situation is serious because cavitated vessels are unable to conduct water. Cavitation of the xylem appears to be at least partly reversible in the lateral shoots of moderately stressed trees, but it is irreversible under more severe stress. Cavitation commences first in the lower crown — another reason why lower leaves yellow and abscise first under drought — then progresses to the upper stem, while the growing apex is the last to cavitate. This mechanism preserves the dominance of the main-stem over the lateral branches under drought and ensures the survival of the growing apex. Tyree et al. (1994) found that 1-year-old stem segments of *P. deltoides*, *P. balsamifera*, and *P. angustifolia* collected from riparian sites in Alberta were more vulnerable to cavitation than any other tree species measured up to that time. Drought-induced cavitation was thought to be the cause of the widespread decline in riparian cottonwoods observed downstream of dams throughout western North America. Variation in cavitation can occur among genotypes of the same poplar species. Black cottonwood populations from hot, dry environments on the east side of the Cascade Mountains in the state of Washington were more resistant to drought-induced cavitation than those from humid coastal areas (Sparks and Black 1999). This variation was due primarily to better stomatal regulation of transpiratory water loss in the east-side genotypes. Tyree et al. (1994) also found similar variation in cavitation among poplar genotypes from dry or wet habitats.

Photosynthesis declines in the lower leaves of a poplar crown, along with the more rapid development of water stress, because of poorer stomatal control of transpiration in the lower crown compared to leaves in the upper crown (Tschaplinski and Blake 1985). Clonal variation in resistance to water loss by leaves also occurs — clones that have large stomata and that lack hairs on the lower leaf surface lose more water (Blake et al. 1984). In experiments where root volume was progressively diminished, there was a decline in leaf water potentials, arrested shoot growth, and ultimately leaf senescence, starting in the lower crown and then moving upwards in the tree. Removing water stress by irrigation or thinning promotes growth by increasing leaf retention and allowing a more continuous growth. These treatments could remove many of the negative effects of intense competition in high-density plantings.

Slower-growing inbred and hybrid poplar clones lack drought tolerance adaptations found in faster-growing hybrid clones (Tschaplinski and Blake 1989). In tolerant clones, soil drought and nutrient deficiency cause carbon to be directed away from shoots and towards root growth (Fig. 13). The resulting increase in water uptake from the soil causes a subsequent increase in shoot growth after drought stress is alleviated (Kramer 1983). So the temporary increase in carbon allocation towards the roots is compensated for by a later increase in shoot growth. Although adjustments to root and shoot stresses both increase leaf growth, root-level stresses facilitate the capture of critical soil resources. By alleviating water deficits in the leaves, the investment in root biomass results in an increase in shoot growth and higher rates of photosynthesis under drought. By contrast, foliar stress (e.g., herbivory, rust defoliation, damage from ozone and other air pollutants, etc.) increases carbon allocation towards the shoot tips, which is required for regeneration of foliage after the stress is relieved. Although this response maximizes carbon gain by the shoot, it occurs at the expense of root growth. The relative increase in shoot growth will eventually be interrupted by an increasing water deficit, which tends to balance root and shoot growth (Fig. 13).

Adjustments in cell water relations

Cessation of growth is one of the earliest responses to water stress. Fast-growing meristematic cells in leaf and shoot apices show the most rapid response to turgor loss (Bradford and Hsiao 1982). The small leaves produced by poplars during a drought period are an indication of the sensitivity of growth processes to water stress. Changes in cell wall elasticity and the accumulation of the inhibitor ABA during drought ensure that growth does not immediately resume after plant water deficits are eliminated. When growth ceases, photosynthesis continues and cells in leaf and branch apices continue to accumulate sugars and other soluble organic compounds in cell vacuoles. This process, called osmotic adjustment, lowers

Fig. 13. Stress that affects the soil (drought, nutrient deficiencies, etc.) leads to carbon allocation to roots because roots are less sensitive to limiting soil resources. Stress on leaves (low humidity, air pollutants, browsing, etc.) leads to carbon allocation back to leaves at the expense of roots.

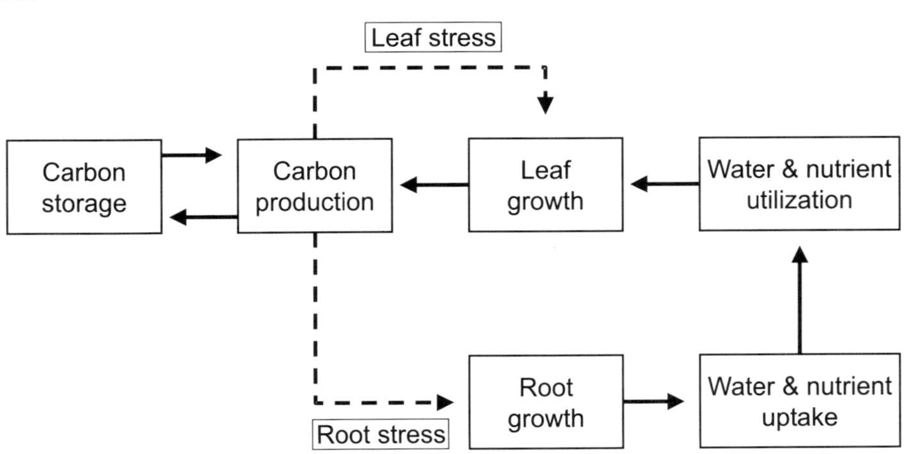

plant water potentials, which allows water uptake to continue and helps to maintain cell turgor during tissue dehydration. In several studies conducted on poplar clones, the ability to osmotically adjust was correlated with faster growth rate (Gebre et al. 1997; Tschaplinski and Blake 1989). In contrast to less vigorous clones, which were unable to adjust, osmoregulation promoted turgor maintenance in faster-growing poplar clones when soil water potential declined under drought. These findings provide further evidence that water stress tolerance is an important prerequisite for faster growth rates. Although the accumulation of soluble compounds in leaves during osmotic adjustment could slow leaf growth immediately after drought is relieved, they also provide energy and carbon skeletons for later growth.

There is much debate as to whether hydraulic or biochemical signals close stomata in drought-stressed plants. Stomatal opening and closing under low humidity result from alterations in guard cell turgor in the absence of any accumulation of the growth inhibitor ABA (Darlington et al. 1997). ABA transport to leaves in the sap stream of trees exposed to soil drought suggests ABA accumulation may be an indicator of water stress. Stomata of *P. trichocarpa* failed to close under drought, despite an accumulation of ABA (Schulte and Hinckley 1987). Application of ABA in this study closed stomata in young, expanding leaves, but not in older, fully expanded foliage. As with most physiological responses, poplar clones differ in their sensitivity to applied ABA (Chen et al. 1997; Ridolfi et al. 1996). Previous exposure of plants to drought can improve stomatal control. For example, stomata of unacclimated *P. trichocarpa* were unable to close under drought, but they exerted greater control of water loss when acclimated by repeated drought (Schulte et al. 1987). ABA may act as a drought signal to close stomata and inhibit shoot growth when soil drought becomes severe; however, ABA appears to be less important for the normal, diurnal stomatal movements that are controlled by guard cell turgor.

Adjustments to flooding

Although standing water is detrimental to tree survival and growth, adaptations in some poplars allow them to thrive in riparian zones, wetlands, and intertidal regions. Root growth is optimal at atmospheric levels of oxygen (20%), but it quickly declines under flooded conditions when oxygen drops below 10%. Oxygen has a very low solubility in water; thus it is poor aeration, rather than excess water, that damages flooded plants. The decline in oxygen to low levels is called hypoxia. Tree vigor declines whenever soil aeration is impaired, e.g., in soils that are compacted, poorly drained, impermeable, fine textured, or with a prolific growth of grass. When partial pressure of oxygen declines to zero — e.g., in stagnant water — plants are exposed to the deleterious effects of anoxia or anaerobic conditions.

Some poplars can survive weeks or even months of flooding. North American riparian poplars tolerant of some flooding include *P. angustifolia*, *P. balsamifera*, and *P. trichocarpa* in section *Tacamahaca* and *P. deltoides* and *P. fremontii* in section *Aigeiros* (Braatne et al. 1996). *Populus heterophylla* in section *Leucoides* may be the most flood-tolerant of all North American poplars. Hybrid poplars and aspens, in general, are less flood-tolerant than their riparian counterparts in the genus, although quaking aspen often grows in habitats where the water table is close to the surface. Tree survival declines with increased duration and depth of flooding (Westhaus 1986). Although moving water is more aerated, rate of water flow apparently has no effect on success of *P. nigra* (Dudek et al. 1998). Specialized aerating tissues are found on flood-tolerant poplars exposed to inundation. For example, callused (hypertrophied) lenticels on the stem, aerenchyma tissue within the bark, and water roots near the water surface ventilate toxic gases produced under anoxic conditions and aerate stems and roots.

Although 16 *P. deltoides* clones all survived 42 days of flooding, there was an early decline in leaf growth, stomatal conductance, and photosynthesis, which started after only 1 day of flooding. A number of detrimental effects of flooding were observed in leaves of these cottonwood clones, including the inhibition of leaf initiation, and chlorosis and abscission of leaves. In addition, leaf size, leaf area, and number of leaves were all reduced when soil was flooded, compared to leaves of plants grown in well-drained soil. Flooding also inhibited root growth and caused the original root system to deteriorate (Cao and Conner 1999). Root resistance to water uptake increased in a *P. trichocarpa* × *P. deltoides* poplar clone, starting 8 h after roots were exposed to anoxic conditions; this treatment reduced water movement by one third and one half, respectively, after 24 and 48 h of anoxia. Despite the increased resistance to water movement in flooded plants, xylem water potentials did not decline (Smit and Stachowiak 1988). Although some flooding symptoms resemble those of drought — e.g., stomatal closure, decline in photosynthesis, and leaf shedding — these symptoms result from ethylene production, rather than the "physiological drought" induced by low water potentials (Blake and Reid 1981). Flooded trees also take much longer to recover after stress is relieved than trees exposed to drought.

The decline in hydraulic conductivity in flooded roots reduces the flow of essential growth factors, including water, nutrients, and plant growth regulators. Additionally, during anaerobic metabolism toxic compounds such as ethanol and lactic acid form and accumulate in the roots of flooded plants. Flood-tolerant plants accumulate higher amounts of alcohol than those that are flood sensitive (Crawford 1989), indicating that tolerance may involve a detoxifying mechanism for alcohol. Ions of manganese and iron, and the gases ethylene and CO_2, also have been found at relatively high concentrations in the roots of flooded plants. Leaves of plants gassed with ethylene develop characteristic symptoms of flooding, including yellowing, drooping, and abscission. Symptoms of ethylene toxicity in shoots result from the movement of an ethylene precursor from the roots of flooded plants. The aerating structures in the stems of flood-tolerant trees are produced largely to vent toxic ethylene.

Mineral nutrient relations

Poplars have elemental nutrient requirements that must be met if they are to grow and thrive. The chemical elements required in the largest amounts are the macronutrients carbon, hydrogen, oxygen, phosphorus, potassium, nitrogen, sulfur, calcium, iron, and magnesium (C, H, O, P, K, N, S, Ca, Fe, and Mg). A mnemonic phrase for this list is "C. Hopkin's Café Mighty Good." Carbon, hydrogen, and oxygen, acquired in the form of CO_2, H_2O, and O_2, however, are not considered mineral nutrients. Micronutrients, although required in minute quantities, also are essential for growth. They include boron (B), chlorine (Cl), copper (Cu), manganese (Mn), molybdenum (Mo), zinc (Zn), and possibly silicon (Si). Because the role of mineral nutrients is similar in all plants, general texts can be consulted for an understanding of mineral nutrient physiology of poplars (Marschner 1995; Epstein 1972).

Trees absorb essential mineral nutrients from the soil solution via the fine roots. Both limitation and toxicity of nutrients on a site can limit tree growth potential. When nutrient limitations reduce tree growth, fertilization can be used to improve tree productivity (see Chap. 5). Problems with nutrient toxicity, although rare, are more difficult to eliminate. For example, nutrient toxicity can occur due to evaporative concentration of salts in irrigation water. Over-fertilization with nitrogen can lead to a "burning" of foliage. Manipulations such as adjusting soil pH by adding lime, managing soil aeration by improving drainage, and leaching out excess salts from the soil all may play a role in improving soil nutrient balance and availability.

Mineral nutrient acquisition depends on root activity. The resurgence of fine-root growth during early spring, for example, is a necessary prelude to the burst of nutrient-demanding growth by buds and the cambium that begins the aboveground growing season. Because minerals can only be absorbed from the soil solution, nutrient acquisition also requires adequate soil water levels. Conditions that decrease root viability can lead to nutrient limitations, even in fertile soils. For example, root rots or grazing by soil nematodes can decrease fine-root mass enough to lead to nutrient and water stress. Factors such as soil compaction or the presence of hardpans in the soil profile can impede root penetration and, thereby, a tree's capacity to acquire water and nutrients.

Symbiotic mycorrhizal fungi are essential for adequate nutrient absorption, especially on sites with low to moderate soil fertility or when fertilization is not employed (Fig. 11). The hyphae of endomycorrhizae, which are more common on young poplars, actually penetrate into the cells of the roots. In contrast, ectomycorrhizae penetrate between the cell walls of the root, in some cases completely enveloping the root tips in a mantle and changing their morphology. The thread-like hyphae of mycorrhizae — much smaller than the finest poplar roots — grow into the soil where they acquire nutrients from minute soil pores inaccessible to tree roots. These fungal associations are most important for uptake of immobile nutrients like P, but mycorrhizae also aid in absorption of N, K, Ca, S, and Zn. Mineral-nutrient uptake rates of mycorrhizal roots can be up to six times those of uninfected roots (Harley and Smith 1983).

Mineral nutrient characteristics

Plant–soil–nutrient interactions are determined by a wide range of environmental and plant characteristics. Nutrient availability to the plant depends not just on the quantity of the element in the soil, but whether it is in a form the plant can use, and whether the nutrient is mobile or immobile in the soil. In general, immobile nutrient ions are cations (positively charged), and interact strongly with the numerous negatively charged sites on soil clay minerals.

The nature of nutrient deficiency symptoms depends on the mobility of the element in the plant. Some nutrients, such as N, are readily translocated in the phloem from senescing leaves, stored over winter as protein in the bark of twigs, and then transferred to young, expanding leaves. Symptoms of N deficiency include yellowing of old leaves, but not young leaves. Other nutrients, particularly Ca, are not mobile in the phloem, and so tend to accumulate in older leaves; deficiency symptoms typically occur first in younger leaves for phloem immobile nutrients.

All nutrients are not equal when expressed in terms of the amount demanded by plants. Macronutrients are needed in larger quantity than micronutrients, but other differences exist between these two categories. Regardless of the amount required, a deficiency or excess in any nutrient can substantially disrupt physiological processes in a plant.

Nitrogen is the mineral element most often limiting plant growth and the one required in largest quantity. Plants combine N with carbohydrates to form amino acids, the building blocks of proteins (plants can manufacture all essential amino acids). Soil N is primarily derived from fixation of atmospheric N by nodulated plants or free-living blue–green algae and from mineralization (decomposition) of organic matter. Soil N pools are very dynamic — N can be immobilized or released by microbes, and it is quite mobile, especially in anionic form (NO_3^-). Because poplars contain such large pools of N (Ericsson et al. 1992), soil N can be strongly depleted by harvesting and removal of the entire aboveground part of trees.

Trees acquire N from the soil in two forms: ammonium (NH_4^+) or nitrate (NO_3^-). Poplars can take up either form of N, showing a slight preference for NH_4^+ (Min et al. 1999). Production of amino acids requires N in the form of NH_4^+, so NO_3^- must be reduced to NH_4^+ in the plant by the enzyme nitrate reductase. Transport of N through the xylem from roots to leaves is in the form of glutamine, an amino acid containing two N molecules. In fact, glutamine is the primary amino acid found in the xylem of poplars (Dickson 1979). The N in glutamine can readily be passed on to other molecules, forming the whole array of amino acids and other N-containing molecules required by the plant. Mycorrhizae also are able to utilize organic N sources in the soil, but the importance of this form of N acquisition to tree N budgets is not yet clear.

Phosphorus is derived from weathering of P-containing minerals and plays a central role in biological energy metabolism, primarily in the form of ATP (adenosine triphosphate). ATP is the main energy currency produced during respiratory metabolism of carbohydrates and during photosynthesis. Plants acquire P from the soil solution in the form of phosphate (PO_4^{++}). Phosphorus is relatively immobile in the soil, but mycorrhizal infections can greatly improve tree P status in low-P soils.

Potassium is derived from the weathering of K-containing minerals. K is essential for charge balance, osmotic regulation, and functioning of many enzymes. Potassium is quite mobile in the soil and very mobile in the phloem.

Sulfur is primarily used in formation of the amino acids cysteine and methionine. Sulfur is acquired from the soil primarily in the form of sulfate (SO_4^{--}).

Calcium is important to cell membrane integrity and functions in hormonal signaling. Calcium may be deficient in very low-pH soils, as well as on high-pH soils.

Magnesium is a required cofactor for many enzymes (particularly those essential for ion transport), and it also is part of the structure of the chlorophyll molecule.

Iron is a constituent of critical enzymes of the electron transport chain, therefore it is essential for photosynthesis and respiration. Iron also is needed for synthesis of chlorophyll. High soil pH can lead to Fe deficiency, with characteristic interveinal leaf yellowing (e.g., lime-induced chlorosis).

Micronutrients. Descriptions of functions for each of the micronutrients can be found in Marschner (1995) or Epstein (1972). Many function as enzyme cofactors; for example, molybdenum is essential for nitrate reduction. Because they are needed in such low quantities, foliar application of micronutrient fertilizers often can successfully relieve a deficiency.

Seasonal patterns and nutrient cycling

Uptake of nutrients occurs throughout the growing season, although it is most active when soil water levels are high; e.g., in the spring. However, the concentration of N in leaves usually declines through the season because increases in structural constituents such as cellulose, lignin, and waxes dilute the N. If early season N concentrations are 3–4% in fully expanded poplar leaves, N is not limiting growth. Nutrient concentrations in plant parts other than leaves also are highest in young tissue and then show a steady decline with age (Ericsson et al. 1992). Internal nutrient cycling is a common feature of deciduous trees like poplar; i.e., there is a regular seasonal pattern of nutrient movement within the tree. We will focus here on N, because it most often limits plant growth and has been studied more than any other nutrient.

During the growing season, most of the tree N is found in the leaves, where the largest portion is in a single leaf protein critical for photosynthesis — ribulose bisphosphate carboxylase oxygenase, commonly called "Rubisco." This protein functions as the enzyme responsible for the initial "fixation" of CO_2 in photosynthesis or O_2 in photorespiration. The approach of autumn is signaled by shorter days and cooler temperatures, triggering internal chemical adjustments in the tree. As leaves senesce, leaf proteins and other cell constituents are broken down, and amino acids (containing the leaf N) and other mobile nutrients are transported out of the leaf and into twigs and branches where they are stored over the winter. In poplar plantations, more than 50% of the N and 30% of the P that are used in growth can come from nutrients internally cycled during the previous autumn (Ericsson et al. 1992). When transport amino acids reach the sites of storage, they are converted back to proteins (Sauter et al. 1989). After winter chilling requirements are met and days once again lengthen in spring, these stores of N are broken down and amino acids are transported in the xylem to the growing points to supply N for leaf construction and branch growth.

The role of roots in the internal cycling of nutrients also is important. Roots continually absorb N and other nutrients from the soil solution, although little N is acquired during the winter. Some of these nutrients are used in the roots for growth and storage, while the major portion is translocated via the xylem to the aboveground parts of the tree. Much of the transported N is in the form of amino acids (Dickson 1979). Root reserves of N, again in the form of protein, are used locally, but in the spring these reserves are broken down and the constituent amino acids are transported upward to the growing branches and stem cambium. Two lines of evidence support this conclusion. First, root N concentrations decline sharply in the spring, building back up through the summer and fall. Second, concentrations of amino acids in the upward-moving xylem reach a peak in the early spring (Dickmann and Pregitzer 1992).

Internal recycling of nutrients is an important conservation mechanism for trees, but the nutrients contained in fallen leaves, twigs, catkins, etc., are not lost to the system. In fact, the contribution of nutrients reclaimed by trees from decomposing organic material in soil litter can be significant — one- to two-thirds of the annual nutrient demands of established plantations can be met from litter mineralization and reabsorption by roots (Ericsson et al. 1992). Mineralization is accomplished by soil microbes and fungi, and their activity is highly dependent on biotic and environmental factors that can vary considerably from site to site. In general, poplar litter is very easily decomposed. The net result of recycling in an established plantation is very little loss of mobile nutrients such as N to the ground water, unless excessive soluble fertilizer is applied or if it is applied during late fall or winter when roots do not actively absorb mineral nutrients.

Physiology of yield

Physiologists with forestry, agronomy, or horticulture backgrounds have long sought to identify the morphological and physiological basis of yield regulation. Interest in this topic, which is also called "production physiology" (Heilman et al. 1996), is especially active in poplars. One of the main questions asked by production physiologists is "Why do poplar trees grow so fast?" The influence of sunlight, water, and mineral nutrients on yield also has long been of interest to physiologists and geneticists. In particular, the main growth-limiting effects of genotype, site, and environment have been studied to optimize tree growth and yield. Absorption of soil resources and environmental or biotic limitations were discussed above. Genotypic (clonal) variation in traits — physiological, anatomical, or morphological — always must be assumed in poplars, and this variation interacts with all the other variables influencing yield. If a particular trait is singled out as desirable, current breeding or gene transformation techniques can probably produce a poplar clone with this characteristic (see Chaps. 2 and 14). Silvicultural manipulations are discussed in Chaps. 4 and 5.

Expressions of yield

The term "yield" has different meanings to different people, depending upon their perspective. Basically, yield means the amount of a useful commodity produced by a stand of trees, expressed as a volume or weight of product produced per unit of land area over some unit of time — usually yearly or at the end of a rotation (see Appendix). In the past, emphasis was placed on wood, the primary economic commodity produced by poplar trees and stands, expressed in terms of m^3 per ha, board feet per acre, or ft^3 per acre. In recent years, there has been more emphasis on the biological productivity of poplars expressed in terms of Mg per ha of biomass and, more recently, energy content of the tree or stand (e.g., J per Mg). Biological productivity includes the entire (whole) above- and below-ground production of the stand or ecosystem and is a useful way of expressing yield for all economic and environmental benefits of poplar culture. Biological productivity, or yield, is a result of the complex interaction of the genetic constitution of the plant and the environment, mediated through physiological processes.

Biological productivity is difficult to estimate, so usually we use a model developed from weights and mensurational parameters measured on a small sample of trees. Equation [2] above can be used to estimate total tree biomass from DBH and height. Because root weights are so difficult to quantify, biomass yields often are expressed on an aboveground basis. The following model can estimate the biomass of the stem and branches of a hybrid poplar (Riemenschneider et al. 2001):

[3] $$B = 6.16 - 2.23(\text{DBH}) + 0.3353(\text{DBH})^2$$

Where B is aboveground tree biomass in kg and DBH is in cm.

Economic yield is that portion of the biological productivity that is harvested, also called harvest yield. Economic yield of a stand varies with the nature of the product, which usually is stems for fiber or timber. Economic yield of poplars for environmental benefits (see Chap. 6), on the other hand, is the overall ecosystem biological productivity above- and below-ground. Other economic products could be foliage in an agroforestry system, cuttings in a nursery, or a certain biochemical in a pharmaceutical planting. Yield of intangible products such as wind or sound amelioration, erosion control, remediation of a polluted site, or landscape beauty are more problematic to quantify, and usually relate more to individual trees than stands. Although these non-traditional commodity yields also have a physiological basis, our ensuing discussion will focus on the yield of wood, biomass, or energy.

Physiological determinants of yield

The biological productivity of a poplar stand is related to its ability to intercept solar radiation and to efficiently convert it to biomass production through the process of photosynthesis. Thus, overall production of an economic commodity is intimately related to leaf production over the course of the growing season throughout the rotation. Several studies have shown a direct linear correlation between capture of solar radiation, as measured by the difference in light above and below a plant canopy, and stand biomass production (Heilman et al. 1996). The more light that is captured, the more biomass is produced. While this model provides a general explanation for production, it raises a number of questions. What plant canopy variables affect interception of solar radiation and its conversion via photosynthesis to sugars? How does light interception vary over the course of a growing season or a tree rotation? How do environmental variables affect light interception?

Leaf area index (LAI) is used to express the status of the leafy canopy of a stand. LAI is the amount of leaf surface displayed by a plant canopy over a unit of ground area, in units of m^2/m^2 but often expressed simply as a number without units. The more leaves that are present over an area of ground, the more incoming solar radiation can be absorbed. The LAI of a stand is under strong genetic control, but also is influenced greatly by the environment. Stands of certain highly productive hybrid poplars reach high LAI levels in the range of 10–12 or above (Heilman et al. 1996), although natural stands may only reach one-third to one-half of these levels. At high LAIs, very little incoming sunlight reaches the soil surface. LAIs are lowest in the year of establishment, then increase through canopy closure, and finally decline to a stable level as lower branches begin to die and abscise (Fig. 14). At close spacings, this pattern of LAI with age is compressed; maximum LAIs develop within a few years but then fall off sharply.

Fig. 14. Changes in leaf area index (m^2/m^2) with stand age for poplar clones 11-11 and 44-136 (*Populus trichocarpa* × *P. deltoides*), Illinois 5 (*P. deltoides*), and 1-12 (*P. trichocarpa*) planted at a 1 × 1 m spacing in western Washington. Redrawn from Heilman et al. (1996).

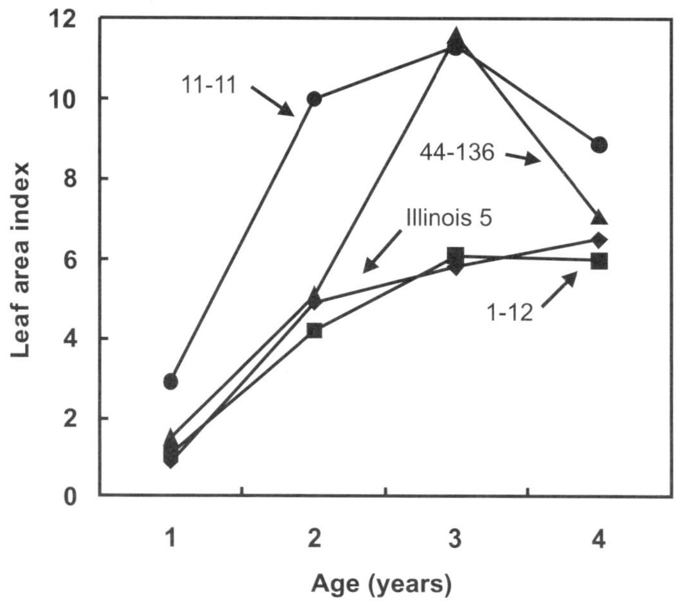

Conversely, this pattern is stretched out over time at wide spacings. Maximization of LAIs occurs only on high quality sites or under growing conditions where stress is minimized. Soil water deficits during a drought or O_3 stress curtail leaf growth and can cause poplars to shed leaves in their lower canopy, decreasing LAI and potential productivity (Dickmann et al. 1992; Isebrands et al. 2000).

Leaf area index of poplar clones is determined by leaf and branch morphology. Because both factors vary greatly, they can be manipulated by altering genotype and silvicultural practices. The angle of the leaf blades from the horizontal determines the depth of light penetration into a canopy; highly angled leaves in the upper crown allow more light to reach the lower leaf layers. Poplar taxa differ considerably in this characteristic. For example, *P. deltoides* leaves are highly angled from the horizontal, *P. trichocarpa* leaves are displayed in a relatively horizontal position, and interspecific hybrids between the two have intermediate characteristics (Ceulemans and Isebrands 1996). Highly angled leaves are more frequent in the upper canopy, where this trait helps reduce heating of leaf blades by the sun, but leaves become more horizontally oriented at deeper canopy levels. This gradual decrease in leaf angles from the top to the bottom of the canopy results in highly efficient vertical profiles of light interception and, therefore, is desirable for high productivity. The dim light at ground level under stands of fast-growing *P. trichocarpa* × *P. deltoides* hybrids in Washington and Oregon indicates that most light is being absorbed by the canopy (Fig. 15). For reasons that

Fig. 15. Poplar stands with high leaf area indexes efficiently capture solar energy and produce high yields. The dim light at ground level in this *Populus trichocarpa* × *P. deltoides* stand in western Oregon denotes its high productivity.

are not fully understood, natural poplar stands or hybrid plantations in other regions of North America seldom achieve the same degree of light utilization.

Leaves are borne on branches, so branching characteristics also affect light interception. Canopy architecture in poplars has been well studied by physiologists. Because poplar trees show a wide range of branching characteristics, they have been model subjects for study. All other yield-related factors being equal, a coarse-branched tree with few sylleptic branches is less desirable than trees with many small sylleptic branches. The leaves on these small branches "fill in the gaps," and allow more light interception in the middle and lower parts of the canopy. Usually these sylleptic branches abscise and, therefore, do not adversely affect wood quality. A long narrow crown, formed by upright branches also improves light interception in a high-density stand. The well-known poplar cultivar 'Lombardy' represents the extreme of this crown shape, although this cultivar is used primarily for amenity planting and not for plantation culture in North America.

Climate — solar radiation, temperature, rainfall, humidity, and wind — probably is the most important determinant of biological productivity of poplar clones. Over millennia, clones and species of *Populus* have become adapted to their native environment, enabling poplars to grow from the tropics to latitudes north of the Arctic Circle. Thus, when one chooses a clone for one particular area, local sources are the best choice until the adaptability of other material has been tested.

Matching a poplar clone to environment in which it will be grown is one of the most fundamentally important principles of poplar culture (Dickmann and Isebrands 1999).

Among climatic variables, several are very significant in controlling yield. The degree of cloudiness greatly affects both temperature and the light available for photosynthesis. Growing degree days — the temperature sum above 4°C over a growing season — is a common expression of the available heat at any particular location. Other things being equal, warm sunny areas are more productive than those cool and cloudy; poplars thrive in the sun. The amazing productivity of poplars grown in irrigated plantations in the high desert of eastern Oregon and Washington illustrates how well poplars respond to these conditions. The ratio between the amount of precipitation (P) and the rate at which water evaporates (E) — the P/E ratio — also is a significant factor determining yield. If this ratio falls below 1 over the course of a growing season, trees will be stressed and productivity will suffer. Over all, water limits productivity more than any other environmental factor in non-irrigated poplar stands.

Optimum soil nutrition and other soil properties also are very important determinants of yield. Soil heterogeneity can, in some regions of North America, even influence the productivity of a given clone in the same field. Other yield-limiting factors include biological stressors such as insect pests, diseases, and mammalian herbivores. Air pollution also affects the complex genetic × environment interactions that determine biological production of poplars (see Chaps. 6–8).

Phenology refers to the relation between climate and the periodic biological activity of plants. The shoot-growth period — the time between bud flushing in the spring and bud set in the fall — is a crucial phenological event relating to yield; a tree cannot grow in height if its buds are set. To attain high yield, this period should be maximized by choosing a clone with an indeterminate growth habit and by using silviculture to optimize the growing environment. But the constraints of growing season length set by the local environment must be considered. For example, a clone that flushes too early in the spring might be damaged by late-spring frosts. Phenology interacts with LAI to determine yield, so the time of leaf abscission in the autumn also is important. The period between bud flush and leaf abscission is called leaf area duration; the longer leaves stay on a tree, the longer they are absorbing solar energy and producing carbohydrates through photosynthesis.

A poplar ideotype

The Australian agronomist C.M. Donald (1968) proposed the ideotype as "a plant model which is expected to yield a greater quantity or quality of ... useful product when developed as a cultivar." In other words, an ideotype comprises the traits of an ideal plant for a particular cultural system and end product. It can serve as a target or guideline for programs that improve trees through selection and breeding or advanced molecular techniques. Donald's ideotype concept has been utilized to good effect in crop improvement, and it also has been proposed for use with trees, including poplars (Dickmann and Keathley 1996; Martin et al. 2001).

A complete ideotype for poplars is presented in Table 2. This ideotype outlines the yield-related morphological and physiological traits currently thought to contribute to high biomass or wood productivity, but it does not yet contain any molecular markers associated with yield. We must emphasize three points about this poplar ideotype. First, it is dynamic. Ideotypes are continually modified as new information becomes available through research. Second, it is unlikely that high-yielding poplar clones will show all of these traits; fast growth can be achieved in a number of different ways. And third, this ideotype in total does not represent a specific breeding goal; the practical limitations of tree breeding set a limit of not more than five or six on the number of traits that can be included in a working ideotype. In fact, some of the traits in this ideotype may be negatively correlated with one another genetically. We need to learn more about the compatibility of traits, especially when they are forced to the extreme end of their range of variability. But these limitations do not negate the importance of the ideotype approach; a completely formulated goal should always be kept in sight.

The ideotype presented in Table 2 is largely morphological because these traits are most commonly linked to yield. Although a few physiological processes are included, they are unlikely to limit yield in and of themselves, except when they become extremely limiting (Dickmann and Keathley 1996). Furthermore, physiological processes are controlled by many enzyme-catalyzed biochemical reactions, each regulated by one or more genes. Selecting for one genetically complex trait in an ideotype to indirectly improve yield — itself a complex genetic trait — is highly inefficient.

The inherent rate of physiological processes seldom limits growth and yield because environmental limitations are so pervasive in plantations. Sunlight, temperature, water availability, wind, or some combination of them nearly always limits first. Many of the differences in leaf photosynthesis rates reported among poplar clones are less a reflection of genetic differences in photosynthesis, per se, than they are a reflection of the influence of environment, morphology, and other physiological processes. For example, the sunlight that is not intercepted by a tree canopy with a low LAI cannot be used in photosynthesis, regardless of how efficiently this process operates in leaves. Careful selection of clones for a particular silvicultural system and site can minimize environmental limitations, with a consequent saving in silvicultural inputs.

There also is a reciprocal interpretation when considering the meaning of a physiological–growth correlation. For example, it is as legitimate to argue that some poplar clones have greater rates of leaf photosynthesis because they grow faster as it is to argue that some clones grow faster because they have inherently greater rates of leaf photosynthesis. There is considerable evidence that photosynthetic

Table 2. An ideotype for poplar trees grown for wood fiber or energy in a high-density, unirrigated, intensive silvicultural system (modified from Dickmann and Keathley 1996).

Phenology:
Early bud flushing but late enough to avoid spring frost injury
Indeterminate shoot growth with bud set just prior to first autumn frosts
Long leaf area duration; i.e., leaf retention well into the autumn
Cambium active until late in the growing season
Physiology:
Thick leaves with a high photosynthetic rate per unit leaf area and a high ratio of net photosynthesis to dark respiration
High water-use efficiency (WUE — CO_2 fixed in photosynthesis per unit of water lost in transpiration)
Stomata closing at moderate levels of water stress
Leaves, cambium, and fine roots osmotically adjusting to gradual dehydration
Effective remobilization of nitrogen and other mobile nutrients into stems and roots prior to leaf abscission
High nutrient-use efficiency (stemwood biomass produced per unit of stemwood nutrient)
Crown morphology:
Upturned branches forming a long, narrow crown
Abundant sylleptic branching
High foliage density on branches
Relatively large, vertically oriented leaves in upper crown grading to small horizontally oriented leaves in the lower crown
High ratio of long (indeterminate) to short (determinate) shoots in the upper crown
Rapid natural pruning of dead branches
Male producing few catkins or sterile
Root morphology:
Hardwood cuttings producing abundant adventitious roots
Long taproot and sinker roots for anchorage and exploitation of deep soil nutrients and water
Many highly branched lateral and fine roots near the soil surface
Slow turnover of fine roots
Colonized by ecto- or endo-mycorrhizal fungi, except when highly fertilized
High concentrations of starch, sugar, and nitrogen reserves during the dormant season
Ecological characteristics:
Weak competitor; i.e., all stems show equal vigor
Resistant to wind, snow, and ice breakage
Tolerant of winter minimum temperatures
Tolerant of common post- and pre-emergent herbicides
Highly resistant to major pathogens, especially *Septoria* canker and *Melampsora* leaf rust
Highly resistant to major insect pests, especially defoliators and borers
Unpalatable to mammals; not rubbed by Cervids in rut

rates of many crop plants, including poplars, increase when demand by rapidly growing or metabolizing tissues is high (Tschaplinski and Blake 1989).

We do not, however, want to imply by the preceding discussion that a certain biochemical step or steps in the overall photosynthetic process, or any other physiological process, might in fact be limiting to yield, with possible genotypic-based variation in the magnitude of this limitation. If such rate-limiting biochemical steps could be identified — and none have been at this writing — a genetic solution might be found, with accompanying increases in yield. The moderating influence of environmental stresses or other limiting factors, however, still might render such gains negligible.

One final caveat is necessary concerning poplar ideotypes. The intent of studying poplar yield physiology and formulating ideotypes is not to find a one-clone solution to a particular cultural system. As we mentioned above, there are many different combinations of traits that can produce an ideal clone for a particular situation. These traits could be expressed in selections from pure species, in various hybrids produced through controlled breeding, or in genetically transformed plants. Each would look different, function in a slightly different way, and occupy a somewhat different ecological niche. Most experts agree that the most reasonable and safest approach to poplar culture regardless of the system employed is to use a suite of suitable clones planted in a mosaic of plantation blocks or individual trees. To use an old but very appropriate aphorism, "never put all your eggs in one basket." Unknown susceptibilities of a particular clone to biotic or environmental factors, which may not have shown up in the testing phase for that clone, may cause a breakdown before the trees can be harvested or before their anticipated life expectancy.

Conclusion

In this chapter, we have provided an overview of how poplars grow and how different genotypes and environmental factors affect physiological growth processes. The final section on yield physiology outlined the traits of poplars that are most closely associated with wood or biomass yields. Although there is much more to be learned, a fairly detailed picture of the way poplar trees function in a complex environment has grown out of research conducted by physiological ecologists. This information can serve as a valuable aid for poplar silviculture and the development of improved clones, regardless of the management objective (Fig. 1). We firmly believe that the continued success of poplar culture in a world where demands for wood and tree planting will continue to intensify depends on deepening our understanding of the physiological ecology of the subject. To paraphrase Paul Kramer, the better we get to know poplars, the better we will be able to grow them.

Acknowledgements

Some of the research cited in this chapter was funded by the U.S. Department of Energy through the Biofuels Feedstock Development Program at Oak Ridge National Laboratory. Support by the Michigan Agricultural Experiment Station and the USDA Forest Service, North Central Research Station, also is acknowledged. Thanks to Gary Hogan and Jon Johnson for critical reviews of an early draft of this chapter.

References

Barnes, B.V. 1966. The clonal habit of American aspens. Ecology, **47**: 439–447.

Bassman, J.H. 2000. Factors affecting water use in poplar culture. *In* Hybrid poplars in the Pacific Northwest. *Edited by* K.A. Blatner, J.D. Johnson, and D.M. Baumgartner. Wash. State Univ. Coop. Ext. Pub. MISC0272, Pullman, WA. pp. 69–75.

Bevilacqua, E., and Blake, T.J. 1997. Modelling growth and ecophysiological responses of *Eucalyptus grandis* clones in Minas Gerais, Brazil. J. Trop. For. Sci. **9**: 505–513.

Blake, T.J. 1983. Coppice systems for short-rotation intensive forestry: the influence of cultural, seasonal and plant factors. Aust. For. Res. **13**: 279–291.

Blake, T.J., and Reid, D.M. 1981. Ethylene, water relations and tolerance to waterlogging of three eucalypt species. Aust. J. Plant Physiol. **8**: 497–505.

Blake, T.J., Tschaplinski, T.J., and Eastham, A. 1984. Stomatal control of water use efficiency in poplar clones and hybrids. Can. J. Bot. **62**: 1344–1351.

Blake, T.J., Bevilacqua, E., and Suiter-Filho, W. 1995. Early selection of *Eucalyptus grandis* clones in central Brazil. J. Trop. For. Sci. **8**: 33–43.

Boysen, B., and Strobl, S. (*Editors*). 1991. A grower's guide to hybrid poplar. Ontario Ministry of Natural Resources, Brockville, ON. 148 pp.

Braatne, J. 1999. Biological aspects of hybrid poplar cultivation on floodplains in Western North America: A review. U.S. Environ. Prot. Agency Doc. No. 910-R-99-002.

Braatne, J.H., Rood, S.B., and Heilman, P.E. 1996. Life history, ecology, and conservation of riparian cottonwoods in North America. *In* Biology of *Populus* and its implications for management and conservation. *Edited by* R.F. Stettler, H.D. Bradshaw, Jr., P.E. Heilman, and T.M. Hinckley. NRC Research Press, Ottawa, ON. pp. 57–85.

Bradford, K.J., and Hsiao, T.C. 1982. Physiological response to moderate water stress. Encycl. Plant Physiol. New Ser. 12B: 263–324.

Cao, F.L., and Conner, W.H. 1999. Selection of flood tolerant *Populus deltoides* clones for reforestation projects in China. For. Ecol. Manage. **117**: 211–220.

Ceulemans, R., and Isebrands, J.G. 1996. Carbon acquisition and allocation. *In* Biology of *Populus* and its implications for management and conservation. *Edited by* R.F. Stettler, H.D. Bradshaw, Jr., P.E. Heilman, and T.M. Hinckley. NRC Research Press, Ottawa, ON. pp. 355–399.

Chen, S.L., Wang, S.S., Altman, A., and Hutterman, A. 1997. Genotypic variation in drought tolerance of poplar in relation to abscisic acid. Tree Physiol. **17**: 797–803.

Crawford, R.M.M. 1989. Studies in plant survival. Blackwell Scientific Publications, Oxford, UK.

Darlington, A.B., Halinska, A., Dat, J.F., and Blake, T.J. 1997. Effects of increasing saturation vapour pressure deficit on growth and ABA levels in black spruce and jack pine. Trees, **11**: 223–228.

Dickmann, D.I., and Isebrands, J.G. 1999. *Caveat emptor*. Am. Nurseryman, **189** (issue 5, Mar. 1): 60–65.

Dickmann, D.I., and Keathley, D.E. 1996. Linking physiology, molecular genetics, and the *Populus* ideotype. *In* Biology of *Populus* and its implications for management and conservation.

Edited by R.F. Stettler, H.D. Bradshaw, Jr., P.E. Heilman, and T.M. Hinckley. NRC Research Press, Ottawa, ON. pp. 491–514.

Dickmann, D.I., and Pregitzer, K.S. 1992. The structure and dynamics of woody plant root systems. *In* Ecophysiology of short-rotation forest crops. *Edited by* C.P. Mitchell, J.B. Ford-Robertson, T. Hinckley, and L. Sennerby-Forsse. Elsevier Applied Science, New York, NY. pp. 95–123.

Dickmann, D.I., Liu, Z., Nyugen, P.V., and Pregitzer, K.S. 1992. Photosynthesis, water relations, and growth of two hybrid *Populus* genotypes during a severe drought. Can. J. For. Res. **22**: 1094–1106.

Dickmann, D.I., Nguyen, P.V., and Pregitzer, K.S. 1996. Effects of irrigation and coppicing on above-ground growth, physiology, and fine-root dynamics of two field-grown hybrid poplar clones. For. Ecol. Manage. **80**: 163–174.

Dickson, R.E. 1979. Xylem translocation of amino acids from roots to shoots in cottonwood plants. Can. J. For. Res. **9**: 374–378.

Dickson, R.E. 1989. Carbon and nitrogen allocation in trees. Ann. Sci. For. **46** (suppl.): 631s–647s.

Dickson R.E., Coleman, M.D., Riemenschneider, D.E., Isebrands, J.G., Hogan, G.D., and Karnosky, D.F. 1998. Growth of five hybrid poplar genotypes exposed to interacting elevated CO_2 and O_3. Can. J. For. Res. **28**: 1706–1716.

Donald. C.M. 1968. The breeding of crop ideotypes. Euphytica, **17**: 385–403.

Dudek, D.M., McLenahen, J.R., and Mitsch, W.J. 1998. Tree growth responses of *Populus deltoides* and *Juglans nigra* to streamflow and climate in a bottomland hardwood forest in central Ohio. Am. Midl. Nat. **140**: 233–244.

Epstein, E. 1972. Mineral nutrition of plants: principles and perspectives. John Wiley & Sons, New York, NY.

Ericsson, T., Rytter, L., and Linder, S. 1992. Nutritional dynamics and requirements of short-rotation forests. *In* Ecophysiology of short-rotation forest crops. *Edited by* C.P. Mitchell, J.B. Ford-Robertson, T. Hinckley, and L. Sennerby-Forsse. Elsevier Applied Science, New York, NY. pp. 35–65.

Ericsson, T., Rytter, L., and Vapaavuori, E. 1996. Physiology of carbon allocation in trees. Biomass Bioenergy **11**: 115–127.

Faulkner, H.G. 1976. Root distribution, amount, and development from 5-year-old *Populus* × *euramericana* (Dode) Guinier. M.S.F. Thesis, Univ. of Toronto, Toronto, ON.

Farquhar, G.D., and Sharkey, T.D. 1982. Stomatal conductance and photosynthesis. Ann. Rev. Plant Physiol. **33**: 317–345.

Friend, A.L., Scarascia-Mugnozza, G., Isebrands, J.G., and Heilman, P.E. 1991. Quantification of two-year-old hybrid poplar root systems: morphology, biomass, and ^{14}C distribution. Tree Physiol. **8**: 109–119.

Gebre, G.M., Brandle, J.R., and Kuhns, M.R. 1997. Influence of rewatering and time of sampling on solute accumulation of two *Populus deltoides* clones. Tree Physiol. **17**: 341–346.

Harley, J.L., and Smith, S.E. 1983. Mycorrhizal symbiosis. Academic Press, New York, NY.

Heilman, P.E., Ekuan, G., and Fogle, D.B. 1994. First-order root development from cuttings of *Populus trichocarpa* × *P. deltoides* hybrids. Tree Physiol. **14**: 911–920.

Heilman, P.E., Hinckley, T.M., Roberts, D.A., and Ceulemans, R. 1996. Production physiology. *In* Biology of *Populus* and its implications for management and conservation. *Edited by* R.F. Stettler, H.D. Bradshaw, Jr., P.E. Heilman, and T.M. Hinckley. NRC Research Press, Ottawa, ON. pp. 459–489.

Hinckley, T., Brooks, J.R., Cermak, J., Ceulemans, R., Kucera, J., Meinzer, F.C., and Roberts, D.A. 1994. Water flux in a hybrid poplar stand. Tree Physiol. **14**: 1005–1018.

Isebrands, J.G., Nelson, N.D., Dickmann, D.I., and Michael, D.A. 1983. Yield physiology of short rotation intensively cultured poplars. *In* Intensive plantation culture: 12 years research. USDA For. Serv. Gen. Tech. Rep. NC-91. pp. 77–93.

Isebrands, J.G., Dickson, R.E., Rebbeck, J., and Karnosky, D.F. 2000. Interacting effects of multiple stresses on growth and physiological processes in northern forest trees. *In* Responses of northern U.S. forests to environmental change. *Edited by* R.A. Mickler, R.A. Birdsey, and J. Hom. Springer-Verlag, New York, NY. pp. 149–180.

Kosola, K.R., Dickmann, D.I., Paul, E.A., and Parry, D. 2001. Repeated insect defoliation effects on growth, nitrogen acquisition, carbohydrates, and root demography of poplars. Oecologia. In press.

Kozlowski, T.T., and Pallardy, S.G. 1997*a*. Physiology of woody plants. 2nd ed. Academic Press, San Diego, CA. 411 pp.

Kozlowski, T.T., and Pallardy, S.G.. 1997*b*. Growth control of woody plants. Academic Press, San Diego, CA. 641 pp.

Kozlowski, T.T., Kramer, P.J., and Pallardy, S.G. 1991. The physiological ecology of woody plants. Academic Press, San Diego, CA. 657 pp.

Kramer, P.J. 1983. Water relations of plants. Academic Press. San Diego, CA. 489 pp.

Kramer, P.J. 1986. The role of physiology in forestry. Tree Physiol. **2**: 1–16.

Lambers, H., Chapin, F.S., III, and Pons, T.L. 1998. Plant physiological ecology. Springer-Verlag, New York, NY.

Larson, P.R. 1994. The vascular cambium. Springer-Verlag, New York, NY. 725 pp.

Larson, P.R., and Isebrands, J.G. 1971. The plastochron index as applied to developmental studies of cottonwood. Can. J. For. Res. **1**: 1–11.

Mahoney, J.M., and Rood, S.B. 1991. A device for studying the influence of declining water table on poplar growth and survival. Tree Physiol. **8**: 305–314.

Marschner, H. 1995. Mineral nutrition of higher plants. Academic Press, San Diego, CA.

Martin, T.A., Johnsen, K.H., and White, T.L. 2001. Ideotype development in southern pines: rationale and strategies for overcoming scale-related obstacles. For. Sci. **47**: 21–28.

Min, X, Siddiqi, M.Y., Guy, R.D., Glass, A.D.M., and Kronzucker, H.J. 1999. A comparative study of fluxes and compartmentation of nitrate and ammonium in early-successional tree species. Plant Cell Envir. **22**: 821–830.

Mitchell, C.P., Ford-Robertson, J.B., Hinckley, T., and Sennerby-Forsse, L. (*Editors*). 1992. Ecophysiology of Short-rotation forest crops. Elsevier Applied Science, New York, NY. 308 pp.

Nelson, N.D. 1984. Woody plants are not inherently low in photosynthetic capacity. Photosynthetica, **18**: 600–605.

Nguyen, P.U., Dickmann, D.I., Pregitzer, K.S., and Hendrick, R. 1990. Late-season changes in allocation of starch and sugar to shoots, coarse roots, and fine roots in two hybrid poplar clones. Tree Physiol. **7**: 95–105.

Pregitzer, K.S., and Friend, A.L. 1996. The structure and function of *Populus* root systems. *In* Biology of *Populus* and its implications for management and conservation. *Edited by* R.F. Stettler, H.D. Bradshaw, Jr., P.E. Heilman, and T.M. Hinckley. NRC Research Press, Ottawa, ON. pp. 331–354.

Pregitzer, K.S., Dickmann, D.I., Hendrick, R., and Nguyen, P. 1990. Whole-tree carbon and nitrogen partitioning in young hybrid poplars. Tree Physiol. **7**: 79–93.

Ridolfi, M., Fauveau, M.L., Label, P., Garrec, J.P., and Dreyer, E. 1996. Responses to water stress in an ABA-unresponsive hybrid poplar (*Populus koreana* × *trichocarpa* cv. Peace). I. Stomatal function. New Phytol. **134**: 445–454.

Riemenschneider D., Tuskan, J., Mohn, C., Hall, R., Stanosz, G., Dickmann, D., and Isebrands, J. 2001. Poplar breeding and testing strategies in the North Central U.S.: demonstration of potential yield and consideration of future research needs. For. Chron. **77**: 245–253.

Sauter, J.J., van Cleve, B., and Wellencamp, S. 1989. Ultrastructural and biochemical results on the localization and distribution of storage proteins in a poplar tree and twigs of other tree species. Holzforschung, **43**: 1–6.

Scarascia-Mugnozza, G.E., Hinckley, T.M., Stettler, R.F., Heilman, P.E., and Isebrands, J.G. 1999. Production physiology and morphology of *Populus* species and their hybrids grown under short rotation III. Seasonal carbon allocation patterns from branches. Can. J. For. Res. **29**: 1419–1432.

Schulte, P.J., and Hinckley, T.M. 1987. Abscisic acid relations and the response of *Populus trichocarpa* stomata to leaf water potential. Tree Physiol. **3**: 103–113.

Schulte, P.J., Hinckley, T.M., and Stettler, R.F. 1987. Stomatal responses of *Populus* to leaf water potential. Can. J. Bot. **65**: 255–260.

Smit, B., and Stachowiak, M. 1988. Effects of hypoxia and elevated carbon dioxide concentration on water flux through *Populus* roots. Tree Physiol. **4**: 153–165.
Smith, N.G., and Wareing, P.F. 1974. The distribution of latent root primordia in stems of *Populus* × *robusta*, and factors affecting the emergence of preformed roots from cuttings. Forestry, **45**: 197–209.
Souch, C.A., and Stephens, W. 1998. Growth, productivity and water use in three hybrid poplar clones. Tree Physiol. **18**: 829–835.
Sparks, J.P., and Black, R.A. 1999. Regulation of water loss in populations of *Populus trichocarpa*: the role of stomatal control in preventing xylem cavitation. Tree Physiol. **19**: 453–459.
Stettler, R.F., Bradshaw, H.D., Jr., Heilman, P.E., and Hinckley, T.M. (*Editors*). 1996. Biology of *Populus* and its implications for management and conservation. NRC Research Press, Ottawa, ON. 539 pp.
Tew, R.K., DeByle, N.V., and Schultz, J.D. 1969. Intraclonal root connections among quaking aspen trees. Ecology, **50**: 920–921.
Tschaplinski, T.J., and Blake, T.J. 1985. Effects of root restriction on growth correlations, water relations and senescence of alder seedlings. Physiol. Plant. **64**: 167–176.
Tschaplinski, T.J., and Blake, T.J. 1989. Water relations, photosynthetic capacity and root/shoot partitioning of photosynthate as determinants of productivity in hybrid poplar cultivars. Can. J. Bot. **67**: 1689–1697.
Tyree, M.T., Kolb, K.J., Rood, S.B., and Patino, S. 1994. Vulnerability to drought-induced cavitation of riparian cottonwood in Alberta: a possible factor in the decline of the ecosystem? Tree Physiol. **14**: 455–466.
Westhaus, W. 1986. Observations on the tolerance to flooding of woody plants and some suggestions for planting. Hercynia, **23**: 346–353.
Widin, K.D., and Schipper, A.L., Jr. 1981. Effect of *Melampsora medusae* leaf rust infection on yield of hybrid poplars in the north-central United States. J. For. Pathol. **11**: 438–448.
Zimmerman, M.H. 1978. Hydraulic architecture of some diffuse porous species. Can. J. Bot. **56**: 2286–2295.
Zimmerman, M.H. 1983. Xylem structure and the ascent of sap. Springer-Verlag, New York, NY.
Zsuffa, L., Giordano, E., Pryor, L.D., and Stettler, R.F. 1996. Trends in poplar culture: some global and regional perspectives. *In* Biology of *Populus* and its implications for management and conservation. *Edited by* R.F. Stettler, H.D. Bradshaw, Jr., P.E. Heilman, and T.M. Hinckley. NRC Research Press, Ottawa, ON. pp. 515–539.

CHAPTER 4
Ecology and silviculture of natural stands of *Populus* species

J.C. Zasada, A.J. David, D.W. Gilmore, and S.M. Landhäusser

Introduction

The trees comprising the *Populus* genus in Canada and the United States are a fascinating group of species in ecological and biological terms and are very important to regional economies. In this chapter, we will deal mainly with aspen (*Populus tremuloides*), the most widely distributed tree species in North America (Note: In the discussion that follows, reference to aspen also refers to bigtooth aspen (*P. grandidentata*) where their ranges overlap unless otherwise stated. Although there are differences between these species, they are similar enough to be considered together for our purposes (Graham et al. 1963; Burns and Honkala 1990; Rauscher et al. 1996)). *P. balsamifera*, related to *P. trichocarpa*, and *P. angustifolia*, is the second most widely distributed tree in the genus; it occurs further north and west than any other tree species in North America. The final three species, *P. deltoides*, *P. fremontii*, and *P. heterophylla* are mainly riparian species (Stettler et al. 1996; see Chap. 1 of this book).

There are important differences among poplar species as well as some remarkable similarities relative to management of natural populations. Several examples are given below. In habitats as contrasting as the relatively hot and dry plains of the southwestern U.S. and north of the Arctic Circle, they are often confined to floodplains and riparian zones and are commonly the largest and the only trees for many kilometers. In the north temperate and boreal forest, they are common in

J.C. Zasada. USDA Forest Service, Forestry Sciences Laboratory, North Central Research Station, Grand Rapids, MN 55744, U.S.A.
A.J. David. Aspen/Larch Genetics Cooperative, University of Minnesota, North Central Research and Outreach Center, 1861 Highway 169 East, Grand Rapids, MN 55744, U.S.A.
D.W. Gilmore. Department of Forest Resources, University of Minnesota, St. Paul, MN 55108, U.S.A.
S.M. Landhäusser. Department of Renewable Resources, University of Alberta, Edmonton, AB T6G 2M7, Canada.
Correct citation: Zasada, J.C., David, A.J., Gilmore, D.W., and Landhäusser, S.M. 2001. Ecology and silviculture of natural stands of *Populus* species. *In* Poplar Culture in North America. Part A, Chapter 4. *Edited by* D.I. Dickmann, J.G. Isebrands, J.E. Eckenwalder, and J. Richardson. NRC Research Press, National Research Council of Canada, Ottawa, ON K1A 0R6, Canada. pp. 119–151.

pure and mixed species stands. Some species manage to exist at the northern or elevational limits of tree growth, while others occur in prime habitats and exhibit very high growth rates in natural and managed stands. As a group, poplars exhibit the broadest variety of vegetative reproduction of any genus of North American trees; but each species has its own unique potential for vegetative reproduction. Relative to other genera, the sexual reproduction of poplars, particularly their seed biology, is very similar; a very short seed life and narrow range of suitable seedbed conditions for germination are key features of this process. All species are important components of wildlife habitat for a variety of game and non-game species in early and mid-succession forests. There are other contrasts and similarities among poplars, but those mentioned serve to provide an introduction to important diversity that is relevant to silviculture and management of this important genus of North American trees.

Measured as volume or biomass, the use of *Populus* spp. as a raw material for forest industry is, at present and for the foreseeable future, dominated by the harvest and utilization of aspen for pulp and paper in the northern Great Lakes region (Minnesota, Wisconsin, and Michigan) and the boreal mixedwood region of Canada. It is interesting to note the contrast between the forests in these two areas of high utilization. The dominance of the northern Lake States forests by aspen resulted mainly from activities associated with settlement of the region by western Europeans and subsequent human disturbance that tends to maintain the forest type. The boreal mixedwood of Canada is a result of natural disturbance, mainly fire, but harvesting is increasing and changing the ratio of natural to harvested forest.

The volume of aspen harvested in the Lake States (Minnesota, Wisconsin, and Michigan) in 1996 was approximately 10.4 million m^3 and accounted for 44% of the pulpwood harvest (Piva 1998). The volume of aspen harvested in Alberta in 1997 was 10.0 million m^3 and accounted for 43% of the total harvest (Natural Resources Canada 2000). For British Columbia, 2.9 million m^3 were harvested in 1998, accounting for 4.3% of the total harvest for the province (Natural Resources Canada 2000). Trends in utilization between these two areas have been generally similar but differ in magnitude. Both have experienced increases, but in Alberta, for example, over the last 15 years utilization has increased by 800% (Natural Resources Canada 2000). In all areas there is concern that there is little room for additional industrial expansion that depends primarily on the availability of aspen. Accurate estimates of area harvested are difficult to derive but have numbered well into the tens of thousands of hectares annually for the last few years.

The history of and potential uses for poplars are much more varied than the current utilization suggests. There is an interesting series of publications from the 1940's that presents a thorough analysis of the properties and potential uses of aspen in the northern Lake States (e.g., Zasada 1947). This series described uses such as house logs, veneer, lumber, chemicals, furniture core stock, and others.

At the time these publications were written, forests were recovering from severe overexploitation and fire during the previous 50–75 years. Foresters and wood products specialists were searching for the highest and best use of this plentiful, but largely unused, resource. There are numerous historical accounts of the importance of other *Populus* spp. for logs and lumber for home construction and other purposes. In the plains regions of western North America, *Populus* spp. were often the only trees of a sufficient size to use for building materials.

The value of *Populus* species goes far beyond the forest products mentioned above. In all parts of their range they are important, in some cases critical, elements in the landscape for their role in reducing erosion on upland and floodplain sites, providing habitat for wildlife and fish, and aesthetics, to name only a few high profile values. Historically these species have been important to Native Americans as a source of food, medicine, and building materials (e.g., Marles et al. 2000). Today parts of some trees, for example, the bark of older *P. trichocarpa* and *P. balsamifera*, are used by carvers and artists for creation of art objects with potentially high value.

The biology, ecology, silviculture, and management of the native *Populus* spp. in Canada and the U.S. have been the topics of a number of excellent reviews in the past two decades (DeByle and Winokur 1985; Peterson and Peterson 1995; Peterson et al. 1996; Burns and Honkala 1990; Rauscher et al. 1995; Stettler et al. 1996). However, during the 1990's concepts that had previously received relatively little attention, e.g., ecosystem management, biodiversity, restoration of riparian forests, and management of mixed species forests, have created new management objectives and altered silvicultural systems for several *Populus* species. This review will cover what we believe are important basic concepts for developing silvicultural systems, in particular the genetic variation in the genus, reproductive characteristics, and site productivity. Following this, we will discuss several trends that are bringing change to the practice of aspen silviculture. This review will be biased toward aspen because there is generally more information available for these species.

Regeneration in *Populus*

Vegetative regeneration

The silvicultural systems applied to aspen stands are centered around the potential for their vegetative reproduction by suckering (Figs. 1, 2). This mode of reproduction creates clones (see below for more detailed information), i.e., stands comprised of genetically identical trees that have all developed from a single seedling that established decades to centuries in the past (Fig. 3). Information on sexual or vegetative regeneration mechanisms of the other *Populus* species is critical whether for the restoration of riparian habitats or for regeneration following natural or human-caused disturbances. Thus, it is essential to understand the

Fig. 1. Aspen root segment with numerous suckers developing from adventitious buds. The suckers can utilize and expand the root system of the parent tree and (or) develop an independent root system. (Photo courtesy of Aspen/Larch Genetics Cooperative, Department of Forest Resources, University of Minnesota.)

Fig. 2. Dense 6-year-old stand of aspen suckers. The density of suckers in the year following harvest commonly ranges from 30 000 to 100 000 suckers/ha. Mortality resulting from self-thinning during the first decade of stand development is high. (Photo courtesy of Aspen/Larch Genetics Cooperative, Department of Forest Resources, University of Minnesota.)

Fig. 3. The high capacity for suckering by trembling aspen results in the development of clones that cover areas of variable size. Three easily identified clones are circled, but there are at least 7–8 other clones in the aspen stands shown in this scene. The other broadleaved species occupying areas around the aspen in this upland area in interior Alaska is paper birch. (Photo courtesy USDA Forest Service.)

process of vegetative reproduction that is critical for the management of these species. Historically, aspen regeneration in most of its range was strongly influenced by wildfire. Regeneration of *P. deltoides*, *P. heterophylla*, *P. trichocarpa*, *P. angustifolia*, and *P. fremontii* is mostly influenced by riverine disturbances, such as flooding, erosion, and fluvial deposits. *P. balsamifera*, which may occur on sites where fire was prevalent or in riparian zones, has a wide potential for vegetative reproduction. This variety of sites and disturbance regimes very much reflects the wide range of possibilities for vegetative reproduction in the different *Populus* species. Although the basic physiological controls that regulate vegetative reproduction are likely the same throughout the range of a species, the location of the roots (in particular depth) and the relative importance of environmental controls (e.g., soil water and temperature, plant competition) most likely differs among geographic regions and sites within regions. These differences are a major reason for some of the variation in response to disturbance summarized below.

The species that are most common on sites historically subjected to periodic fire have developed a vegetative regeneration strategy that relies heavily on the root system, which is more likely to survive fire than are the above-ground parts. Root suckers develop from adventitious primordia (meristematic cells) in the cork cambium laid down during secondary growth of the roots (Fig. 1). These primordia may be present before the stand replacement disturbance or develop after the disturbance (Schier 1973); clone history and genetics are important factors determining their presence. It is well-documented that the removal of apical dominance, i.e., a change in the balance between auxins produced in the shoots and cytokinins in the roots, and increasing soil temperatures stimulate the flushing and expansion of these primordia into suckers. However, the magnitude of sucker suppression via apical dominance is not quite clear, since suckering has also been observed within stands with intact canopies (Betters and Woods 1981; Lieffers and Campbell 1983; Jones and DeByle 1985; Doucet 1989). Suckering within stands may also be related to insect defoliation or decline in individual tree vigor that alters hormone balance and allows the forest floor to receive more sunlight. Soil temperatures increase due to the removal of the canopy and other vegetation. In addition, further increases in soil temperatures are caused by disturbance or removal of organic soil layers and by the blackening of the soil surface by fire. Suckering is normally completed after 2 years; at that time hormone levels are thought to have stabilized and the closing canopy leads to cooler soil temperatures (Hungerford 1988; Burns and Honkala 1990).

There has been much discussion on the persistence and maintenance of the parental root systems in aspen clones (Jones and DeByle 1985) and this characteristic most likely varies among and within regions. Some authors suggest that root connections persist only until the death of one of the connected saplings or trees interrupts the connection (Sandberg 1951). On the other hand, Shepperd (1993*a*) found that root systems in regenerating aspen stands in the Rocky Mountains were intact for at least 14 years, while DeByle (1964) found live connections in aspen after 50 years. New research in central Alberta has shown that the parent root system in boreal aspen clones can stay intact and functioning throughout the life of the new stand (DesRochers and Lieffers 2000) while new roots are established to complement and replace the parent root system over time (Shepperd 1993*a*; DesRochers 2000). DesRochers and Lieffers (2000) also found that the root systems of dead trees within a clone were maintained and utilized by the remainder of the clone throughout the life of the stand and that clonal connections were further increased by the development of root grafts. In contrast to earlier studies, root grafts were commonly found in these northern aspen clones.

Suckers will also form a new independent root system through the growth of adventitious roots from the base of the stem. New root initiation seems to vary among geographic areas and sites within regions. While Sandberg (1951) found that aspen quickly produced a new root system after suckering, Schier and Campbell (1978) found that deteriorating clones in the Rocky Mountain region of Utah were slow in developing their own new root system. DesRochers (2000) found a negative relationship between the development of new roots in regenerating stands and the mass of the parent root system, suggesting that the size of the parent root system strongly affects sucker performance and new root initiation. If the arising suckers do not have to spend energy on the development of a new root system, they are able to invest more energy into height and leaf area growth (DesRochers 2000). The initial number of suckers and corresponding development of leaf area has to reflect the respiration needs of the existing root system, since root to shoot ratios in these young regenerating stands are extremely high (Shepperd and Smith 1993). An inadequate number of suckers or leaf area might not be sufficient to maintain the parent root system, resulting in the death of portions of it. DesRochers (2000) and Shepperd (1993*a*) reported leaf area indices (leaf area per unit of ground area) of 2–4 m^2 per m^2 in stands 4–8 years old, which are comparable or even higher than what has been reported for mature stands (Delong et al. 1997).

Aspen stands normally regenerate well after clearcutting (Fig. 2); however, aspen suckering can be negatively affected by a combination of factors relating to stand age, timing and type of harvest, site conditions, soil disturbance, and plant competition (Navratil and Bella 1990). Soil temperature, root carbohydrate reserves, clonal variability, and herbivory are among other factors that can be important for suckering and growth of suckers (Maini 1967; Zasada and Schier 1973; Shepperd and Fairweather 1994). Armillaria root rot infection increases with time since harvest in aspen stands that are of sucker origin. Armillaria root rot, therefore, may limit rotation length and the number of times that aspen stands can be successfully regenerated vegetatively (Stanosz and Patton 1987).

Currently it is recommended that sites be clearcut and left with minimal soil disturbance to promote aspen suckering. But in northern Alberta there have been numerous examples where this technique has resulted in sparse and sporadic sucker initiation (Darrah 1991). There are several reasons why the above recommendation may not be appropriate for aspen regeneration in the boreal forest. Cold soil temperatures during the growing season are one of the major factors determining poor suckering and growth performance after harvesting in the boreal forests. Low soil temperature is a major limiting factor in the boreal forest, especially in soils with thick organic layers (Hogg and Lieffers 1991) or high slash loads (Shepperd 1996), and it has a strong effect on suckering (Zasada and Schier 1973; Lavertu et al. 1994) and growth (Landhäusser and Lieffers 1998). A threshold soil temperature of 15°C has been suggested as necessary for successful aspen sucker regeneration (Hungerford 1988; Maini 1967). Results from growth chamber studies suggest that new fine roots, essential for water and nutrient uptake, are lacking at soil temperatures below 6°C (Landhäusser and Lieffers 1998; Wan et al. 1999).

Other effects of harvesting on sucker regeneration of *Populus* species are soil compaction, physical damage to the root system, and increase in soil moisture after harvesting (Navratil et al. 1996). The amount of damage to the root system is related to both soil conditions and the amount of traffic (Shepperd 1993*b*). The effect of harvesting and skidding traffic on the suckering ability is known to be significant (Kabzems 1996; Bates et al. 1993; Stone et al. 2000). The effect of wounding of aspen roots on suckering is not clear. While there have been suggestions that wounding of aspen roots can increase suckering, there have also been general statements that wounding is detrimental to suckering or the long-term growth of the stands due to the potential of increased stain and decay (Steneker 1973; Navratil and Bella 1990). For example, Stanosz and Patton (1987) found that over 70% of the trees in naturally regenerating aspen stands in Wisconsin were infected by *Armillaria* 15 years after harvest.

While aspen usually shows prolific suckering after a disturbance has killed or removed the above ground parts, *P. balsamifera* suckering is generally not as abundant (Navratil 1996; Peterson et al. 1996). However, compared to the other *Populus* species from the *Tacamahaca* and *Aigeiros* sections, it suckers well; e.g., 3000 stems/ha after cutting in the western boreal forest of Canada (Navratil and Bella 1990). *Populus balsamifera* exhibits the largest variety of vegetative regeneration strategies of any of the North American poplars and perhaps any other tree species in northern forests. These regeneration strategies include sprouting from existing or adventitious buds at the stump and (or) the root collar, and the ability to form roots on buried branch fragments. All have been observed in harvested areas (Burns and Honkala 1990). Rooting of branch segments of *P. balsamifera* is probably more common in wet riparian or seepage habitats than on the upland sites.

The tendency for *P. balsamifera* to occur in seasonally wet depressions on upland sites suggests a greater tolerance to elevated water levels in the soil than aspen. In fact, *P. balsamifera* is considered to be very flood tolerant (Viereck 1970; Braatne et al. 1996; Krasny et al. 1988*b*). The characteristically riparian *P. deltoides* has been found to have large clonal variation in flood tolerance (Cao and Conner 1999). One component of flood tolerance is the ability of stems covered by silt deposited during flooding to send out new roots into the silt. *P. balsamifera* readily expands into newly deposited silt layers, developing a distinctly layered root system; aspen apparently does not have this capability. Although suckering has been observed in most of the riparian poplars, especially in the *Tacamahaca* section (Rood et al. 1994), it does not seem to be the prevalent method of vegetative reproduction in other riparian cottonwoods (Braatne et al. 1996). Clonal expansion of seedling-origin *P. balsamifera* through the production of suckers from the root system can occur relatively rapidly on floodplains (Krasny et al. 1988*a*). Vegetative regeneration includes mostly sprouting of pre-existing or adventitious buds from stump and root collar and the rooting of buried branches.

In *P. trichocarpa*, regeneration via branches is even further advanced. Small shootlets that abscise during the growing season have the capacity to root and establish if they fall on or drift in water to a suitable microsite, a process called cladoptosis. In an extensive study along the floodplains of a southern Alberta river system, Rood et al. (1994) found that overall 52% of the regeneration of *P. angustifolia*, *P. balsamifera*, and *P. deltoides* was via seedlings, 30% through root suckers, and 18% as shoot resprouts. Regeneration via cladoptosis in these three species was less than 1%. Root suckers occurred only in *P. angustifolia* and *P. balsamifera*; they generally occurred in drier microsites away from the river; wetter microsites closer to the river were occupied by seedlings.

Coppicing from basal stem and root collar buds is possible in all poplar species at some time in their life. In aspen, basal sprouting may be common in trees less than 15–20 years old but is rare in older plants. In other species, basal sprouting is more common in trees that are greater than 40–50 years old. Sprouting has been observed in harvested *P. balsamifera* that were more than 100 years old.

Sexual regeneration

Regeneration from seed has been of essentially no importance in developing silvicultural systems for aspen, but it is very significant in the development of floodplain forests. A key feature in the biology of *Populus* seeds is their short seed life, generally less than 2–3 weeks following seed dispersal. Seed dispersal, occurring primarily in late spring and early to mid summer, must be closely coupled with the presence of adequate safe sites for successful seedling establishment. Although regeneration from seed is not an important consideration in aspen silviculture, it has been an important part of the ecology of the species. Seedlings are believed to have played a large role in development of Lake States aspen forests after exploitation of the pre-settlement forests, subsequent burning and clearing for agriculture, and then abandonment.

Two examples help to illustrate how the balance may be tipped to favor seed germination and seedling establishment. First, regeneration from seed was believed to be a rarity in the western U.S., although seed production was apparently not a limitation. The bottleneck seemed to be in germination and establishment. The fires in Yellowstone Park in 1988 resulted in large areas of exposed or ash-covered mineral soil seedbeds and aspen seedlings established abundantly on these microsites (Kay 1993; Romme et al. 1997), suggesting that at least, in part, fire exclusion policies may have been limiting seed regeneration. A second example is from floodplain sites in interior Alaska. On these sites, aspen is rare in primary succession (Viereck 1970; Krasny et al. 1988*a*, *b*), yet sites receive annual seed rain even though the nearest seed sources are several kilometers distant. However, disturbance caused by harvesting and prescribed burning of mature to over-mature white spruce stands altered these site conditions to allow aspen seed regeneration as part of the mix of colonizing species (Dyrness et al. 1988). Aspen stocking was as high as 90% 2 years after burning, and small, heavily browsed, saplings were still common 15 years later. No seedlings were present in unburned areas except where mineral soil had been exposed by logging or mechanical site preparation. These and other observations suggest that with seedbed treatments timed specifically to coincide with *P. tremuloides* seed rain, it is possible to favor the establishment of seedlings and introduce aspen to sites where it did not previously exist.

The floodplains of North American rivers are special places for the *Populus* species adapted to these habitats (Fig. 4). Seed regeneration has been commonly observed on sites from dry southwestern areas to the northern- and western-most occurrence of trees on the continent. Braatne et al. (1996) provided an excellent

Fig. 4. (*a*) Floodplain of the Tanana River, Alaska, showing the diversity of habitats in which *Populus balsamifera* is found. The circled areas denote stands in which *P. balsamifera* is clearly the dominant species. Other sites have various *Salix* and *Alnus* species as well as *P. balsamifera*. The older stands will have a white spruce component in the understory. (*b*) Initial stages of primary succession on the Susitna River, Alaska. Although seed regeneration is common under these conditions, development of clones by suckering can be rapid and contribute significantly to stand formation. This site will be flooded periodically, resulting in the deposition of silt and the development of a layered root system as new roots develop and occupy the newly deposited sediment. (*c*) A 60-year-old *P. balsamifera* stand on the Tanana River. Stands located on the active river channel are subjected to continual erosion. Peak flood events can destroy large parts of a stand. (Photos courtesy of John Shaw, Department of Forest Resources, Utah State University, Logan, UT.)

(*a*)

(*b*)

Fig. 4 (*concluded*).

account of the intricacies of the entire process of seedling establishment on floodplains. As on upland areas, seeds are usually available, most commonly by wind dispersal. However, seeds are carried long distances in the water and can be deposited on suitable microsites as water levels recede, following periods of high water. In fact, germination often occurs when seeds are in water. Mineral soil surfaces are usually common in association with free-flowing rivers. However, there are a wide variety of conditions such as water level, location in the floodplain relative to the main channel, substrate texture, and surface chemistry that determine the availability of suitable microsites (Braatne et al. 1996; Krasny et al. 1988*a*). As on upland areas, opportunities for seedling establishment are likely more closely tied to substrate condition than to seed availability, although the latter is definitely important. Across the span of river conditions in North America there are many factors related to human development that are having a negative effect on the poplars that inhabit these sites. Although this development affects all aspects of the ecology of these species, dams and levees that alter the normal seasonal flow of the river limit regeneration and renewal of populations (Braatne et al. 1996; Rood et al. 1995).

Genetics of natural populations of *Populus*

Genetic variability in Populus *populations*

The eight species of *Populus* native to North America belong to four different sections and all are wind-pollinated species with male and female flowers borne on individual trees. This method of reproduction is associated with high levels of genetic diversity at the gene, individual tree and population levels, and provides for a high level of gene flow within species. The large natural ranges for *P. deltoides*, *P. tremuloides*, *P. balsamifera*, and *P. trichocarpa* cover a wide variety of environmental conditions, suggesting that genetic variation for adaptive traits is extensive. Furthermore, because many of these species have similar site requirements they can coexist in stands. This physical proximity combined with a common chromosome number ($2N = 38$) and overlapping flowering phenologies provides an opportunity for natural hybridization and gene flow among several species. Collectively these facts predict a large level of genetic variation in *Populus* species native to North America.

Genetic structure

Although a comprehensive investigation into the range-wide genetic structure of trembling aspen has not been attempted, investigations from smaller geographic areas using molecular markers indicate a high level of genetic diversity at the population level (Yeh et al. 1995). This is undoubtedly due to extensive gene flow and little inbreeding associated with a species that is both wind pollinated and dioecious.

Relative to *inter*clonal genetic variation, the investigation of *intra*clonal (or somaclonal) genetic variation in aspen has been all but ignored. On first consideration all stems of the same clone should be genetically identical because all stems trace their lineage back to the original seedling. However, due to the size and age of extremely large clones, there are potentially several generations of stems with different lineages back to the same original seedling. One study focusing on mature aspen clones in Yellowstone National Park found unexpected incremental genetic variation among stems in each of 10 mature clonal stands of *P. tremuloides* (Tuskan et al. 1996). Because these results have not been duplicated, it is difficult to ascertain the frequency or the cause of somaclonal variation. The potential impact of this genetic variation on the ecological function of natural stands or the performance of clonally propagated individuals is unknown.

Clonal attributes

In the subarctic or mountainous West, individual clones may be easy to identify because of the physical distance between clones or the phenology of individual clones. In the Lake States and in the eastern portion of the continent, intermingled clones make identification of individuals difficult (Fig. 3). Traits such as sex of the clone, date of bud break or leaf senescence, leaf shape, and branch angle among others can be used to identify individual clones and estimate their size. Average clone size in the Lake States and Alaska range from 0.05 to 0.2 acres, but these are dwarfed by clonal sizes of aspen in the Rocky Mountain region. In south central Utah, an aspen clone exists that may be one of the most massive organisms in the world, weighing an estimated 6 million kg and covering 43 ha with 47 000 total stems (Grant et al. 1992).

Because they are dioecious, clones are either male or female, although examples of hermaphroditism do exist. In aspen, the ratio of males to females has been debated, but it is now generally accepted that the sex ratio in a population is balanced. The increased frequency of male versus female flowering and the higher number of male versus female flowers on a given clone may have contributed to the estimation of a male bias in sex ratio. This decreased frequency of flowering in female clones and the lower number of flowers per clone may explain why female clones have a faster growth rate and larger clone sizes than males. These findings are consistent with the widely held notion that energy expenditures for flowering in female clones exceed that of males.

Occasionally, triploid or even tetraploid clones are found in natural populations or as the result of controlled crosses in breeding programs. There is evidence to suggest that triploid clones possess faster growth rates and better fiber quality than diploid clones, but their usefulness in a breeding program is questionable due to chromosomal mispairings at meiosis.

Clonal variation

Because aspen can vegetatively propagate, clones function at the level of the individual. Therefore clonal performance is akin to individual performance in most other tree species, the difference being that clones function on a larger spatial scale. Variation among clones exists for a wide variety of traits, including growth rate, wood quality traits such as density and fiber length, insect and disease susceptibility, and drought stress to name a few. Clearly, clonal variation in aspen exists for a variety of traits, but extrapolating the results of individual experiments to a functional clonal organism in the field is difficult. Clone size also confounds the ability to predict community responses because interclonal variation will be felt on a spatial scale larger than an individual tree. Thus when placed in the matrix of a boreal or northern forest the impact of clonal variation on the complexity of the forest community becomes difficult to forecast.

Natural hybridization

Natural hybridization among native *Populus* spp. is common, and both intra- and intersectional crosses are represented. High levels of natural hybridization occur because of a common diploid state within *Populus* ($2N = 38$), the close physical proximity of different species owing to similar site requirements, and the occurrence of overlapping flowering phenologies. One of the best-studied *Populus* hybrid zones occurs in Weber Canyon near Salt Lake City, Utah, between *P. fremontii* and *P. angustifolia*. In this hybrid zone, nuclear and mitochondrial genetic markers have been used to identify parental species, F1 hybrids, and backcross generations where the hybrids mate with a parental species. Analysis of genotypes from throughout the hybrid zone indicates that F1 hybrids are backcrossing only with *P. angustifolia* and the complexity of the backcrosses is clinal, increasing with elevation up the drainage. Thus unidirectional introgression is occurring between these two species, and it appears that the hybrid zone is dynamic, spreading down the canyon at the expense of the *P. fremontii* population (Keim et al. 1989; Paige et al. 1991).

Other poplar hybrid zones are described in the literature mainly from an increased incidence of pests or pathogens on the hybrids relative to the parental species. This occurrence of hybrid breakdown, where hybrid performance is below that of the pure parental species, is known to occur in the hybrid zones of other plant species. In several cases, the ability of insects to detect hybrid individuals is so strong that insect bioassays have been proposed as a fast, reliable method of supplementing morphological traits in identifying parental species and hybrids in the field (Floate and Whitham 1995).

From a biodiversity standpoint, the heightened level of herbivory on hybrids contributes to increased levels of species richness in riparian ecosystems. This species richness created by the genetic differences of parental and hybrid individuals undoubtedly plays an important role in the ecology of riparian communities by serving as a focal point for host–pathogen and predator–prey relationships, among others.

Effects of silvicultural systems on Populus *genetics*

Because *Populus* species are obligate outcrossers and wind pollinated, they maintain high levels of gene flow among populations, which results in high levels of genetic variation within populations. With high gene flow levels and seed capable of traveling long distances, we predict that harvesting would have little impact on loss of genetic variation from pre-harvest to post-harvest forests. In the case of aspen, with its strong suckering response to harvesting, the assemblage of genotypes in the post-harvest forest should be identical to the pre-harvest genotypes, although the area occupied by each clone may vary according to clonal suckering ability. New genotypes may arise if conditions for seedling germination are met, but competition with root suckers for available light and other resources would be severe.

In contrast to aspen, other poplars are usually not as numerous in post-harvest forests. This is due to their decreased ability to clonally propagate and the lack of adequate microsites for seedling establishment in a post-harvest environment. This increased dependence on seedling recruitment means that the post-harvest forests of non-aspen poplar species will be a mixture of old and new genotypes. This mixture will be determined by a species' propensity for vegetative propagation and the ability of individuals in the stand to clonally propagate. By comparison to post-harvest aspen sites, these sites will be skewed towards new genotypes.

The evolution of genetic markers provides the potential to answer some fundamental questions about aspen regeneration after harvesting and how it might be manipulated to increase stand productivity. Both *P. tremuloides* and *P. grandidentata* respond dramatically to harvest by strong root suckering. However, the individual clonal response to harvesting is largely a matter of conjecture. Morphological characteristics have been used to identify individual clones in pure aspen stands, but identifying individual clones on the basis of root suckers becomes problematic due to the distortion of morphological features associated with the fast growth of root suckers. Simple sequence repeats, or SSRs, are a relatively new class of genetic markers that detect large levels of genetic variation and would allow for quick and accurate identification of clones and clonal boundaries regardless of growth rate (Dayanandan et al. 1998). Short rotation ages for coppice stands of native aspen in the Lake States (currently 40–55 years), coupled with the fact that aspen is harvested primarily for pulp fiber and wood chips, indicates that shorter rotations may be possible in native stands as improvements in harvesting machinery and processing technology allow smaller diameter material to be utilized. If harvesting favors clones that root sprout heavily, then genotype losses and the restructuring of genetic variation may occur at a faster rate as rotation ages decrease with improvements in technology.

If there is a clonal response to harvesting and rotation ages are shortened, the concept of plus clone selection (Perala 1977) may gain favor as a silvicultural treatment. Ideally, clonal response to harvesting would be correlated with some useful phenotypic trait with moderate to high heritability that would make the identification of clones superior for growth traits and suckering ability relatively easy. Sucker formation by less desirable clones could be discouraged through the application of herbicides, thinning during the rotation, or simply by allowing these clones to break up from old age.

Productivity

Site quality

We now briefly describe the developmental sequence of an aspen stand development in general terms. Immediately after harvest or burning, sucker density ranges from 10 000 to 100 000 stems/ha. Within 8–10 years, density dependent mortality reduces this to 5000–10 000 stems/ha. Mortality tends to be much greater on more productive sites during this period. At maturity, tree density ranges from 500 to 700 stems/ha. Mortality during the first 30 years of stand development tends to occur in waves rather than through a gradual loss of individual stems (Burns and Honkala 1990; Perala et al. 1996, 1999; Doucet 1989). The range of heights attained at various ages are: 1–2 m at 1 year; 6–23 m at 20–25 years; and 20–25 m at maturity, with heights sometimes exceeding 30 m. The large geographic range of aspen means variability will exist among areas, but in some cases there is as much variation within a specific area as among them. One important feature of stand development is longevity, which tends to increase for higher latitudes and with altitude in the Rocky Mountains, where trees are commonly 100 years old and the oldest near 200. Perala et al. (1996, 1999) have developed models for predicting stand development based on the self-thinning concept. These models include independent variables like average July temperature and precipitation so that they can be adapted to local conditions.

The clonal habit of aspen has many important implications for site quality assessment and growth and yield. In almost all cases, assessment of productivity in natural stands is based on measurement of one or more clones comprised of varying numbers of individual, but possibly connected, trees. Although clones from all areas originated from seed at some time in the past, most commercially harvested stands are likely several generations removed from the original establishment event. Consequently, once established on a site, an aspen clone or group of clones may propagate themselves indefinitely. The size of the clone provides some idea of the age since establishment — the larger it is the older it is.

Where seedling and sucker growth have been measured, suckers grow faster than plants of seedling origin by a factor of at least 3–5 times during the first decade. Generally there have been few comparisons between suckers and seedlings in terms of age to achieve specified vegetative and reproductive growth stages. For aspen seedling regeneration, competition from herbaceous species and vegetative reproduction from other woody species slows growth and causes significant early mortality.

Estimated or direct measures of site index are the fundamental bases for height growth equations used in the majority of growth and yield models currently available. Two assumptions in the use of site index as an indicator of site quality are: (*i*) tree height is an indicator of site potential and is not influenced by non-site factors such as stand density or site preparation treatments, and (*ii*) site index is constant over time. The above assumptions are compromised if aspen trees of the same clone are used in the assessment of site quality. Ramets (individual stems within clones) are genetically identical. Therefore, tree height may be influenced either positively or negatively through the clone's genetic makeup as opposed to site factors. Second, because clones of aspen share a common root system, a ramet could be growing on a poor microsite that would not support its growth if the entire root system were confined to this area.

Numerous site index curves have been published for *P. tremuloides* across its natural range. A common caveat published with site index equations is that their applicability be confined to the region where the study was conducted due to changes in site productivity associated with geographic location (Carmean et al. 1989). Differences in predicted values of site index may be more an artifact of the prediction equation than an actual difference in site index. Common site index curves for *P. tremuloides*, however, may be applicable across the natural range of this species. We graphically compared site index curves for trembling aspen developed for different regions of North America (Fig. 5). A common model form with separate parameters was used for three of the sets of site index curves

(Carmean et al. 1989) and a similar model form was used for a fourth (Chen et al. 1998*a*). Two of these studies used an index age based on a total tree age of 50 years (Gevorkiantz 1956; Carmean 1978) and two used an index age based on a breast height (bh, 1.3 m) age of 50 years (Deschamps 1991; Chen et al. 1998*a*). Carmean et al. (1989) suggested a correction factor of 4 years for an aspen tree to reach bh. We used this published correction factor, but acknowledge that aspen suckers and seedling can reach a height of 4.5 ft (1 ft = 0.305 m) in their first growing season.

There was less than a 3-m difference in predicted values of site index at any given age among these four sets of site index equations. Differences in predictability were greatest at the lowest and highest site indices in the 15–30-year range; differences were least around the index age of 50 years. Above the index age, differences in predicted site index were generally less than 1 m up to an age of 80 years. While limited in scope, this comparison implies that a common set of site index curves may be applicable across the natural range of aspen. This hypothesis should be tested, however, as our review of an additional set of site index curves from the Central Rocky Mountains (Edminster et al. 1985) suggests lower productivity of aspen in that region of North America. It was not possible to graphically compare the site index curves developed by Edminster et al. (1985) because they used an index age of 80 years at bh. From a statistical perspective, the selection of an appropriate model form is important to minimize biases in height prediction (Chen et al. 1998*a*). From a pragmatic perspective, the construction of separate site index curves for different regions to account for nuances in prediction patterns is difficult to justify (Gilmore et al. 1993). The site index curves compared were derived from aspen stands that developed in the absence of silvicultural treatments following natural or human-caused disturbances. Our comparison of site index equations, while limited, corroborates Perala et al. (1996, 1999) who detected little difference in stand level characteristics (growth and yield, self-thinning) of *P. tremuloides* and *P. tremula*, which are considered by some to be a single circumpolar species.

Although there are differences in productivity with increasing latitude, one of the features that is not captured in equations and stand level analysis is presence of aspen at the landscape level of resolution. For example, in the Lake States region aspen grows on a wide variety of sites, as indicated by site indices of pure stands between 50 and 90. At the north end of the distribution, the range of sites on which aspen occurs seems to be much smaller. It mainly occurs on warmer south-facing sites and is generally absent from large areas because of the presence of cold soils and permafrost.

The above site quality and growth and yield work have been retrospective and concentrated in undisturbed or second-growth aspen stands. Similarities in the growth and yield of aspen stands, with the notable exception of *P. grandidentata* (Perala et al. 1996), suggests that future work should focus on height growth and volume patterns following disturbance from harvesting that can reduce both suckering density and early growth.

The effect of insects and diseases on growth and yield of aspen and other poplars is beyond the scope of our paper; these topics are covered in detail in Chaps. 7 and 8 of this book. One general statement that we believe important is that there does not appear to be any single insect or disease that is of overriding importance to growth throughout the entire range of a *Populus* species. For example, the generally high incidence of Hypoxylon stem canker on trembling aspen in the Lake States does not occur in the northern parts of the range in Alaska and Canada or in the Rocky Mountains. In these parts of the range other diseases become relatively more important.

The *P. balsamifera – P. trichocarpa* complex in Alaska provides an example of the need for multiple sets of site index curves to adequately assess growth and productivity. The geographic area of interest, divided by the Alaska Range, contains large rivers along which the species grow, and is a well-documented area of range overlap and hybridization between the two species. In their analysis, Shaw and Packee (1998) recommend four sets of polymorphix site index curves. One covers the area north of the Alaska Range where balsam poplar occurs almost exclusively on floodplain sites. South of the Alaska Range, three sets of curves are recommended, one for what they refer to as "typical" floodplain sites, a second for floodplain sites in the upper reaches of the Susitna River valley, and a third covers stands on well-drained upland sites south of the range. It is interesting to note that the upper Susitna area coincides to a large degree with the area of hybridization between the two species.

Soil-site studies

Numerous soil-site studies for quaking aspen have been conducted over the past 50 years. The depth to the ground water table and percent organic matter in the upper 20 cm of the soil profile were found to have a strong relationship between site index, yields at age 40, and the annual rate of height growth in native aspen stands in central Wisconsin (Wilde and Pronin 1949; Carmean and Li 1998) (Fig. 5). A summary of variables found to have a correlation with aspen site index in the Lake States, Ontario, and British Columbia are provided in Table 1. These studies were conducted independently and lack a common experimental design. Consequently, the relative importance of variables to site quality from one region to another can only be inferred. In general, site quality of aspen is influenced by the fertility of the surface soil horizons, as determined through chemical analyses or inferred through soil texture analyses, soil drainage class, and depth to a root-restricting layer. In British Columbia, a positive relationship was found between site index and latitude, and a negative relationship between site index and elevation (Chen et al. 1998*a*, *b*).

Using Minnesota Cooperative Soil Survey, Forest Inventory Analysis, and Cooperative Stand Analysis data, Prettyman (1992) found the most important predictive relationships of site index for quaking aspen to be potassium (K) in the upper soil horizons, phosphorus (P) in the middle and upper soil horizons, and

Fig. 5. Height growth curves generated from site index studies of quaking aspen in the Lake States (Gevorkiantz 1956, solid line), Wisconsin (Carmean 1978, dash–dot line), Ontario (Deschamps 1991, dashed line), and British Columbia (Chen et al. 1998*a*, dotted line).

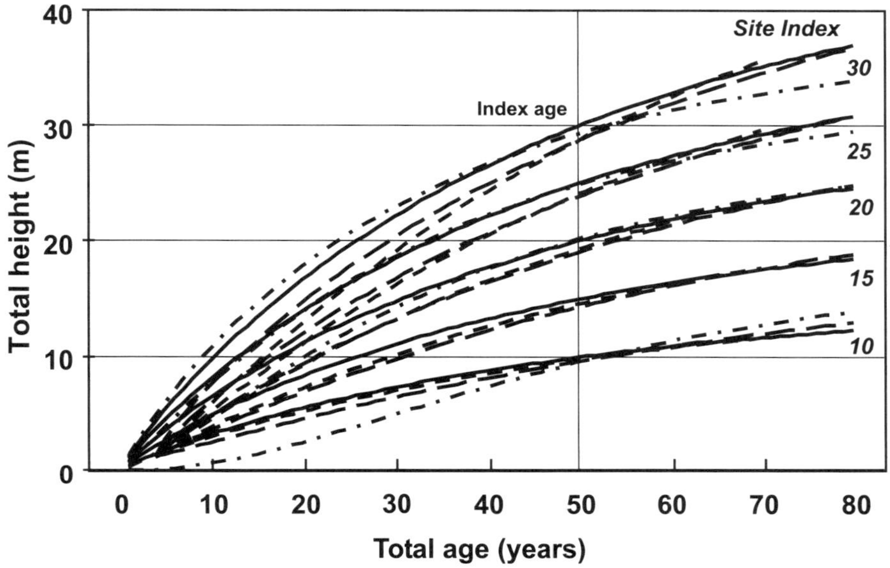

calcium (Ca) and cation exchange capacity (CEC) in the upper soil horizons. In small-scale harvesting studies on the effect of aspen rotation length on productivity, rotation lengths of less then 10 years have been considered unsustainable due to a drain of nutrient capital from a site (Berry and Stiell 1978).

Soil-site studies, while valuable, provide little information on sites where trees will not grow because such sites are in effect excluded by default from a sampling design. Thus, sites not suitable for a *Populus* species may exhibit some attributes of high quality but in combination with other variables create a poor environment for tree growth. For example, a site having adequate moisture and acceptable percentages of silt and clay in the mineral soil may have high percentages of coarse fragments in the C horizon that effectively create a poor environment for aspen growth.

Published optimal nutrient requirements for *Populus* are based more on anecdotal studies than experimentation. Fertilizer recommendations are often based on a single nutrient analysis for a site that is used as a relative gauge of soil fertility. Additional information that would contribute to the scientific underpinnings of fertilizer recommendations includes results of in situ and laboratory nutrient mineralization rates for a given site to determine the seasonal timing of nutrient release for a given soil. This information combined with seasonal plant nutrient demands would contribute to a greater efficiency in fertilizer application.

Another limitation of the soil-site studies included in this review is that they were all conducted on stands that developed in the absence of human-caused

Table 1. Correlations between site index and edaphic variables from selected soil-site studies in the Lake States, Ontario, and British Columbia.

	Forest floor	Mineral soil	Mineral soil	Morainal soils	Glacio-fluvial soils	Lacu-strine soils
Reference	Chen et al. (1998*b*)	Stoeckeler (1960)	Chen et al. (1998*b*)	Carmean and Li (1998)		
% silt + clay		0.64				
% clay		0.72				
% coarse fragments in B horizon				–0.39		
% coarse fragments in C horizon				–0.70		
% sand in A horizon				–0.35		
% silt in A horizon				0.40		
% silt + clay in A horizon				0.65		
% clay in BC horizon						0.48
% clay in C horizon					0.50	0.77
Moisture equivalent		0.71				
Conductivity		0.60				
Bulk density (g/cm^3)	–0.45					
pH	0.70	0.56	ns[a]			
Cation exchange capacity (CEC)		0.63				
N		0.54				
Total C (%)	ns		0.50			
Total N (%)	0.57		ns			
Mineral N (mg/g)			0.43			
Available P (mg/g)			–0.67			
Extractable K (mg/g)	0.49		ns			
Extractable Ca (mg/g)	0.66	0.50	ns			
Extractable Mg (mg/g)	0.63		ns			
Extractable S (mg/g)			0.51			
A horizon thickness		0.47				
BC horizon thickness				0.42		
Thickness	0.66					

Table 1 (*concluded*).

	Forest floor	Mineral soil	Mineral soil	Morainal soils	Glacio-fluvial soils	Lacu-strine soils
Depth to root restricting layer				0.45	0.68	
Depth to drainage mottles				0.45		0.65
Drainage class					0.51	

[a]ns = not significant.

disturbances. The effects of soil compaction and nutrient drain, while not quantified for aspen, will likely reduce individual tree productivity through restricted root growth and a lack of soil oxygen. Future work in this arena should focus on the effects of soil compaction and changes in water table levels through timber harvesting on aspen productivity.

Silvicultural systems for aspen

The silvicultural prescription for management of aspen that has evolved over the last 50 years is relatively uncomplicated. This species was called the "phoenix tree" by Graham et al. (1963) for good reason. It can overwhelm the site and capture most of the growing space within 1–2 years after disturbance because of its potential for root suckering (Figs. 1, 2). The most common silvicultural prescription allows the vegetatively regenerated stand to develop naturally with no intermediate treatments between final harvest cuts. Rapid self-thinning reduces stem density. There is no report of stagnation or other severe effects of density; in fact dense stands may be less susceptible to some insects and diseases (Perala 1977; Peterson and Peterson 1995).

As with all aspects of forestry, the past decade has signaled changes in diverse areas of aspen silviculture. We will discuss three topics — ecosystem management, harvesting technology, and thinning — that seem to us to be currently important aspects of aspen silviculture.

Ecosystem management

Clearcutting in the strictest sense will continue to be the dominant silvicultural system over most of the aspen type in North America. However, with the evolution of ecosystem management, retention of live trees on harvest units is becoming increasingly common (Fig. 6). The methods of application and immediate and long-term goals driving these practices differ by region. In Alberta, the Alberta–Pacific Corporation is retaining approximately 5% of aspen volume in cutting units primarily to meet wildlife habitat objectives (Stelfox 1995). Also, in

Alberta, partial cutting is prescribed in stands with varying mixtures of aspen and conifers. In these mixed stands, aspen is removed in the first entry leaving the conifers to occupy the site, but the goal is to maintain a mixed species stand with both aspen and conifer outputs over the long term (Fig. 7; Navratil et al. 1994). In the northern Great Lakes states, there is a tendency on all ownerships, but particularly on lands managed by the federal government, to retain both aspen and other species. The objectives of these practices include type conversion to longer-lived species (retained trees provide seed sources and suppress suckering), protection and restoration of riparian habitat, and wildlife habitat, with particular emphasis on providing conditions for species that require later succession forests.

Fig. 6. (*a*) One method used to meet ecosystem management objectives is to retain reserve trees in areas that are otherwise clearcut. The purposes of the reserve trees are to suppress suckering, maintain large trees for animal habitat, and provide large woody debris for the future. (Photo courtesy of Dr. Doug Stone, USDA Forest Service, Grand Rapids, MN.)
(*b*) Over-mature aspen are very susceptible to damage by wind storms. This susceptibility should be considered in selecting stands in which retention is being considered. This area was damaged in a severe wind storm, affecting more than 150 000 ha in the Boundary Waters Canoe Area, Superior National Forest, MN. (Photo courtesy USDA Forest Service.)

Fig. 6 (*concluded*).

Fig. 7. The area in the photo was harvested using cut-to-length processors. In this stand, an aspen overstory was removed, leaving the white spruce understory. The pattern of strips with all trees removed for access, alternating with uncut or partially cut strips, is the standard pattern for all cut-to-length harvesting and use of forwarders to move products to the landing. (Photo courtesy USDA Forest Service.)

Retention of trees on a harvest unit will reduce aspen suckering and early growth, but the effect has not been adequately quantified over a range of sites and tree densities. Perala (1977) indicated that 2.4–3.6 m^2/ha of residual overstory will reduce sucker growth by 35–40%. Doucet (1989), however, cited examples of adequate stocking with residual basal areas as high as 14 m^2/ha. In a case study conducted on the Superior National Forest in Minnesota, Stone et al. (2000), reported that leaving 75 uniformly dispersed reserve trees/ha on a harvest unit reduced aspen sucker density 33–41%, but sucker growth was not affected. Similar effects were observed by Navratil (1996) in the vicinity of residual groups of *P. balsamifera*. In areas where aspen management remains a major management objective, yet retention of some overstory is desirable, the question becomes "How can trees be retained and have the least impact on aspen regeneration and growth?"

The distribution of retained trees is a key consideration determining the effects on growth. Trees can be uniformly distributed or aggregated within a harvest unit. Each type of distribution has a different effect on development of the regeneration. If it is necessary to meet specific retention objectives on a harvest unit, it is possible to have the same average tree density (e.g., 50 trees/ha) over an entire harvest unit but distributed in spatial patterns that provide different amounts of growing space. For example, if an average of 50 trees/ha were required in a 10-ha unit (a total of 500 trees) one option is to have them uniformly distributed at an average spacing of 14 m. There are various options for aggregating the trees, e.g., twenty 25-tree groups or ten 50-tree groups. In the aggregated options, parts of the harvest unit would likely not differ significantly from a clearcut with no residuals. In addition to the effect on growing space for suckers, each of these distribution patterns could affect mortality, logging damage, animal habitat, and susceptibility to insect and disease.

The importance of aspen forests to numerous animal species is well known and an important aspect of ecosystem management. A primary reason for harvesting and managing aspen is to promote important game species like white-tailed deer and moose. Regenerating suckers are an important source of browse. The silvicultural system prescribed to improve and maintain ruffed grouse habitat in the Lake States is centered around treatment of aspen stands. This system recommends the use of 4–5-ha clearcuts dispersed among older stands in the landscape. Ruffed grouse require stands of different ages and density to meet their needs. Such diverse concerns as downed logs for drumming sites and mature male aspen clones for a supply of flower buds also need to be addressed in managing aspen for ruffed grouse habitat. It is believed that use of these recommendations can help to stabilize populations and reduce the large fluctuations that are characteristic of grouse (Gullion 1984).

Although nongame animals also utilize early successional aspen stands, they also utilize mature to over-mature trees and stands. For example, in the mixed-species boreal forest of western Canada, *Populus* species serve a different ecological function for woodpeckers than coniferous species. The use of *Populus* species by woodpeckers as nest trees is a function of tree size — hairy woodpeckers nested in trees that were on average 37 cm in diameter whereas downy woodpeckers preferred trees about half that diameter. *Populus* trees of adequate size are used primarily for nesting, while larger, decaying coniferous trees often contain ant nests, an important food source during winter (Caza 1993). These examples illustrate the need to consider retention of older, larger trees within a landscape managed for aspen if the goal is to provide habitat for a range of species (Stelfox 1995).

Harvesting technology

The development of cut-to-length, whole tree processors and the associated forwarding machines is creating new opportunities for planning and conducting harvest operations. The maneuverability of this equipment and the skill of the machine operators provides the lowest impact machine harvesting that has ever been available. This technology is used for both clearcutting and partial cutting (Figs. 7, 8; Sauder 1994). Protection of an established understory of desirable trees during clearcutting or of crop trees in a thinning operation is a common practice with these machines. In addition to protecting desirable trees, the branch and unmerchantable stem wood can be distributed such that machine travel occurs over this material, reducing the amount of direct contact with the forest floor and mineral soil.

The harvesting pattern in stands to be partially harvested or where protection of regeneration is necessary usually consists of alternating narrow clearcut strips where the machines travel with uncut or partially cut strips with no machine travel (Figs. 7, 8). Depending on the machine and application, the clearcut travel corridors are 2.5–4.5 m wide. The boom on the processor can reach 6–7 m on either side of the strip, making it possible to have the partially cut strips 12–14 m wide when travel corridors are on each side.

An extremely important aspect in the use of these machines is the skill and training of the machine operator. In many cases, the operator is given a density or basal area guide to follow and no trees are marked prior to harvest. The machine operator applies the silvicultural treatment, checking as the operation progresses with prism plots to assure harvesting is within the recommended guides. Sauder (1994) discusses the training needs necessary to assure successful execution of the silvicultural prescription.

Thinning

Although not a common practice throughout the range of aspen, thinning illustrates how interest in a given silvicultural practice varies through time and why it is important to maintain a broad basic and applied research program to meet the changing information needs for forest management.

Fig. 8. Cut-to-length processor, harvesting small diameter aspen from partially cut strip; processor is in the strip from which all trees are harvested. Merchantable logs are piled to the left side for removal by forwarder. Unmerchantable material is placed in the total removal strip to provide protection to the soil surface and reduce compaction. (Photo courtesy of Jim Marshall, Blandin–UPM Paper Company, Grand Rapids, MN.)

Thinning in aspen has been a topic of research for nearly 60 years. Although there has been research in other parts of the range of aspen, the majority of the information on thinning is from studies in Minnesota. Renewal of interest in thinning has occurred because of the imbalance in age classes and predicted wood shortages, and the need to obtain wood from younger stands because of its higher quality for some uses. This is particularly the case in Minnesota. Changes in harvesting technology have also been responsible for a renewed interest in thinning. Several thinning regimes have been proposed including both precommercial operations at about 10 years and commercial thinning at about 30 years (Perala 1977). Precommercial thinning has been conducted on a relatively large-scale basis. Blandin Paper Company in Grand Rapids, Minnesota, has, for example, mechanically thinned about 6500 ha in the past decade. The standard practice consisted of alternating strips (2–2.5 m wide) of flattened, but not severed, 8–10-year-old aspen saplings with 2–3 m wide untreated strips. More recently chipping has been tried to reduce the tangle created by just pushing stems over to improve access for hunters. Significant growth responses in the residual trees have been observed. In growth models, thinning is assumed to remove the smaller trees, i.e., thinning from below (Perala et al. 1996). The currently prescribed methods, basically row thinning, depart from the methods used in past research, so models need to be tested and likely modified to accommodate these new methods.

Commercial thinning is recommended for stands growing on good sites (site index 24.5+ m at 50 years) that are 25–30 years old and which have basal areas of 25–30 m^2/ha (Perala 1977). The introduction of cut-to-length harvesters could greatly increase the occurrence of commercial thinning. The maneuverability of these machines reduces damage to the residual stand compared to older mechanical harvesting (Fig. 8). In addition, their ability to distribute the unmerchantable materials to use as a travel surface reduces impact to the forest floor and has the potential for reducing soil compaction and root damage. As with precommercial thinning, commercial thinning is basically row thinning with some selection from the strips between the clearcut travel corridors.

Silvicultural systems for other *Populus* species

In aspen silviculture, managers are dealing with secondary succession; i.e., aspen is regenerated on sites where it occurred in the previous rotation and often for several previous rotations. For other poplars, the situation may be similar to that for aspen. However, stands of some species, e.g., *P. balsamifera* and *P. deltoides*, develop on sites that were formed relatively recently on active river floodplains (Fig. 3). Forest development under these circumstances can be described as primary succession. Changes in composition and soil conditions occurring during this process have been described (e.g., Viereck 1970; Braatne et al. 1996). In these situations, the poplars generally develop with other shade-intolerant, early colonizers such as willows and alders. They are eventually replaced by more tolerant hardwoods and conifers. Replacement of cottonwoods following harvesting is not, generally, as automatic as described for aspen. The species remains a part of the next stand, but they usually do not dominate the site. There is certainly an attractiveness to the potential for managing these species wherever they occur on floodplains because of their rapid growth — frequently described as the most rapid for the given region. The following two examples for *P. balsamifera* and *P. deltoides* provide two very different silvicultural scenarios and represent the ends of the continuum from extensive to intensive silviculture of natural stands. In many cases, the natural stands on these sites are planted with the same species or hybrid poplars or are converted to other species. The culture of planted poplars is discussed in detail in Chap. 5 of this volume.

P. balsamifera occurs in two general situations, as essentially pure stands on the floodplains of rivers in Canada and Alaska and on upland sites as a component of mixed stands with aspen, white spruce, and other northern species. In mixed stands, it may comprise about one third of the stand but usually less. It is frequently associated with poorly drained microsites in these stands. In some cases, it is harvested along with the other species, and in other situations it is not harvested (Peterson and Peterson 1995; Navratil 1996).

Much of the discussion regarding the silviculture of stands in which *P. balsamifera* occurs deals with the consequences of leaving it, particularly as it affects aspen suckering. Navratil (1996) concluded that aspen suckering within groups of uncut trees was reduced, but there was not a significant effect in the clearcut area around the reserved trees. Following harvest, *P. balsamifera* regeneneration occurs from all possible sources (root suckering, stump sprouting, detached branches, and seeds), but none of these sources alone or in combination provide adequate stocking.

The other end of the management continuum for natural poplar stands is summarized by Johnson and Shropshire (1983) for early successional stands along rivers like the Mississippi, Ohio, and Missouri. These stands, like those of their northern counterparts, are comprised of a poplar — *P. deltoides* — and other intolerant species such as black willow (*Salix nigra*). The growth potential in these stands is very high, and rotations are generally in the range of 30–40 years. When possible, light thinnings may begin as early as age 5 years with commercial thinnings at 15–20 years old. Recommended minimum basal areas are 9–21 m^2/ha. These species develop on recently deposited alluvial soils and do not replace themselves but succeed to more tolerant hardwood species. Regeneration of these stands does not occur without some type of fairly severe site preparation. The success of these cultural practices is not guaranteed unless post-treatment surface soil moisture is adequate for germination. Thus the tendency is to plant these sites if *P. deltoides* is desired in the next stand. This requirement of a close coupling between seed dispersal and seedbed conditions is a recurring theme regardless of the type of site, or whether regeneration is a planned silvicultural event or part of the natural process of colonization.

Summary

The *Populus* species native to Canada and the U.S. are at once a varied, yet similar, group. All tend to be early successional and thus are dependent on disturbance for regeneration and maintenance of populations at the stand and landscape level. Species, mainly aspen, that occur primarily on upland sites are well-adapted to periodic stand level disturbances. Historically these conditions resulted from fire, but presently forest harvesting is a major contributor to disturbance regimes. The riparian species depend on the floodplain processes of erosion and deposition to create conditions for stand establishment — both by sexual and asexual reproduction — and maintenance. The alteration of flow regimes in North American rivers has greatly influenced these species. In many areas, restoration procedures are needed to increase the extent of riparian poplar ecosystems.

The *Populus* species considered here are all wind-pollinated obligate outcrossers, which results in high levels of genetic variation at the population level and large amounts of gene flow between populations. Combined with long-range seed dispersal this means that loss of genetic variation as a result of harvesting or natural disturbance will be minimized. Loss of genetic variation is particularly unlikely with aspen because the same genotypes will be present on the site after a stand disturbance event due to its strong suckering ability.

The aspens are the most important poplars economically because they are a major source of fiber for the manufactured wood products industry (e.g., paper and oriented strandboard). Aspens also have high noncommodity values, e.g., soil stabilization, wildlife habitat, and nontimber forest products. These values vary by geographical region and cannot be overlooked in management and silvicultural planning. Aspen is a very resilient species because of its ability to rapidly dominate disturbed sites by suckering from the root system. The silviculture of this species depends on the regeneration potential of the root system. Although adequate stands are usually established following harvesting, there are a significant number of examples of poor stocking in harvested areas. A main contributing factor is soil compaction that changes the physical properties of the soil and may damage the root system. Other factors associated with less than full regeneration are competition from residual overstory, cold soils, and waterlogged soils. In short, the stocking and growth of aspen in the next rotation on a given site will be, to a significant degree, dependent on the vitality of the root system remaining after harvest of the parent stand. In the long term — i.e., multiple rotations of the species on the same site — the nutrient status of the soil must be considered. Although we did not discuss this issue in detail in our chapter, the current view is that retaining leaves and unmerchantable bolewood and branches on the harvested site will minimize effects on site nutrient capital. However, the final word on this topic has yet to be written.

The level of management intensity for other poplars varies significantly by region and by the value of the species to that region. Natural stands of *P. deltoides*, *P. trichocarpa*, and *P. balsamifera* have all provided important wood and fiber products for local use and export. Silvicultural systems for natural stands of these species do not appear to be well-developed. The main silvicultural problem seems to be stand replacement through natural regeneration because vegetative reproduction is not as reliable as it is for aspen. Seed regeneration in harvested areas is not readily achieved even with site preparation. The poplar species and their hybrids that occur in riparian areas and active floodplains play very important roles in river dynamics and provide habitat for many animal species. Silvicultural systems and restoration procedures need to be developed to meet all management objectives for these riparian poplars.

Acknowledgements

Research supported by the USDA Forest Service, North Central Research Station; College of Natural Resources and Aspen/Larch Cooperative, University of Minnesota; and the University of Alberta. Published as paper no. 004420028 of the Minnesota Agricultural Experiment Station. We thank Wayne Shepperd, Richard Kabzems, and Rob Farmer for reviewing an early draft of this chapter. Thanks also to Doug Stone, Jim Marshall, and John Shaw for providing photographs.

References

Bates, P.C., Blinn, C.R., and Alm, A.A. 1993. Harvesting impacts on quaking aspen regeneration in northern Minnesota. Can. J. For. Res. **23**: 2403–2412.

Berry, A.B., and Stiell, W.M. 1978. Effect of rotation length on productivity of aspen sucker stands. For. Chron. **54**: 265–267.

Betters, D.R., and Woods, R.F. 1981. Uneven-aged stand structure and growth of Rocky Mountain aspen. J. For. **79**: 673–676.

Braatne, J.H., Rood, S.B., and Heilman, P.E. 1996. Life history, ecology, and conservation of riparian cottonwoods in North America. *In* Biology of *Populus* and its implications for management and conservation. *Edited by* R.F. Stettler, H.D. Bradshaw, Jr., P.E. Heilmann, and T.M. Hinckley. NRC Research Press, Ottawa. pp. 57–85.

Burns, R.M., and Honkala, B.H. (*Technical Coordinators*). 1990. Silvics of North America. Volume 2, hardwoods. USDA For. Serv. Agric. Handb. 654.

Cao, F.L., and Conner, W.H. 1999. Selection of flood-tolerant *Populus* deltoides clones for reforestation projects in China. For. Ecol. Manage. **117**: 211–220.

Carmean, W.H. 1978. Site index curves for northern hardwoods in northern Wisconsin and Upper Michigan. USDA For. Serv. Res. Pap. NC-160.

Carmean, W.H., and Li, J. 1998. Soil-site relations for trembling aspen in northwest Ontario. North. J. Appl. For. **15**: 146–153.

Carmean, W.H., Hahn J.T., and Jacobs, R.D. 1989. Site index curves for forest tree species in the eastern United States. USDA For. Serv. Gen. Tech. Rep. NC-128.

Caza, C.L. 1993. Woody debris in the forests of British Columbia: a review of the literature and current research. B.C. Min. For. Land Manage. Rep. 78.

Chen, H.Y.A., Klinka, K., and Kabzems, R.D. 1998*a*. Height growth and site index models for trembling aspen (*Populus tremuloides* Michx.) in northern British Columbia. For. Ecol. Manage. **102**: 157–65.

Chen, H.Y.A., Klinka, K., and Kabzems, R.D. 1998*b*. Site index, site quality, and foliar nutrients of trembling aspen: relationships and prediction. Can. J. For. Res. **28**: 1743–1755.

Darrah, D.W. 1991. Aspen harvesting: a government perspective. *In* Aspen management for the 21st century. *Edited by* S. Navratil and P.B. Chapman. Forestry Canada and Poplar Council of Canada, Edmonton. pp. 61–66.

Dayanandan, S., Rajora, O.P., and Bawa, K.S. 1998. Isolation and characterization of microsatellites in trembling aspen (*Populus tremuloides*). Theor. Appl. Genet. **96**: 950–956.

DeByle, N.B., and Winokur, R.P. (*Editors*). 1985. Aspen ecology and management in the western United States. USDA For. Serv. Gen. Tech. Rep. RM-119.

DeByle, N.V. 1964. Detection of functional interclonal root connections by tracers and excavation. For. Sci. **10**: 386–396.

Delong, H.B., Lieffers, V.J., and Blenis, P.V. 1997. Microsite effects on first-year establishment and over-winter survival of white spruce in aspen-dominated boreal mixedwoods. Can. J. For. Res. **27**: 1452–1457.

Deschamps, K.C. 1991. Polymorphic site index curves for trembling aspen in north central Ontario. Lakehead University, Thunder Bay.

DesRochers, A. 2000. Aspen (*Populus tremuloides* Michx) clonal root dynamics and respiration. Ph.D. thesis, University of Alberta, Edmonton.

DesRochers, A., and Lieffers, V.J. 2001. Structural root system of mature aspen (*Populus tremuloides*) in declining stands in Alberta, Canada. J. Veg. Sci. In press.

Doucet, R. 1989. Regeneration silviculture of aspen. For. Chron. **65**: 23–27.

Dyrness, C.T., Viereck, L.A., Foote, M.J., and Zasada, J.C. 1988. The effect on vegetation and soil temperature of logging flood-plain white spruce. USDA For. Serv. Res. Pap. PNW-RP-392.

Edminster, C.B., Mowrer, H.T., and Shepperd, W.D. 1985. Site index curves for aspen in the central Rocky Mountains. USDA For. Serv. Res. Note RM-453. 4 pp.

Floate, K.D., and Whitham, T.G. 1995. Insects as traits in plant systematics: their use in discriminating between hybrid cottonwoods. Can. J. Bot. **73**(1): 1–13.

Gevorkiantz, S.R. 1956. Site index curves for aspen in the Lake States. USDA For. Serv. Lake States For. Exp. Stn. Tech. Note 464.
Gilmore, D.W., Briggs, R.D., and Seymour, R.S. 1993. Stem volume and site index equations for European larch in Maine. North. J. Appl. For. **10**: 70–74.
Graham, S.A., Harrison, R.P., Jr., Westell, C.E., Jr. 1963. Aspens: phoenix trees of the Great Lakes region. University of Michigan Press, Ann Arbor.
Grant, M.C., Mitton, J.B., and Linhart, Y.P. 1992. Even larger organisms. Nature (Lond.), **360**: 212.
Gullion, G.W. 1984. Ruffed grouse management — where do we stand in the eighties. *In* Ruffed grouse management: state of the art in the early 1980's. Proceedings of the Symposium of the 45th Midwest Fish and Wildlife Conference, St. Louis, Missouri, December 5–7, 1983. *Edited by* W.L. Robinson. Book Crafters, Chelsea. pp. 168–181.
Hogg, E.H., and Lieffers, V.J. 1991. The impact of *Calamagrostis canadensis* on soil thermal regimes after logging in northern Alberta. Can. J. For. Res. **21**: 382–394.
Hungerford, R.D. 1988. Soil temperature and suckering in burned and unburned aspen stands in Idaho. USDA For. Serv. Res. Note INT-378.
Johnson, R.L., and Shropshire, F.W. 1983. Bottomland hardwoods. *In* Silvicultural systems for the major forest types of the United States. USDA For. Serv. Agric. Handbk. 445. pp. 175–179.
Jones, J.R., and DeByle, N.B. 1985. Morphology. *In* Aspen ecology and management in the western United States. USDA For. Serv. Gen. Tech. Rep. RM-119. *Edited by* N.V. DeByle and R.P. Winokur. pp. 11–18.
Kabzems, R. 1996. Boreal long-term soil productivity study. B.C. Min. For. Res. Note Pg-06.
Kay, C.E. 1993. Aspen seedlings in recently burned areas of Grand Teton and Yellowstone National Parks. Northwest Sci. **67**: 94–104.
Keim, P., Paige, K., Whitham, T.G., and Lark, K.G. 1989. Genetic analysis of an interspecific hybrid swarm of *Populus*: occurrence of unidirectional introgression. Genetics, **123**: 557–565.
Krasny, M.E., Vogt, K.A., and Zasada, J.C. 1988*a*. Establishment of four Salicaceae species on river bars in interior Alaska. Holarct. Ecol. **11**: 210–219.
Krasny, M.E., Zasada, J.C., and Vogt, K.A. 1988*b*. Adventitious rooting of four Salicaceae species in response to a flood event. Can. J. Bot. **66**: 2597–2598.
Landhäusser, S.M., and Lieffers, V.J. 1998. Growth of *Populus tremuloides* in association with *Calamagrostis canadensis*. Can. J. For. Res. **28**: 396–401.
Lavertu, D., Mauffette, Y., and Bergeron, Y. 1994. Effects of stand age and litter removal on the regeneration of *Populus tremuloides*. J. Veg. Sci. **5**: 561–568.
Lieffers, V.J., and Campbell, J.S. 1983. Biomass and growth of *Populus tremuloides* in northeastern Alberta: estimates using hierarchy in tree size. Can. J. For. Res. **14**: 610–616.
Maini, J.S. 1967. Variation in the vegetative propagation of *Populus* in natural populations. Bull. Ecol. Soc. **48**: 75–76.
Marles, R.J., Clavelle, C., Monteleone, L., Tays, N., Burns, D. 2000. Aboriginal plant use in Canada's northwest boreal forest. UBC Press, Vancouver.
Natural Resources Canada, Canada Forest Service. 2000. National forestry data base. <http://nfdp.ccfm.org> (accessed 6 March 2000).
Navratil, S. 1996. Sustained aspen productivity on hardwood and mixedwood sites. *In* Ecology and management of B.C. hardwoods. Canada – British Columbia Partnership Agreement FRDA Report No. 255. *Edited by* P.G. Comeau, G.J. Harper, M.E. Blache, J.O. Boateng, and K.D. Thomas. pp. 53–64.
Navratil, S., and Bella, I.E. 1990. Regeneration, development and density management in aspen stands. *In* Management and utilization of Alberta's poplar. Proceedings of the Poplar Council of Canada 10th Annual Meeting, Edmonton, Alberta, October 1988. Forestry Canada, Northern Forestry Centre, Edmonton, and Alberta Dept. Forestry, Lands and Wildlife, Edmonton. pp. 19–37.
Navratil, S., Brace, L.G., Sauder, A., and Lux, S. 1994. Silvicultural and harvesting options to favor immature spruce and aspen regeneration in boreal mixedwoods. Can. For. Serv. Inf. Rep. NOR-X-337.
Navratil, S., Hayward, R., and Brace, L. 1996. Management of aspen regeneration density on boreal mixedwood sites. Nat. Res. Can. Can. For. Serv. North. For. Cent. and Alberta Prot. Land For. Serv. Can-Alberta Partner. Agree. For. Rep. 139. 43 pp.

Paige, N.K., Capman, W.C., and Jennetten, P. 1991. Mitochondrial inheritance pattens across a cottonwood hybrid zone: cytonuclear disequilibria and hybrid zone dynamics. Evolution, **45**(6): 1360–1369.

Perala, D.A. 1977. Manager's handbook for aspen in the north central states. USDA For. Serv. Gen. Tech. Rep. NC-36.

Perala, D.A., Host, G.E., Jordan, J.K., and Cieszewski, C.J. 1996. A multiproduct growth and yield model for the circumboreal aspens. North. J. Appl. For. **13**: 164–170.

Perala, D.A., Leary, R.A., and Cieszewski, C.J. 1999. Self-thinning and stockability of circumboreal aspens (*Populus tremuloides* Michx., and *P. tremula* L.). USDA For. Serv. Res. Pap. NC-335. 16 pp.

Peterson, E.B., and Peterson, N.M. 1995. Aspen managers' handbook for British Columbia. Canada – British Columbia Partnership Agreement. FRDA II. FRDA Rep. 230, Victoria.

Peterson, E.B., Peterson, N.M., and McLennan, D.S. 1996. Black cottonwood and balsam polar managers' handbook for British Columbia. Canada – British Columbia Partnership Agreement. FRDA II. FRDA Rep. 250, Victoria.

Piva, R. 1998. Pulpwood production in North-Central region, 1996. USDA For. Serv. Gen. Tech. Rep. NC-190.

Prettyman, D.H. 1992. Forest soils – climate – site index relationships for Minnesota. Minn. Rep. 228 (Item No. AD-MR-6062-D), University of Minnesota, St. Paul.

Rauscher, H.M., Perala, D.A., and Worth, C.V. 1995. The ecology and management of aspen. AI Applic. **9**: 59 (includes 2 disks with text and references).

Romme, W.H., Turner, M.G., Gardner, R.H., Hargrove, W.W., Tuskan, G.A., Despain, D.G., and Renkin, R.A. 1997. A rare episode of sexual reproduction in aspen (*Populus tremuloides* Michx.) following the 1988 Yellowstone fires. Nat. Areas J. **17**(1): 17–25.

Rood, S.B., Hillman, C., Sanche, T., and Mahoney, J.B. 1994. Clonal reproduction of riparian cottonwoods in southern Alberta. Can. J. Bot. **72**: 1766–1774.

Rood, S.B., Mahoney, J.B., Reid, D.E., and Zilm, L. 1995. Instream flows and the decline of riparian cottonwoods along the St. Mary River, Alberta. Can. J. Bot. **73**: 1250–1260.

Sandberg, D. 1951. The regeneration of quaking aspen by root suckering. Masters thesis, School of Forestry, University of Minnesota, St. Paul.

Sauder, E.A. 1994. Harvesting practices for alternative silvicultural systems in the Canadian boreal forest. *In* Proceedings of the Innovative Silvicultural Systems in the Boreal Forests, Edmonton, Alberta, October 2–8, 1994. *Edited by* C.R. Bamsey. Clear Lake Ltd., Edmonton. pp. 56–60.

Schier, G.A. 1973. Origin and development of root suckers in aspen. Can. J. For. Res. **3**: 45–53.

Schier, G.A., and Campbell, R.B. 1978. Aspen sucker regeneration following burning and clearcutting on two sites in the Rocky Mountains. For. Sci. **24**: 303–308.

Shaw, J.D., and Packee, E.C. 1998. Site index of balsam poplar/western cottonwood in interior and southcentral Alaska. North. J. Appl. For. **15**: 174–181.

Shepperd, W.D. 1993*a*. Initial growth, development, and clonal dynamics of regenerated aspen in the Rocky Mountains. USDA For. Serv. Res. Pap. RM-312.

Shepperd, W.D. 1993*b*. The effect of commercial harvest activities on root compaction and suckering of aspen. West. J. Appl. For. **8**: 62–66.

Shepperd, W.D. 1996. Response of aspen root suckers to regeneration methods and post-harvest protection. USDA For. Serv. Res. Pap. RM-324.

Shepperd, W.D., and Fairweather, M.L. 1994. Impact of large ungulates in restoration of aspen communities in a Southwestern ponderosa pine ecosystem. USDA For. Serv. Gen. Tech. Rep. RM-247. pp. 344–347.

Shepperd, W.D., and Smith, F.W. 1993. The role of near-surface lateral roots in the life cycle of aspen in the central Rocky mountains. For. Ecol. Manage. **61**: 157–170.

Stanosz, G.R., and Patton, R.F. 1987. Armillaria rot in Wisconsin sucker stands. Can. J. For. Res. **17**: 995–1000.

Stelfox, J.B. (*Editor*). 1995. Relationships between stand age, stand structure, and biodiversity in aspen mixedwood forests in Alberta. Alberta Environmental Centre (AECV95-R1), Vegreville, and Canadian Forest Service, Edmonton. p. 308.

Steneker, G.A. 1973. The size of trembling aspen (*Populus tremuloides* Michx.) clones in Manitoba. Can. J. For. Res. **3**: 472–478.

Stettler, R.F., Bradshaw, H.D., Jr., Heilmann, P.E., and Hinckley, T.M. (*Editors*). 1996. Biology of *Populus* and its implications for management and conservation. NRC Research Press, Ottawa. 539 pp.

Stoeckeler, J.H. 1960. Soil factors affecting the growth of quaking aspen forests in the Lake States. Tech. Bull. 233, University of Minnesota Agricultural Experiment Station.

Stone, D.M., Elioff, J.D., Potter, D.V., Peterson, D.B., and Wagner, R. 2000. Restoration of aspen-dominated ecosystems in the Lake States. *In* Sustaining aspen in western landscapes. Proceedings Symposium. Grand Junction, Colorado, July 13–15, 2000. USDA For. Serv. Gen. Tech. Rep. In press.

Tuskan, G.A., Francis, K.E., Russ, S.L., Romme, W.H., and Turner, M.G. 1996. RAPD markers reveal diversity within and among clonal and seedling stands of aspen in Yellowstone National Park, USA. Can. J. For. Res. **26**: 2088–2098.

Viereck, L.A. 1970. Forest succession and soil development adjacent to the Chena River in interior Alaska. Arct. Alp. Res. **2**: 1–26.

Wan, X., Landhäusser, S.M., Zwiazek, J.J., and Lieffers, V.J. 1999. Root water flow and growth of aspen (*Populus tremuloides*) at low root temperatures. Tree Phys. **19**: 879–884.

Wilde, S.A., and Pronin, D.T. 1949. Growth of trembling aspen in relation to ground water and soil organic matter. Soil Sci. Soc. Am. Proc. **13**: 345–347.

Yeh, F.C., Chong, D.K.X., and Yang, R.Y. 1995. RAPD variation within and among natural populations of trembling aspen (*Populus tremuloides* Michx.) from Alberta. J. Hered. **86**: 454–460.

Zasada, J.C., and Schier, G.A. 1973. Aspen root suckering in Alaska: effect of clone, collection time, and temperature. North. Sci. **47**: 100–104.

Zasada, Z.A. 1947. Aspen properties and uses. USDA For. Serv. Lake States Aspen Rep. 1.

CHAPTER 5
Ecology and silviculture of poplar plantations

John A. Stanturf, Cees van Oosten, Daniel A. Netzer, Mark D. Coleman, and C. Jeffrey Portwood

Introduction

Poplars are some of the fastest growing trees in North America and foresters have sought to capitalize on this potential since the 1940s. Interest in growing poplars has fluctuated, and objectives have shifted between producing sawlogs, pulpwood, or more densely spaced "woodgrass" or biofuels. Currently, most poplar plantations are established for pulpwood or chip production on rotations of 10 years or less, but interest in sawlog production is increasing. Sid McKnight (1970) characterized cottonwood as a *prima donna* species: under ideal conditions, growth rates are just short of spectacular. Just as this can be applied to all poplars, it is equally true that all poplars are demanding of good sites and careful establishment. Growing poplars in plantations is challenging, and good establishment the first year is critical to long-term success. If a grower lacks the commitment or resources to provide needed treatments at critical times, then species other than poplars should be considered. Successful poplar culture can be illustrated by the triangle in Fig. 1 — plant proven clones on good sites and provide timely, appropriate cultural treatments. Our objective in this chapter is to provide

J.A. Stanturf.[1] USDA Forest Service, Center for Bottomland Hardwoods Research, P.O. Box 227, Stoneville, MS 38776, U.S.A.
C. van Oosten. SilviConsult Woody Crops Technology Inc., 2356 York Crescent, Nanaimo, BC V9T 4N3, Canada.
D.A. Netzer. USDA Forest Service, North Central Research Station, Rhinelander, WI 54501, U.S.A.
M.D. Coleman. USDA Forest Service, Savannah River Institute, P.O. Box 700, Building 760-15G, New Ellenton, SC 29809, U.S.A.
C.J. Portwood.[2] Crown Vantage Corp., 5925 N. Washington St., Vicksburg, MS 39180, U.S.A.
Correct citation: Stanturf, J.A., van Oosten, C., Netzer, D.A., Coleman, M.D., and Portwood, C.J. 2001. Ecology and silviculture of poplar plantations. *In* Poplar Culture in North America. Part A, Chapter 5. *Edited by* D.I. Dickmann, J.G. Isebrands, J.E. Eckenwalder, and J. Richardson. NRC Research Press, National Research Council of Canada, Ottawa, ON K1A 0R6, Canada. pp. 153–206.

[1]Present address: Forestry Sciences Lab, 320 Green Street, Athens, GA 30602, U.S.A.
[2]Present address: Temple-Inland Forest, 207 N. Temple Drive, Diboll, TX 75941, U.S.A.

Fig. 1. Poplar plantation culture depends on three things: planting the best quality stock on high-quality sites and providing timely and appropriate cultural treatments.

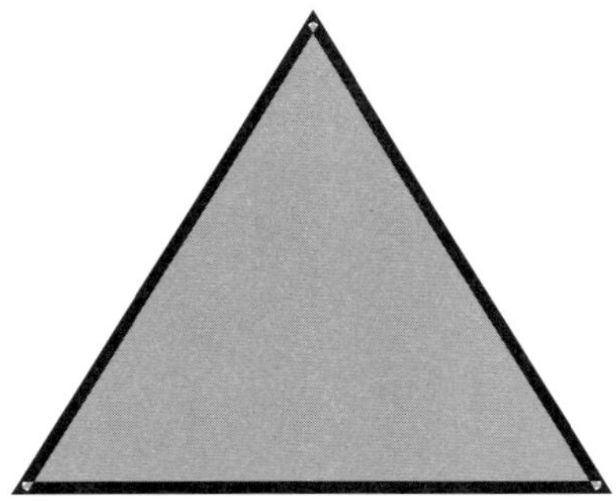

growers with current information for establishing and tending poplar plantations, as practiced in North America. Where we have sufficient information, differences between the poplar-growing regions of the United States and Canada will be noted. Mostly information is available on eastern and black cottonwood, and their hybrids.

Propagation and production of planting stock

Great strides have been made in selecting and breeding superior poplar genotypes. One advantage of poplars is that superior material is quickly available for operational use because species of the *Aigeiros* and *Tacamahaca* sections used in North America are easy to propagate through asexual means, usually by vegetative propagation of unrooted dormant stem cuttings or sets (also called whips). This lends itself well to mass-propagation of selected varieties for operational use, but poor rooting ability may disqualify some genotypes. Eastern cottonwood (*Aigeiros* section) displays great variability in rooting ability. Interspecific hybrids within and between the *Aigeiros* and *Tacamahaca* sections usually root well. Poplars in section *Populus* (the aspens) are difficult to propagate from stem cuttings, as are the interspecific hybrids between *P. tremuloides* and *P. tremula*. Two methods used in Canada for mass propagation of aspen are dormant root cuttings and seedlings from open-pollinated sources. Both methods are expensive and take longer to deploy superior genotypes.

Planting stock types

Poplar stock can be produced in several different types and is mostly a function of ease of propagation, desired end product, and cost (Table 1). Unrooted dormant cuttings (Fig. 2) are produced from 1-year-old stem material, varying in length

Table 1. Conditions under which certain poplar stock types can be used.

	Unrooted stock		Rooted stock		
			Bare root		
	Cuttings	Sets	Small	Sets	Container
Density of plantation (stems ha^{-1})	>700	<400	>700	<400	>700
Plantation purpose	Fibre and solid wood	Solid wood	Fibre and solid wood	Solid wood	Research trials; new stoolbeds; extreme drought conditions at planting
Soil moisture conditions	Good	Excellent	Good	Good	Needs irrigation if planted in full leaf
Weed control	Excellent	Reasonable	Excellent	Reasonable	Excellent
Threat of browsers	High[a]–low	High	Low	High	Low
Timing of planting	U.S. South: early winter				
	U.S. Midwest, Pacific Northwest, and Canada: late winter to early spring	Late winter to early spring	Late winter to late spring	Late winter to late spring	Late winter to late spring (irrigated)

[a]With a high threat of browsers, deer fencing may be necessary.

from mini (2–3 cm) to regular cuttings (15 cm to a maximum of about 1 m long). When planted in soil, adventitious roots grow from stem pieces, but viable buds must be present for stems to form. Unrooted dormant sets can be cut from 1- or 2-year-old dormant material, but roots develop better from 1-year-old material. Sets vary in length from 1.5 m to as long as 5 or 6 m. As with cuttings, buds are necessary for new stems to develop.

Planting unrooted dormant cuttings or sets in a nursery bed and allowing them to grow a viable root system produces rooted cuttings. Rooted cuttings (also called barbatelles) can be out-planted as dormant bareroot cuttings, equivalent to a 1–0 seedling. Container plants are produced from seed, root cuttings (aspen), or small single-bud hardwood or greenwood cuttings. These plants are usually dormant

Fig. 2. Dormant, 1-year-old, unrooted hardwood cuttings of poplar. The large cutting on the top is optimum; the cutting on the bottom is less likely to give an established plant. Photo by Don Dickmann.

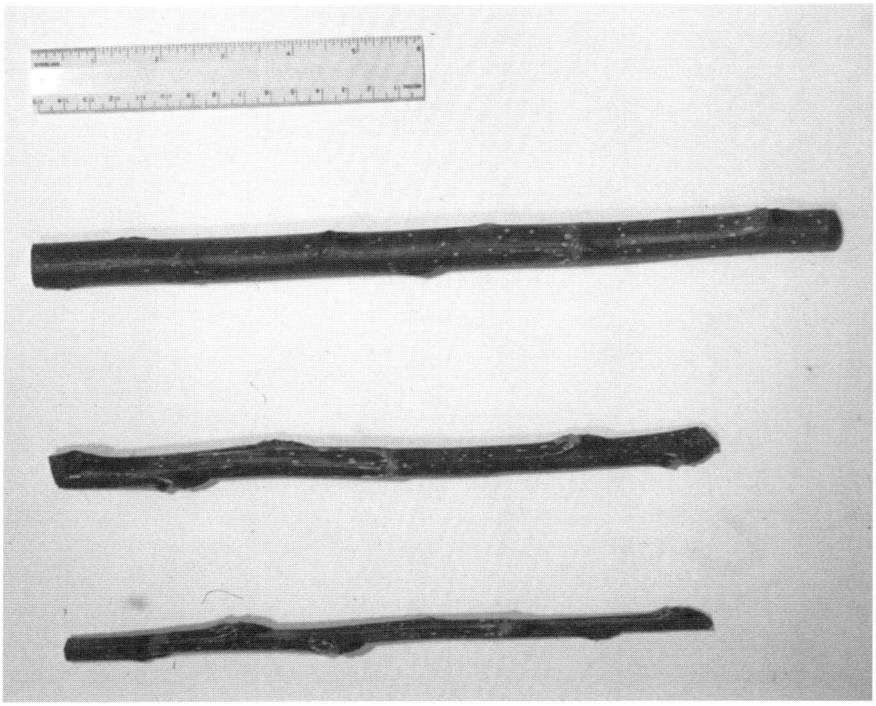

when planted, but they can be planted after breaking dormancy in the same growing season, if done immediately and there is sufficient time remaining in the season to develop an adequate root system.

Stock production systems

Unrooted dormant cuttings and sets

Most production of poplar planting stock takes place in stoolbed nurseries (Fig. 3). A stool is a stump from which new sprouts emerge. Stools may be started from any stock type, but normally dormant cuttings are used. Stools are cut back annually to a height of 5–15 cm in winter, thus producing 1-year-old sprouts every year. When very tall planting stock is required, the stools are cut back every other year to produce a 2-year-old set. Harvested sprouts are sawn into cuttings or sets in early winter in the southern U.S. and late winter or early spring elsewhere in North America. Stock must be refrigerated and remain dormant waiting out-planting. Storage is in coolers or freezers, depending on the length of storage. For the best production of the healthiest stock, the lifespan of a stoolbed should be limited to 3–7 years.

Fig. 3. Eastern cottonwood nursery in the Lower Mississippi Alluvial Valley. Fitler Plantation, Fitler, MS. Note irrigation system. Photo by Jeff Portwood.

The density of the stools in beds is typically 0.3 × 0.3 m, or slightly less than 0.1 m^2 per stool. The density of the stools determines the caliper of the sprouts and controls the number of viable buds. The grower wants a uniform sprout with a basal caliper that only slightly exceeds the maximum set by the customer, thereby minimizing waste. Each cutting or set must have dormant viable buds. When the stools are planted too widely, sunlight that penetrates the canopy stimulates buds to develop into sylleptic branches, rendering the sprouts useless for cuttings. Varieties vary enormously in their tendency to form sylleptic branches. For instance, *deltoides* × *nigra* (D×N) hybrids are usually not a problem, whereas many *trichocarpa* × *deltoides* (T×D) or *trichocarpa* × *nigra* (T×N) hybrids grow prolific amounts of sylleptic branches.

Weed control strategies

Competition from weeds is a serious threat during establishment of new stoolbeds. Herbicides provide the most effective control of weeds (Table 2). Mulching can be used to control weeds, but they re-establish over time and the mulch can create habitat for rodents. Sawdust has been used as mulch, but it will tie up available nitrogen and can acidify the soil. During site preparation, grasses and broadleaved weeds can be effectively controlled with a tank mix of glyphosate (various formulations as Roundup®, Accord®, Vision®) and 2,4-dimethylamine (2,4-D). Repeated applications of glyphosate may be needed for control of perennial grasses (e.g., quack grass, reed canary grass) that spread by rhizomes. After cuttings are planted, a pre-emergent or pre-bud-break herbicide application is advisable. Choice will vary by location, soil texture and pH, and weed species.

Table 2. Partial list of herbicides for use in poplar plantations. Always check the label for current registration, rates, and application timing. Labels are available online at websites such as <http://picol.cahe.wsu.edu>. Most chemical companies provide downloads of their latest herbicide labels at their websites. Where rates of application have not been listed, the EPA# is provided under the Region column.

Active ingredient	Product name	Manufacturer	Application	Timing	Rates (imperial units)[a]	Rates (metric units)[b]	Region
2,4-D Dimethylamine	Various	Various	Post-emergent weed control	Apply as a shielded spray to kill old stumps	1–2 pints	1.2–2.3 L	U.S.
Azafenidin	Milestone	DuPont	Pre-emergent weed control	Apply prior to bud flush	5–10 oz	0.35–0.7 kg	U.S. (label pending)
Clopyralid	Transline	DowAgro Sciences	Selective post-emergent weed control	Apply as a broadcast foliar spray over trees or banded or directed	1/3 to 2/3 pints not to exceed 1 1/3 pints/year	0.39–0.77 L not to exceed 1.56 L/year	U.S.
Clopyralid	Stinger	DowAgro Sciences	Selective post-emergent weed control	Apply as a broadcast foliar spray over trees or banded or directed	1/3 to 2/3 pints not to exceed 1 1/3 pints/year	0.13–0.28 kg (active ingredient)	U.S.
Clopyralid	Transline	DowAgro Sciences	Selective post-emergent weed control	See label	See label	See label	Oregon and Washington EPA# 62719-259
Dichlobenil	Casoron 4G	Uniroyal	Pre- and post-emergent	Early spring and late fall	98–150 lb	110–174 kg	Canada
Diuron	Karmex DF	Griffin	Pre-emergent weed control	Apply to trees 1 year old and older	1–3 lb	1.12–3.36 kg	U.S. Prairie States — CO, MT, NE, SD, ND, ID, OR, WA
Diuron	Direx 4L	Griffin	Pre-emergent weed control	Apply to trees 1 year old and older	2–4 qt	4.7–9.4 L	U.S. Prairie States — CO, MT, NE, SD, ND

Table 2 (*continued*).

Active ingredient	Product name	Manufacturer	Application	Timing	Rates (imperial units)[a]	Rates (metric units)[b]	Region
Diuron	Diuron 4L	Drexel	Pre-emergent weed control	Apply preplant or dormant post-plant or as a shielded application	2–4 qt	0.56–1.68 kg (active ingredient)	Western Washington
Diuron	Diuron 80 DF	DowAgro Sciences	Pre-emergent weed control	Apply to trees 1 year old and older	2.5–5 lb	2.8–5.6 kg	U.S. Prairie States — CO, MT, NE, SD, ND
Fluazifop-*p*-butyl	Fusilade DX	Zeneca	Post-emergent grass control	Apply over actively growing trees to control grass	Split application (12 fl oz followed by 8 fl oz) Application timing is critical	Split application (0.88 followed by 0.58 L) Application timing is critical	U.S.
Fluazifop-*p*-butyl	Venture L	Zeneca Agro	Post-emergent grass control	Apply over actively growing trees to control grass		Up to maximum 2 L/ha & only one application per year	Canada
Glyphosate	Various	Various	Preplant site preparation, directed spray in older trees	Apply when trees are completely dormant or as a careful directed spray	3/4 to 3 qt	1.75–7 L	U.S., Canada
Imazaquin	Scepter 70 DG	BASF	Pre- and post-emergent weed control	Broadcast before and after bud break	2.8 oz	0.2 kg	U.S. (30 states)
Imazaquin/ pendimethalin	Squadron	BASF	Pre-emergent weed control	Preplant incorporated or pre-emergent	3–6 pints	3.5–7 L	U.S. (30 states)

Table 2 (*continued*).

Active ingredient	Product name	Manufacturer	Application	Timing	Rates (imperial units)[a]	Rates (metric units)[b]	Region
Linuron	Lorox DF	Griffin	Pre- and early post-emergent weed control	Broadcast before bud break or directed spray after bud break	2–4 lb Use less on light soils	2.25–4.5 kg Use less on light soils	Midwest U.S.
Linuron	Linex 4L	Griffin	Pre- and early post-emergent weed control	Broadcast before bud break or directed spray after bud break	2–4 pints Use less on light soils	2.3–4.7 L product Use less on light soils	Midwest U.S.
Oryzalin	Surflan A.S.	DowAgro Sciences	Pre-emergent weed control	Apply before weed flush Will not control active weeds	2 qt Not more than 8/year	4.7 L Not more than 18.7 L/year, 3 months between applications	U.S.
Oxyfluorfen	Goal 2XL Plus	Rohm & Haas	Pre-emergent weed control	Broadcast before bud break or directed spray after bud break	64 oz pre-bud break, 32 oz after bud break	4.7 L pre-bud break, 2.3 L after bud break	U.S.
Oxyfluorfen	Goal 1.6E	Rohm & Haas	Pre-emergent weed control	Broadcast before bud break or directed spray after bud break	Not more than 10 pints/year	Not more than 11.7 L/year	U.S.
Oxyfluorfen	Galigan 2E Oxyfluorfen Herbicide	Makhteshim-Agan of North American Inc.	Pre-emergent weed control	Broadcast before bud break or directed spray after bud break	See label	See label	Oregon and Washington EPA# 66222-28

Table 2 (*continued*).

Active ingredient	Product name	Manufacturer	Application	Timing	Rates (imperial units)[a]	Rates (metric units)[b]	Region
Paraquat dichloride	Gramoxone Extra	Zeneca	Post-emergent weed control	Apply dormant postplant in combination with oxyfluorfen or oryzalin	2 pints	2.3 L	Southeast U.S.
Paraquat	Griffin BOA Herbicide	Griffin	Post-emergent weed control	See label	See label	See label	Oregon and Washington EPA# 1812-420
Pendimethalin	Pendulum 3.3 EC	American Cyanamid	Pre-emergent weed control	Broadcast before and after bud break	2.4–4.8 qt	5.6–11.2 L	U.S.
Quizalofop	Assure II	DuPont	Post-emergent grass control	Apply over actively growing trees to control grass	5–10 oz	0.37–0.73 L	MN
Sethyoxydim	Poast, Poast-Plus	BASF	Post-emergent grass control	Apply over actively growing trees to control grass	1–2 pints	1.2–2.3 L	U.S.
Sulfometuron methyl	Oust	DuPont	Pre-emergent weed control	See label restrictions	0.5–2 oz	0.04–0.14 kg	WI, MN, WA, OR
Terbacil	Sinbar	DuPont	Pre-emergent weed control	Apply pre- or post-plant	1–2 lb	1.12–2.24 kg	WA, OR
Trifluralin	TRIAP 4HF	IAP	Soil incorporated	Preplant soil incorporated Older plantation incorporate to depth to not injure tree roots	See label	See label	Oregon and Washington EPA# 71058-1

Table 2 (*concluded*).

Active ingredient	Product name	Manufacturer	Application	Timing	Rates (imperial units)[a]	Rates (metric units)[b]	Region
Trifluralin	Trilin	Griffin	Soil incorporated	Pre-plant soil incorporated Older plantation incorporate to depth to not injure tree roots	1–2 pints, dependant on soil and rainfall	2.3–4.7 L, dependant on soil and rainfall	U.S.
Trifluralin	Trilin 5	Griffin	Soil incorporated	Preplant soil incorporated Older plantation incorporate to depth to not injure tree roots	0.8–3.2 pints, dependant on soil, rainfall, and age of planting	0.9–3.6 L	U.S.
Trifluralin	Trilin 10G	Griffin	Soil incorporated	Preplant soil incorporated Older plantation incorporate to depth to not injure tree roots	5–20 lb, dependant on soil, rainfall, and age of planting	5.6–22.4 kg	U.S.
Trifluralin	Treflan	DowAgro Sciences	Soil incorporated	Preplant soil incorporated Older plantation incorporate to depth to not injure tree roots	1–4 pints, dependant on soil, rainfall, tree age	1.2–4.7 L, dependant on soil, rainfall, tree age	U.S.

[a]In pounds, lb (1 lb = 0.454 kg); ounces, oz (1 oz = 28.35 g; 1 fl oz = 28.41 cm^3); pints (1 pint = 0.568 dm^3); or quarts, qt (1 qt = 1.14 dm^3) product per acre (1 acre = 0.405 ha) unless specified.

[b]In kilograms (kg) or liters (L) product per hectare unless specified.

Several nurseries in the Pacific Northwest use oxyfluorfen to maintain a weed-free stoolbed. In British Columbia, dichlobenil (Casoron®) was used successfully immediately following planting. In the southern U.S., an oxyfluorfen plus paraquat or glyphosate (Goal® plus Gramoxone® or Accord®) tank mix is used to control weeds. Leaves and succulent stems of very small cuttings can be damaged by splash of oxyfluorfen caused by overhead irrigation droplets, but cuttings usually grow out of any damage without lasting effects. In the Midwest, oxyfluorfen is also used to control weeds in newly planted dormant stool beds. Linuron (Lorox®) and oryzalin (Surflan®) also provide good weed control. If cuttings have active leaves, oryzalin may cause less damage than other herbicides.

It is common to use manual labour to hand-weed portions of stoolbeds. Weeding needs decline rapidly when the stock fully occupies the stoolbed and shades out the weeds. During the next fall and winter, leaf litter forms a layer of mulch, which effectively suppresses weeds.

Fertilization and irrigation

Nutrient deficiencies and moisture stress should be avoided in stoolbeds, but fertilization and irrigation schedules are very specific to local conditions (see Diagnosing nutrient deficiencies). Usually a balanced application of nutrients at the start of the growing season is sufficient. Direct foliar applications of nutrients can correct nutrient imbalances that develop during the growing season. An oversupply of nitrogen, however, can cause the crop to grow too fast, promote formation of sylleptic branches, and delay the onset of dormancy (especially when applied after early August). Excess nitrogen can also increase weed competition. Growers must be able to manipulate crop development by supplying or withholding nitrogen at the right times. The same principles apply to irrigation where the aim is to provide just enough water to maintain even growth. Over-irrigation can promote the development of sylleptic branches. Water should be withheld late in the growing season to promote hardening off and avoid frost damage.

Crop health, protection, and hygiene

The three most serious disease and pest problems facing the nursery grower are leaf rusts, blackstem diseases, and the cottonwood leaf beetle. Protection strategies are a combination of chemical control, cultural practices, and use of resistant varieties. High stoolbed densities favour foliage diseases such as Melampsora rusts, especially with overhead irrigation. Varieties with normally low susceptibility in plantations may develop serious problems in stoolbeds. The grower can avoid these varieties or use registered fungicides (e.g., Bayleton®). If Melampsora rust causes early defoliation, cuttings in this physiologically weakened state are more vulnerable to blackstem disease.

Blackstem diseases are caused by a number of organisms (*Cytospora chrysosperma*, *Phomopsis oblonga*, and *Colletotrichum gloeosporioides*) that are opportunistic on stressed plants. Blackstem is often considered a storage disease, and although improper storage can cause the disease to spread, it usually starts in a stressed plant well before it is put into storage. Stress can occur in the stoolbed because of drought, insufficient light or nutrients, frost damage, insect damage, or leaf diseases such as Melamspora rust. Upon out-planting, portions of the bark die off and turn black (hence the name blackstem disease). The disease spreads and usually leads to poor growth and often mortality. Diseased cuttings become a source of inoculum, and inadequate culling worsens the condition.

The cottonwood leaf beetle or CLB (*Chrysomela scripta*) is the most serious insect threat in stoolbeds and is a serious pest in plantations. The CLB defoliates developing leaves and in extreme cases feeds on the woody part of the stem. Monitoring for CLB must continue throughout the growing season as multiple generations are produced at about 1-month intervals. Successful control is achieved with several commercial insecticides registered for use in eastern cottonwood and hybrid poplar in the U.S., including several formulations of carbaryl (Sevin®) and dimethoate (Dimate 4E®). Several Bt (*Bacillus thuringiensis*) products are available and Novodor® is used operationally in Minnesota.

Unrooted dormant branch cuttings

Dormant material can be harvested from branches of young plantations instead of stoolbeds. These are known also as serial cuttings. First-order branches near the top of the tree produce vigorous cuttings of sufficient diameter. A 2-year-old tree can provide 20–30 cuttings, depending on branching characteristics. In plantations of T×D hybrids, sylleptic branches can be used for cuttings. Sylleptic branches from the previous year grow to a reasonable size the second year, but only the 1-year-old portion of these branches is used. This produces small-diameter cuttings, which are marginally suitable for planting in the field but can be used to establish stool beds. Branch cuttings also must be stored in coolers or freezers until planting.

Rooted dormant cuttings

Bareroot dormant cuttings can be used to establish widely spaced plantations for solid wood products (Table 1). This system of plant production is expensive, labour intensive, and is not normally undertaken to merely establish fibre plantations. After 1 year of growth in the nursery, the grower excavates bareroot plants with the root systems intact for out-planting in the field. Root systems may be trimmed to a manageable size at the nursery. Often the tops are also trimmed for easier handling or to balance top and roots. Bareroot stock is lifted in winter or early spring, while the trees are dormant. Large stock cannot be stored easily and must be transported and planted immediately. Large stock can be several meters tall, sometimes 2 years old, with large caliper. It requires machinery for planting; for example, a tractor-mounted auger for digging the planting hole.

Container nursery for rooted plants

Materials that can be produced in a container nursery may be grown from single-bud stem cuttings, root cuttings, or seed. Dormant single-bud hardwood cuttings are used for varieties that are difficult to propagate or if only a limited amount of material is available, such as from a breeding program. Each sprout or branch is divided into very small cuttings containing a single bud and rooted in containers. Single-bud greenwood cuttings are collected from actively growing young, succulent green shoots and planted (with leaves) into a container, usually with rooting hormones in mist beds. This method is expensive and labour intensive, but can be used to quickly multiply a single mother plant into thousands of identical plants. Uses include establishing a new stoolbed with an improved genotype or for experimental purposes.

For the hard-to-propagate aspens (e.g., *P. tremuloides* and its hybrids), dormant root cuttings are placed in containers in a greenhouse in order to produce fully rooted plants with soil for out-planting. The container crop is initiated in the late winter in the greenhouse, and grows during the spring and summer into large plants with well-developed root systems. During the late summer, the containers are placed outdoors. The following winter, dormant seedlings are extracted from the containers, packaged, and stored in a cooler or freezer, pending out-planting the next spring.

Containerized seedlings can be produced for operational planting of aspen. In the Prairie Region of Canada, seed from open-pollinated trees is used to produce planting stock for reforestation. The seed is sown in containers in the late spring. With few exceptions, the new seedlings will be ready for out-planting that fall.

Stock harvesting, processing, and quality control

Although there are good arguments for and against monoclonal plantings, clones must be identified and kept separate in stoolbeds so that only appropriate clones are planted on a site. Harvesting and processing should be done one variety at a time to eliminate the risk of mixing with another. Harvesting can be of individual stems or by mowing or cutting many stems at once (called mass harvesting).

Harvesting

Individual stem harvest requires experienced personnel who can determine the quality of each sprout before its harvest and select against poorly formed, diseased, or undersized sprouts. This is positive selection of good material, as poor quality material is not cut. Individual stem harvesting has few options to mechanize, which is a disadvantage for large stoolbed operations. It is labour intensive and costly, although it could lead to savings at the processing plant. In mass harvesting, sprouts are cut with a hand-held brushing saw or a mower attachment to a tractor. In the southern U.S., a modified sugar cane harvester has proven successful. Mass harvesting achievs higher production levels at lower harvesting costs. Quality control costs will increase and there is more waste to handle because both good and bad material arrives at the processing plant.

Processing

Processing can be accomplished by individual stem or by applying assembly line techniques for mass processing. In the individual stem method, the cutter uses a saw or hand-powered, hydraulic or pneumatic shears and processes each sprout separately (Fig. 4). Advantages of this system are high quality control through better recognition of defects, maximum recovery per sprout (especially when the variety is in short supply), quality control accountability by cutter, and more variety of tasks (packaging, counting, etc.). It also allows recovery of odd stock sizes (such as sets) from material not suitable for cuttings. Cutters work independently, and are not affected by assembly line breakdowns. This system works best for processing small to moderate quantities of cuttings of several varieties, but could be more costly than mass processing of large quantities.

In mass processing, sprouts are cut to size by a set of mechanized cutting saws, usually operated by one or two persons. Other workers sort the resulting cuttings, followed by additional workers packaging the stock. The main advantage is fast processing, which is especially beneficial when processing large numbers of a single variety. Disadvantages are bottlenecks caused by a breakdown of the mechanized saws or frequent changeovers to different stock sizes or other varieties. Cuts are not always at the correct location in relation to buds so there is greater waste.

Fig. 4. Reducing dormant eastern cottonwood whips to cutting length. Fitler Plantation, Fitler, MS. Photo by Jeff Portwood.

Quality control

Quality control encompasses both culling substandard material and properly identifying varieties. Individual cuttings should meet size specifications, be properly formed and free of diseases such as blackstem or lack evidence of stem borers, and be dormant. Mixing or mislabelling varieties occurs frequently at nurseries. Harvesting and processing protocols are critical to minimize these problems. The nursery manager must maintain good records of each variety, by stock source, stoolbed, customer, etc. Advances in DNA technology allow for precise fingerprinting at reasonable costs and can be used by nurseries and customers alike to ensure identity. Variety contamination is a costly problem for the nursery and can lead to loss of customers.

Stock packaging and storage

Processed cutting stock should be packed in sealed plastic bags to prevent moisture loss. Each bag should be labelled with the variety name or number, the amount, packaging date, and nursery name. The bags are then placed in cardboard boxes or larger storage bins.

This makes quality control easier, facilitates the allocation of stock to planting areas, payment of planting contractors, and helps in the overall administration of the planting project. Sets up to 2 m in length can be packaged and stored in a similar fashion. Plastic sleeves can be cut to size to hold the material.

The boxes or bins are stored at +2°C to +4°C for short-term storage of up to 1 month, or at –2°C to –4°C for longer-term storage. Boxes can be stacked on a pallet, 2–4 boxes high and 4 boxes deep. Bins or pallets with boxes can be stacked on top of each other, but must have free air circulating between them to prevent over-heating and sprouting. If cold storage is not possible, stock can be stored in a snowbank (in the north) when temperatures are around freezing or in a shady and cool spot for short periods. In the southern U.S., there is a serious risk of the stock drying out during a dry winter or spring after planting. The stock can be soaked for a day or two in freshwater prior to planting as a preventive measure. Prolonged soaking should be avoided, however, as it promotes premature sprouting and may promote disease.

Site requirements and site selection

For poplar to live up to its reputation as the fastest growing species in North America, the best varieties must be planted on the best sites, with the best crop tending (Fig. 1). There is an unfortunate misconception that poplar likes wet sites; an inexperienced grower with a few acres of swampland who plants poplar and expects it to do well will be disappointed. Other misinformed forest managers have been known to plant poplar on very good quality land and expected fabulous growth without having to spend much effort on tending the crop. This point bears emphasizing: if a grower is not wholly committed to providing the necessary cultural treatments, especially early in the life of a plantation, then the grower should not plant poplars.

Site requirements

The site requirements for optimal performance of poplar can be stated simply as "Best performance can be expected on soils that are a well-aerated, have sufficient moisture and nutrients, are sufficiently deep (>1.0 meter to the water table), have a medium texture (sand/loam) and have a soil pH in the 5.0 to 7.5 range" (Baker and Broadfoot 1979). Poplars thrive under growing season conditions of high light intensity and warm temperatures. The influence of soil texture and drainage condition on site quality for poplar is summarized in Table 3.

Poplars grow well under many site conditions and it may be easier to list some factors that are generally *unfavorable* to poplar growth. The grower can control several of these factors. For example, poplar grows very well in the desert-like conditions of eastern Washington and Oregon, where climatic conditions are perfect — lots of sun and warm temperatures — but soils are mostly sand. Through fertigation (application of nutrients in the irrigation water) the grower can transform this high desert into a poplar forest. In the southeastern U.S., forest industry is investigating the feasibility of establishing short-rotation plantations, including poplar on deep sands. These plantations are located near existing mills and can be logged when most sites are too wet. Soil texture and drainage class determine to a great degree the suitability of a site for poplar (Table 3).

Unfavorable site conditions

Soils that are saturated and waterlogged during the growing season develop anaerobic conditions and starve the root systems of oxygen, leading to drought-like symptoms. The leaves turn yellowish-green and remain very small. The stressed tree exhausts its reserves and slowly dies. Most poplar varieties cannot tolerate anaerobic conditions for very long into the spring months and must have well-aerated soils by the beginning of June to survive and thrive. Younger trees are more vulnerable.

Some varieties do not tolerate saturated soil conditions in the winter very well either. For example, in northwest Washington and southwest British Columbia, hybrids of *Populus trichocarpa* × *P. maximowiczii* suffered significant loss of height and diameter growth the next growing season, whereas hybrid varieties of *P. trichocarpa* × *P. deltoides* do well under these circumstances.

Heavy soils (clay, clay loam, and silty clay loam textures) are considered less favorable for poplar growth than coarser textured soils, but the advent of better chemical weed control has improved the prospect for poplars on these soils.

Table 3. The influence of soil texture and drainage condition on site quality (very good – poor) for poplar. Shaded fields indicate potential to improve suitability through ditching, installing drain tile, subsoiling, or some combination (source: Dickmann and Stuart 1983).

	Natural drainage class		
Dominant profile textures	Well and moderately well drained	Somewhat poorly drained	Poorly and very poorly drained
Fine clay (>60% clay)	Fair	Fair	Poor
Clay (40–60%)	Fair	Fair	Poor
Clay loam and silty clay loam	Good	Poor	Poor
Loam and silt loam	Good – very good	Fair	Poor
Loam and silt loam 25–50 cm over well-decomposed peat	Good – very good	Poor	Poor
Loam and silt loam marbled with well-decomposed peat	Good – very good	Fair–good	Poor
Sandy loam	Very good	Fair–good	Poor
Loamy sand	Very good	Fair–good	Poor
Sand	Poor	Fair	Poor
Sandy loam 35–100 cm over clay	Very good	Fair	Poor
Sandy loam 50–100 cm over loam – clay loam	Very good	Fair	Poor
Sandy loam 50–100 cm over sand	Good	Very good	Poor
Loamy sand 35–100 cm over clay	Very good	Fair	Poor
Sand – loamy sand 50–100 cm over loam – clay loam	Very good	Very good	Poor
Sand – loamy sand 100–150 cm over loam–clay	Good	Very good	Poor
Muck	N/A	N/A	Poor–fair

Because finer textured soils generally have poor aeration and poor drainage, they restrict equipment access during wet periods, making weed control difficult. Survival is reduced and growth during the first few years can be disappointing. The lack of rapid growth and early crown closure leads to an abundance of weed competition, slowing tree growth even more. Recent advances in pre-emergent herbicides and application technology have improved weed control, enabling poplars to be established successfully on these sites. Eastern cottonwood grows better on medium textured soils but performs acceptably on soils with as much as 90% clay as long as weed competition can be controlled.

Saline conditions are not tolerated by the poplar species in North America. *P. trichocarpa* is extremely intolerant of salt and so are its hybrids; *P. deltoides* is slightly less intolerant. Salt damage to the trees resembles desiccation damage. Physiologically the tree suffers from drought stress. Leaves remain small and yellowish-green. Sometimes the leaf edges become necrotic. The condition worsens as summer drought sets in, resulting in tree mortality. Sensitivity to salinity should be a concern to growers who rely on irrigation or fertigation to manage their poplar crop, and adequate drainage must be provided along with sufficient water to flush salts through the rooting zone.

Poplars can perform well on shallow soils, although windthrow may be a problem. Shallowness of the rooting zone can be caused by a high water table that does not retreat during the summer, an impermeable soil layer, bedrock, soils that are naturally very compact, or compaction resulting from heavy machine traffic.

It is commonly thought that peat soils do not support good poplar growth. Peats are usually waterlogged and very acidic, but there are exceptions. Weed control on peat soils can be challenging. Access may be difficult at critical times due to waterlogging, precluding mechanical control. Soils with high organic matter content will bind and render ineffective many pre-emergent herbicides. Artificial drainage may be the key to successful poplar management on these soils. Several sites with a high peat component in northwest Washington and Oregon are reasonably well drained and support good growth of hybrid poplar. Windthrow damage is a real threat especially if water tables are shallow, but some poplar varieties are well suited to these conditions and hardly pose a serious windthrow problem.

Site selection

Despite being armed with the knowledge of site requirements, site selection can still be a daunting task. Sites are never uniform, and multiple combinations of site factors can occur. This is especially true for alluvial sites, where river action adds subsoil variability to a site under a blanket of uniform surface soil. It pays to determine soil texture, drainage conditions, subsoil properties, available nutrients, pH, and organic matter content. Where there is even a remote possibility of salinity, the site should be ruled out. If sites are subject to growing season flooding, historic inundation regimes should be determined.

Table 3 highlights several situations where drainage can be enhanced, leading to more favourable site conditions for poplar. Many otherwise suitable sites may require enhanced drainage. Improving and maintaining ditches, subsoiling, and installing drain tile can accomplish this. To maximize efficiency of planting and subsequent maintenance, block planting of a single variety is often desirable but may not provide maximum yields. A flexible, good performing variety may reduce the complexity of stand establishment but result in lower yields if site conditions vary substantially. Varieties can be easily matched with specific soil characteristics, leading to greater yields. There is also a school of thought that favours mixtures of varieties because site resources are more completely utilized and insects and disease problems are minimized.

Other important aspects of site selection are economic ones. A marginal site close to the mill may be more attractive financially than a good site farther from the mill. Transportation, mobilization, and management costs become prohibitive for sites too far from the mill site. Shape and size of a potential poplar site are also important. A small or odd-shaped area is awkward to manage. The length and orientation of plantation rows often determine the cost of cultivation maintenance. For large commercial operations, sites generally must be larger than 40 ha to be economical, although the concentration of acreage within a management area may be more critical. These factors have nothing to do with suitability of the soil, but everything to do with the suitability of the site.

Site preparation

Proper site preparation for planting is essential to the successful establishment of poplar plantations. Without adequate site preparation, survival and growth of poplars may be drastically diminished. A thorough evaluation to determine specific soil and site conditions of a potential poplar plantation will aid in the selection of appropriate treatments to apply. The main benefits will be reduced planting costs; more effective herbaceous weed control, and reduced damage to young poplars in mechanical cultivation; and disruption of impervious soil layers, which will improve internal drainage and aeration. Bear in mind that the main purpose of site preparation is to get poplars off to a fast start and to provide easy access to the site for essential weed control.

There are many combinations of site prep methods in use today throughout North America, depending mostly on site conditions. Sites one might encounter include open pasture or agricultural land, cutover natural stands, or prior plantations. On prior pasture or farmland, site prep can be very simple. On cutover forest or prior plantations, site prep becomes complex and very expensive due to stumps, logging debris, and heavy vegetation.

Open agricultural land is commonly prepared using combinations of conventional and minimum tillage methods, such as disking, chisel plowing, subsoiling (Fig. 5), and mowing. Many poplar growers have added herbicide treatments to their arsenal of site prep tools in order to reduce early weed competition. Raised beds or bedding is relatively new to poplar culture but has a long history of success in pine plantation culture on poorly drained sites.

Where fertigation is used, site preparation is more complicated and involves heavy construction. An existing center pivot irrigation system may need to be removed. The old irrigation piping system could be utilized to reduce expenses; otherwise, a completely new irrigation system infrastructure must be developed. This involves installing pumping stations, underground water lines, mains, and submains. Following this intensive initial process, conventional site prep methods as described above are used. Finally, drip hose is laid and connected to submains and emitters are installed.

Fig. 5. Tractor with subsoil shank used to break up plow pans and to inject fertilizers prior to planting cottonwood cuttings on former agricultural fields in the Lower Mississippi Alluvial Valley. Photo by Jeff Portwood.

Preparation of sites after timber harvest is also more involved. The longer the previous rotation, the larger, and more troublesome will be the material still on the site. New growth of herbaceous and woody vegetation, stumps, roots, and compaction from logging traffic can further complicate this process. Conventional land clearing methods such as shearing, raking, piling, and burning have not changed much over the years. These are still the preferred methods used in the southern U.S. Poplar growers recognize the need for less intensive, more cost-effective means of clearing harvested plantations. In the West, site prep between existing stumps has been successful, using an orchard flail to reduce woody debris, followed by a rototiller to further grind and incorporate debris into the soil. This leaves stumps intact. The planting bed is prepared between old rows while the soil is still loose from tilling.

Location of rows should be clearly marked according to the selected spacing. The row should be slit or bedded to a depth sufficient for the length of cutting to be planted (Fig. 5). Slitting can be accomplished by modifying conventional farm equipment. Reasonably straight rows are important for cultivating and for spray machines to avoid damage to young plants. It is best to mark in both directions when cross cultivation is planned. When slitting, at least 15 cm of rainfall is required to fill trenches with silt before planting can begin.

Planting

Planting is a crucial phase of plantation establishment, and only quality planting stock should be used. Select genetically improved poplar cultivars, developed by U.S. Forest Service, university, and forest products industry researchers, are available for purchase from government, private, and industry nurseries or through industry–landowner assistance programs. Planting stock varies in length from 15 to 45 cm. Optimum cutting size is from 1.0 to 2.0 cm in diameter. Cuttings larger than 2.0 cm are excellent planting stock but are hard to handle.

Although either seedlings or cuttings can be used, cuttings are preferred planting stock for poplars throughout North America. They survive and grow as well as seedlings and cost less to produce and plant. Additionally, cuttings are more desirable than seedlings because genetically superior varieties can be expanded more rapidly through vegetative propagation.

In drier regions of North America, harvested poplar whips or cuttings should be soaked in fresh water for a minimum of 2 days to prevent them from drying out during storage and planting. Cuttings or even whips should not be exposed to drying conditions during transport to planting sites. Exposure to light for extended periods before planting is also harmful. A tarp will keep the stock in good shape. When planting will be delayed until after the start of the normal growing season, cuttings must be kept in freezer storage.

Planting spacing varies from 2.1 × 3.0 m to 4 × 4 m, depending on poplar species and the desired product size (pulpwood or sawlog). Poplars can be planted by machine or hand, but hand planting is more common (Fig. 6). The cutting must contact soil and be planted as deeply as possible to take full advantage of soil moisture. Depth of planting will vary with cutting length. Shallow planting usually results in poor survival and reduced height growth. Aboveground exposure should be minimized to reduce the likelihood of undesirable multiple sprouts. Nonetheless, about 5 cm should be left above ground so that cuttings are visible to equipment operators during early cultural treatments. Cuttings always should be planted with vegetative buds pointing upward. The tops of cuttings can be spray painted orange to insure proper orientation and speed planting. This also assists in monitoring planting contracts.

Poplars may be planted any time during the dormant season. In the southern U.S. this extends from the first severe frost in the fall until buds begin to open in the spring. In areas of North America that have frozen soil in winter, cuttings are normally planted in the spring when soil temperature reaches 4°C. In the Midwest, planting usually is done when the soil is warm enough to plant corn. Planting material should always be checked for dormancy before using; succulent green tissue of rapid growth may persist for a short time after the first fall frost, but the buds and current season's growth must have stopped growing and be hardened before cuttings are made. In addition, cuttings that have already sprouted are

Fig. 6. Planting dormant cottonwood cuttings in 3.7 × 3.7 m furrows in the Lower Mississippi Alluvial Valley. Photo by Jeff Portwood.

a poor risk. Delaying planting until after bud break of surrounding vegetation in the spring has been successful and affords an opportunity to plant sites that remain wet throughout the winter and flood during the normal planting season. Delayed planting is advantageous on low wet sites and should not be used on drier ridges unless irrigation is available. Cuttings should remain in freezer storage until planting.

Competition control

Competition in any form will affect poplar plantation growth and survival. Poplars must have full sunlight, adequate water, and nutrients for maximum growth potential (Demeritt 1990). Control of competing vegetation is critical to successfully establish poplar plantations (Schuette and Kaiser 1996; Von Althen 1981; Hansen and Netzer 1985). Weeds will compete better than poplars for available water and light, resulting in diminished growth or mortality. In addition to competition from vegetation, browsing by deer and rodents can reduce survival and growth, as will outbreaks of insects such as cottonwood leaf beetle early in the rotation (Ostry et al. 1989).

Weed control

Control of competing vegetation especially during the establishment years will allow poplars to survive and grow to the potential of the site. Competition control strategies vary by region and depend on annual rainfall, soils, and herbicide registration. In the southern U.S. newly planted eastern cottonwoods are sprayed in bands 0.9 m in width as needed directly over the tree rows with oxyfluorfen alone or in combination with herbicides such as imazaquin (Fig. 7). The area between rows is disked as needed to control invading weeds. The strategy in the Midwest is to broadcast apply herbicides such as linuron or imazaquin over entire plantations of dormant newly planted hybrid poplar cuttings (Hansen 1993; Hansen et al. 1993; Netzer and Noste 1978). This is followed by shallow cultivation as the herbicides become ineffective. In the Pacific Northwest weed control strategies vary, depending on the local rainfall patterns. Extremely low rainfall areas east of the Cascade Mountains are often irrigated, and herbicides such as trifluralin are soil-incorporated prior to tree planting. West of the Cascades, weed control strategies are similar to the Midwest, using capping herbicides and cultivation (Heilman et al. 1995).

Poplars typically are grown on sites that were recently in agriculture, and the weed complex is herbaceous broadleaves and grasses, although persistent woody vines are a problem in the southern U.S. Sites that have not been in crop production for several years may have additional woody brush and small trees. Control of all existing weeds can be done by applications of non-residual herbicides such as glyphosate, alone or in combination with 2,4-D. This is usually done the year prior to plantation establishment before mechanical site preparation begins. Sites

Fig. 7. Results of banded herbicide application over a 1-month-old cottonwood plantation in the Lower Mississippi Alluvial Valley. Note top of planted cutting and sprout. Photo by Jeff Portwood.

subject to winter erosion should be planted in fall with a cover crop such as annual rye grass (*Lolium multiflorum*).

Weed competition must be controlled during the first growing season. Poplars are extremely sensitive to herbicide damage (Buhler et al. 1998; Netzer and Hansen 1992; Netzer and Hansen 1994; Netzer et al. 1997; OMNR 1991), but several herbicides have been identified that poplars will tolerate (Table 2). Even labeled herbicides need to be tested by site to insure safe performance. Most herbicides are applied immediately before or after planting while cuttings are dormant. One exception is diuron, which is applied in northwestern Washington the fall prior to spring planting. Herbicides requiring soil incorporation are usually applied prior to planting. Other herbicides are sprayed directly over newly planted dormant stock. These herbicides usually do not provide complete control throughout the growing season.

Grass herbicides such as sethoxydim and fluazifop-*p*-butyl can be used directly over actively growing trees without damage. Clopyralid, imazaquin, and oxyfluorfen are used to control broadleaf weeds while poplars are growing, although leaf injury has been observed in several instances. Local trials need to be performed to insure tree clone tolerance to application timing and chemical rates.

Several types of cultivators including rototillers, discs, and various shovel and spring cultivators are used to control invading weeds during the growing season (Fig. 8). Cultivation equipment must be kept shallow enough to avoid root damage to the poplars, usually no deeper than 5 cm. Cultivators with guide wheels can control the depth of cultivation accurately. Care must also be taken to avoid damage from tool bars or other equipment to the bark and buds of young trees. Shields have been used in the Midwest to protect young trees from covering by displaced soil during cultivation.

Tending

At the end of the first growing season dormant hybrid poplars may be treated after leaf fall and prior to ground freeze up with herbicides such as azafenidin, low rates of sulfometuron, and others to control weed growth the following spring and part way through the succeeding growing season (Table 2). Care must be taken to ensure the trees are completely dormant to avoid herbicide injury. These applications can be reapplied at the end of the second growing season and beyond as needed. As trees grow taller, directed or shielded spray of low rates of glyphosate may effectively control weeds and grasses near the trees during the growing season. In northwest Washington, a shielded application of glyphosate and diuron (tank-mixed) is made between the plant rows during mid October of the first and second year. Fall applications keep the cultivated portion between tree rows free of weeds during the winter in areas without soil frost and avoid having to bring in equipment too early in spring when soils are wet.

Fig. 8. Disc cultivation of a cottonwood plantation in the Lower Mississippi Alluvial Valley during the second growing season. Photo by Jeff Portwood.

Other competitors

Mammals

Poplars are a preferred browse for most cervid species (deer, elk, and moose) and may cause establishment failure, especially of smaller plantations subject to high browsing pressure. Deterrents such as electric fences and repellants may reduce browsing to tolerable levels. Trees may grow out of the reach of deer if browsing pressure is low, by the end of the second growing season (Netzer 1984), but will remain susceptible for several years to bucks rubbing during the rutting season. Large mammal browsing can be so serious that the landowner is left with only two options: fence or forget growing poplar. In cutover forest stands, slash can be bulldozed into brush fences 3 m or higher (McKnight 1970). Electric fencing is another option, but requires continual maintenance while plants are susceptible. A five-strand fence, with the lowest strand 25 cm off the ground and the other strands 30 cm apart above it, has worked in the northeast (Brenneman 1982). Other options are available, including a more expensive woven wire fence (Dickman and Lantagne 1997). To be effective, at least two tiers of 1.2-m woven wire are required. Stay wires (no wider than 15 cm apart), a third tier of fencing, or a strand or two of barbed wire will be needed to keep deer from penetrating.

Periodically high vole (*Microtus* spp.) populations can be a problem. Grass cover in 3- and 4-year-old plantations provide protection from predators, allowing voles to feed on roots and lower stems, which can lead to heavy tree mortality. Serious damage has occurred to plantations in the Pacific Northwest, the Midwest, and southwestern British Columbia. Grass control can prevent this problem, although mice and voles can still cause trouble under snow cover.

Insects and diseases

Major pests of cottonwood plantations include defoliating insects such as cottonwood leaf beetle, poplar tent maker (*Clostera inclusa*); borers such as the cottonwood twig borer (*Gypsonoma haimbachiana*), cottonwood clearwing borer (*Paranthrene dollii*), and cottonwood borer (*Plectrodera scalator*); and aphids, mites, and leafhoppers (Morris et al. 1975; Solomon 1985). A frequent monitoring schedule should be used to control these insects prior to large infestations. Labeled general-purpose insecticides such as carbaryl or *Bacillus thuringiensis* (Bt) may be applied to control these pests (see Chap. 7).

Fertilization

The objective of a nutrient management program is to maximize plantation growth by minimizing nutrient limitations. Nutrient limitations are related to high inherent requirements due to high productivity of poplars, limited availability of native soil nutrients, and imbalance among essential nutrients. Understanding how these factors interact to affect poplar productivity focuses on nitrogen (N) as the main element limiting poplar growth in all regions. Although growth on some sites has been shown to respond to other nutrients, it is most important to provide adequate N supply and keep other nutrients balanced with N to avoid relative deficiencies.

Nitrogen requirements

The amount of N required to support optimum growth is shown in Table 4. These estimates demonstrate the very high N requirement of rapidly growing poplar, especially hybrid poplars, compared with other forest types. The high nutrient demand is due to the young age of intensively managed poplar plantations and

Table 4. Amount of nutrients required to sustain growth of poplar species and their hybrids compared with an average of temperate deciduous and conifer forest types.

Genotype	Age (years)	NPP[a] (tons ha^{-1} $year^{-1}$)	Requirement (kg ha^{-1} $year^{-1}$) N	P	K	Ca	Mg	Reference
P. deltoides	4–6	17	102	11.5	88	151	17.9	Cited by Bernier 1984
P. deltoides	7	17	107	11	91	157	18	Nelson et al. 1987
P. trichocarpa	4	7–18	95–159					Heilman and Stettler 1986
P. trichocarpa P. × deltoides	4	27–28	271–276					Heilman and Stettler 1986
P. × canadensis	4	11	168					Heilman and Stettler 1986
P. × canadensis	1–2	12–24	182–246	20–36	113–171	121–237	38	Cited by Bernier 1984
Temperate deciduous	30–120	10	98	7.2	48	56	10.4	Cited by Bernier 1984
Temperate conifers	15–450	8.3	46	5.5	28	20	4.6	Cited by Bernier 1984

[a]Net Primary Productivity (NPP) includes belowground and aboveground biomass, including foliar mass.

their high productivity. The variation in nutrient requirements among genotypes may be related to efficiency of N use (Blackmon et al. 1979), which has important ramifications for protecting surface and ground water from nitrate contamination.

Nitrogen to meet plant needs is supplied from various sources including internal cycling and N mineralized from soil organic matter and litter decomposition. We are uncertain how much of the annual N requirement is met by these sources, which limits our ability to accurately prescribe cost-effective nutrient additions. We know that the relative importance of internal cycling increases as the stand develops. Many sites with high native soil fertility do not respond to fertilization, indicating the site supply capacity is adequate to meet even the high nutrient requirements of poplar. Nonetheless, nutrients not adequately supplied by the site must be supplemented through fertilization if optimum growth rates are to be maintained. Peak demand occurs by age 5 or 6 years (Nelson et al. 1987).

Diagnosing nutrient deficiencies

Agricultural crop nutrient requirements and common nutrient deficiencies in a region provide some hints to the poplar grower. Even with practical knowledge, however, diagnostic techniques are needed to evaluate nutrient deficiencies and identify imbalances among essential plant nutrients. The effectiveness of both nutrient management approaches must be monitored to achieve maximum growth potential and avoid negative environmental effects caused by over-fertilization. Diagnostic techniques are especially necessary where variation in site and climate may affect nutrient demand, as in the Midwest and eastern North America. Although there is critical need for diagnostic and prescription techniques, few accurate tests are available.

Leaf analysis is the most common diagnostic technique for determining poplar nutrient deficiency. Nutrient concentrations are analyzed on leaves collected from the upper canopy during midsummer. Consistency in timing and canopy position of sample collection is important because variation in either will affect results. Fertilizer recommendations typically focus on N — critical levels below which fertilization is recommended are between 2 and 3% foliar N (Dickmann and Stuart 1983; Hansen 1993). Growth rates are known to increase at higher foliar concentrations (Jia and Ingestad 1984; Coleman et al. 1998), but these levels are difficult to achieve operationally. The critical foliar concentration level may vary with genotype because of differences in N use efficiency (Heilman 1985). More rapid diagnostic techniques such as the SPAD meter (Spectrum Technologies, Plainfield, IL) hold promise because of the good relationship between leaf N and SPAD value ($r^2 > 0.7$) when foliar N levels are greater than about 2.0% (Young and Berguson 2000). The SPAD meter utilizes the absorbance peak of chlorophyll in the red region (400–500 nm) with the lack of transmittance in the near-infrared (500–600 nm) region to calculate a SPAD value, which is proportional to leaf chlorophyll. Chlorophyll and nitrogen contents are highly correlated in many plant leaves. Standardized leaf sampling location or collecting weight per unit area information is especially important with such light transmittance meters because leaf thickness influence values.

Maintaining balance between N and other essential nutrients is critical for achieving optimum production. For example, many poplar stands do not respond to N additions unless accompanied by additions of P, K, or other nutrients (Blackmon 1976). Two diagnostic techniques based on foliar ratios between nutrients are available for evaluating the balance among nutrients — Ingestad's and DRIS. The ratios of several essential plant nutrients to N can be very consistent for high productivity plantations, leading to the use of Ingestad ratios for diagnostic purposes (Ericsson et al. 1992). Recommended ratios for poplar, based on laboratory-grown plants, are 100 N : 11 P : 48 K : 7 Ca : 7 Mg. Luxury consumption of N, however, affects the accuracy of ratios and interpretation of multiple nutrient ratios, making it difficult for plantation managers to determine which nutrient is actually deficient.

Another technique is called DRIS (Diagnostic and Recommendation Integrated System), adapted by Leech and Kim (1981) for use in poplar. DRIS uses all combinations of nutrient ratio means and deviations to calculate balance indices for each nutrient element in the analysis. These indices are easily interpreted and provide a method of identifying which nutrient is deficient relative to the others. For example, fertilization with N alone may result in deficiency of other nutrients and thereby limit growth. DRIS analysis is capable of diagnosing such an imbalance.

Soil testing

Soil testing has been used in forestry for characterizing major differences between soil types and landscapes. In general, routine testing for specific nutrient deficiencies has not been successful, for several reasons. Tree roots access nutrients from multiple layers and often at considerable depth; sampling the total volume utilized by the tree would be prohibitively costly. Further, interpretation of soil test results in terms of tree requirements is difficult because what is extracted by chemical tests may bear little relationship to what the tree can extract. Unless considerable effort is made to calibrate soil test results against tree nutrient status or fertilizer response, interpretation is impossible. Nevertheless, soil testing can play an important role when establishing new plantations on former agricultural land. Rough guidance for tree crops can be obtained from soil test results correlated with the previous row crops; the nutrient-demanding poplars grown on short rotation are not that different from agricultural crops. In areas where severe macro- or micro-nutrient deficiencies of trees have been demonstrated and can be correlated with soil test results, critical levels can be established to guide preventive fertilization. Over time, relationships within a fixed area between soil test results and fertilizer response can be established.

Approaches to fertilizing poplar

The amount of fertilizer to be applied depends not only on the crop nutrient requirement, but also on the application system. Two distinct approaches are used: (1) the dryland approach involves fertilizing the site as little as once per rotation or as often as once per year; (2) the fertigation approach seeks to constantly maintain optimal concentrations in the soil solution during the active growing season (Fig. 9). The dryland approach is suited (1) to non-irrigated plantations where stand entry or over-flights are the only alternatives, and (2) for supplementing micronutrients or relatively immobile nutrients that need to be applied once per rotation (e.g., lime). The dryland approach is economically

Fig. 9. Hybrid cottonwood planting in the first year of growth in western Oregon. Note lead for the fertigation system at the beginning of the row. Photo by John Stanturf.

attractive because of low capital requirements and suitability for contracting out application.

Fertilizing the site with high rates of N (150–500 kg ha^{-1}) only once per rotation assumes that applied nutrients are quickly immobilized in the soil and slowly released to supply tree growth. High rates are expected to produce a long-term fertilizer effect and may not increase growth the first year more than low rates. Nutrients not captured by vegetation or immobilized in soil may contaminate ground and surface water with nitrates.

More frequent fertilization with lower rates (50–150 kg N ha^{-1}) can sustain maximum production and avoid water quality degradation, but application costs increase. The amount of N applied annually can be adjusted to the developmental stage of the stand by ramping the rate up during establishment, reaching a maximum rate at canopy closure, and then either maintaining the maximum rate throughout the rotation or backing down as cycling on the site supplies more of the annual N requirement. Matching developmental requirements involves matching application rates to stand growth patterns and requires more sophisticated diagnostic methods than currently are available. Fertilization prior to canopy closure also risks enhancing growth of competing vegetation. Therefore, a practical approximation of an optimal nutrient regime is annual or biennial applications of constant low fertilizer rates (e.g., 50–100 kg N ha^{-1} $year^{-1}$), beginning when the canopy has closed. This less intensive approach avoids increasing weed control needs during establishment, but it risks missing maximum growth potential by under-fertilizing, or nitrate leaching by over-fertilizing (Table 5).

The fertigation approach provides the greatest flexibility in supplying nutrient uptake requirements but requires high capital costs initially and constant attention to the delivery system. This approach assumes that applied nutrients are available for uptake by poplar roots from soil solution. Nutrients removed from solution through uptake or immobilization are incrementally replenished as often as several times per week so that relatively constant nutrient concentrations are maintained in the soil solution. Such frequent incremental additions are of low concentrations but adequately supply annual growth requirements and minimize risk of nitrate leaching losses. This approach is well suited to drip irrigated plantations for applying mobile or easily fixed nutrients that are required in large quantities such as N, P, or K.

Other nutrients

Nutrients besides N may improve poplar growth, including phosphorus (P), potassium (K), calcium (Ca), and micronutrients such as boron (B), molybdenum (Mo), and zinc (Zn). Other micronutrients may be required to maintain optimum balance on certain sites. These nutrients can be applied separately or with N in fertilizer blends, using the dryland approach or using appropriate concentrates in fertigation systems.

Phosphorous may be limiting on sites such as the coarse-textured well-drained soils used for fertigation systems, highly weathered soils of the southeast U.S., or upland marine and some alluvial soils in the Pacific Northwest. Phosphorus applied at planting will encourage root development. It will persist and become slowly available for several years (possibly even through the rotation) because of mineral fixation with iron, aluminum, and calcium, as well as immobilization in organic matter. Superphosphate can be broadcast along with N, but fertilizer use efficiency can be low if roots have not fully exploited the site, and soluble P exposed to a large reaction surface on soil particles is easily fixed. Granular superphosphate, alone or in a mixture with N, may be banded and incorporated along planting rows or placed in a patch directly below the cutting at establishment

Table 5. Typical fertilizer rates applied in various poplar growing regions of North America.

Region	Cropping system	Nutrients applied	Application frequency	Application rate (elemental) (kg ha^{-1})	Chemical formulations
Lake States	Non-irrigated	N	Every 2nd year after canopy closure	85–185	Urea 45–0–0
Eastern Canada and Northeast U.S.	Non-irrigated	N	Annually	60–100	Urea 45–0–0
		N	Biosolids 55 tons ha^{-1} applied at planting	1008	1.8% N
	Irrigated	N	Every irrigation cycle during 1st year	60	Ammonium polyphosphate 10–34–0
		N	Every irrigation cycle after 1st year	125	Urea – ammonium nitrate solution 28–0–0
Pacific Northwest					
Vancouver Island	Non-irrigated	N	At planting and canopy closure	25–200	Urea 45–0–0
		P	At planting and canopy closure	25–200	Monoammonium phosphate 11–48–0
		S	At planting	8	Copper and zinc sulfate
Lower Columbia River	Non-irrigated	N, P, K	Not required		
Eastside	Irrigated	P	First month of establishment, every irrigation cycle (at least once daily)	12	Ammonium polyphosphate solution 10–34–0
		N	May through July, every irrigation cycle	60 first year, increased by 30 every year to 150 by 4th	Urea – ammonium nitrate solution 32–0–0

Table 5 (*concluded*).

Region	Cropping system	Nutrients applied	Application frequency	Application rate (elemental) (kg ha^{-1})	Chemical formulations
		Zn	1st year	30 (see text)	Zn chelate for rapid response, $ZnSO_4$ for long term
			4th year	2	
Southeast U.S.					
Lower Mississippi Valley	Non-irrigated	N	Subsoil injection at planting	100–120	Urea – ammonium nitrate solution 32–0–0
Coastal Plain	Irrigated	N	Every irrigation cycle, April through October	60–250	8–2–8 or 12–2–8
		Lime	At planting	Achieve pH 6.5	

(van den Driessche 1999). This decreases the contact between fertilizer and soil and improves efficiency of use. Another approach is to inject a mixture of N and P where the base of the cutting will be during the subsoiling / row marking operation. This places the nutrients at an optimal location for tree roots and out of the reach for shallow-rooted competing vegetation.

Fertigation systems can take advantage of slow P availability by adding one or more pulses during the rotation. Much of this pulsed P will become immobilized and slowly mineralized at rates sufficient to meet uptake requirements. Alternatively, fertigation can supply small amounts of P constantly to maintain soil solution concentrations of P adequate to meet requirements. This supplies nutrients directly to waiting roots so there is less opportunity for fixation or immobilization.

Potassium can also increase growth of poplar, usually only if supplied along with N and P (Blackmon 1976). This element can be supplied at planting by broadcast application or banding. On sandy soils, K is easily leached and may require several applications over a rotation. Soils containing expanding lattice (2:1) clay minerals such as smectite are common on slackwater deposits in the southern U.S. and are capable of fixing large quantities of K. On these soils, K should be applied in bands.

Calcium amendments by liming may be needed to raise soil pH. Poplars prefer pH levels of 6.0–6.5, but do well between 5.5 and 7.5. Black cottonwood and its hybrids perform best between pH 5.0 and 6.0. Acid soils may require heavy lime additions, but individual clones may vary in their preference. Lime additions are required for calcium supply if exchangeable Ca is less than 1000 mg kg^{-1}; levels between 3400 and 3800 mg kg^{-1} are optimal. Calcium and Mg deficiencies are rarely observed, probably because poplars occur naturally, and grow in plantations, on good sites. Attempts to grow poplars on sands and other less fertile sites using fertigation may demonstrate the need for calcium fertilization, especially on old, highly weathered parent materials as found in the southern U.S.

Sulfur and micronutrients such as zinc are also known to be limiting on some sites, but a single addition of these nutrient elements will last throughout the rotation. Because of the small amounts of micronutrients required, surface banding along the planting rows or spot treatments at each tree location are adequate. Excess amounts of these micronutrients can become toxic, so care should be taken to add only required amounts.

Regional distinctions

Poplar nutrient management programs in all regions include N additions, although the rates, timing, formulations, and methods of applying amendments vary widely.

Table 5 provides examples of the nutrient amendments used in North America. Dryland approaches to fertilizing poplar occur in regions with sufficient precipitation during the growing season or access to ground water. This area includes much of the eastern continent, as well as the west side of the Cascade Mountains in the Pacific Northwest. Fertigation is used in the arid regions east of the Cascades in the Pacific Northwest and on well-drained sites in the southeastern U.S. where extended periods between summer rains make irrigation necessary.

North Central

Sites in the North Central region range from organic peat soils to coarse glacial tills. Climate varies from moist summers and extreme winter temperatures in Canada to drier summers and milder winters in the Great Plains. Such variation requires diagnostic tools for evaluating fertilizer needs. Productivity is correlated with N levels and response to fertilization is certain when leaf N levels are at or below 2%. Fertilizer response is less certain with leaf N above 2.5%. Typically, urea is applied (85–185 kg N ha^{-1}) after canopy closure in the third or fourth growing season. Applications may continue as often as every second year thereafter, but the effectiveness of multiple applications has not been thoroughly evaluated. Although difficult to predict, growth response to blended NPK fertilizer over N only has been observed but depends on clone and site.

Northeast

Surprisingly few commercial plantations of cottonwood or balsam poplars occur in the Northeast region. Fertilization in eastern Canada consists almost exclusively of mill biosolids applied by Domtar Forest Products. Favorable response is obtained with mill residue mixed from primary and secondary clarifiers. Trees are planted through a biosolids layer applied at 55 Mg ha^{-1}, which contains 1.8% N, 0.26% P; and 0.98% K plus a full complement of micronutrients.

Both dryland and fertigation approaches are used by Mead Fiber Board in Maine. Leaf N concentrations above 3% are maintained by supplying 61 kg ha^{-1} during the first year and then 125 kg ha^{-1} annually, beginning in the second season. These rates provide increased growth over unfertilized plots, but no further effect is seen with greater rates.

Hybrid poplar clones NM-5 and NM-6 are included in the New York coppice system where 100 kg N ha^{-1} of sulfur coated urea is applied at the start of every coppice cycle. Significant productivity increases due to fertilizer application are seen compared with unfertilized controls.

Pacific Northwest — westside

No response has been seen to N application rates up to 370 kg ha^{-1} on the rich alluvial soils along the lower Columbia River even though productivity is high (Fig. 10). Some stands growing on heavy clay soils can be chlorotic and have low Mg, Zn, or Mo concentrations. Such deficiencies are rare and easily remedied by aerial applications of a micronutrient mix. In contrast, marine and alluvial soils on east Vancouver Island respond to N, P, and perhaps S at planting, and at mid-rotation. Significant growth response to N and P is obtained by banding 100–200 kg ha^{-1} of each nutrient with incorporation along the planting row, or by placing fertilizer below the surface at the base of the cutting (25–50 kg ha^{-1}). Poplars growing on marine and alluvial soils on Vancouver Island respond to fertilization just before canopy closure (3–4 years), at rates up to 200 kg N ha^{-1} and 100 kg P ha^{-1}.

Pacific Northwest — eastside

Poplar production using the fertigation approach has reached operational scale east of the Cascade Mountains in the Columbia River basin. This production system depends on fertilizer applied through the irrigation system to meet N, P, and Zn requirements, and all other essential nutrients are supplied by the coarse alluvial soils. Under this regime at the Potlatch Corporation fiber farm in Boardman, Oregon, all nutrients are applied through the irrigation system. During the first month after planting, a 10–34–0 concentrate (12 kg P ha^{-1}) is applied during each irrigation cycle. The high P concentration in the fertigation is used to encourage root growth while at the same time supplying trees with some N for establishment. The P solution is applied until mid May, then N is supplied using a 32–0–0 concentrate starting at a rate of 60 kg ha^{-1} for the first year. This rate is increased by 30 kg ha^{-1} each year until 150 kg ha^{-1} is reached in the fourth year. The N solution is applied during each irrigation cycle, and the total annual treatment is completed by the beginning of August. Zinc chelate is applied (12 kg ha^{-1})

Fig. 10. High-yielding hybrid cottonwood plantation in its fourth year of growth on an alluvial soil in western Oregon. Photo by Don Dickmann.

once trees are 0.75–1 m tall during establishment. An additional 6 kg ha^{-1} is applied mid season; in September Zn is applied as a sulfate. The chelate gives quick response during the growing season, and the sulfate form supplies long-term Zn requirements. An additional 2.5 kg Zn ha^{-1} is applied as zinc sulfate in the fourth growing season.

Foliar samples are collected monthly and analyzed for all essential nutrients. This regular monitoring program provides information on the adequacy of the fertilizer rates as well as a check on the operation of the fertigation system.

Southeast Coastal Plain, irrigated

Although the southeast U.S. has a humid climate, evaporation deficits during the growing season and the possibility of drought years lead to productivity gains for irrigated stands and present an opportunity for the fertigation approach to apply mineral nutrients. Typically, 60–250 kg N ha^{-1} are applied using an 8–2–8 or 12–2–8 liquid fertilizer concentrate. Nutrients are added during every irrigation cycle between April and October. Broadcast additions of lime, micronutrients, and P are common prior to planting. Good response is seen to irrigation, but fertilizer response depends on the site.

Lower Mississippi River Valley, non-irrigated

The Lower Mississippi River Valley contains important commercial poplar plantations on sites within the present active floodplain and on sites protected by levees (Fig. 11). Crown Vantage is the predominate grower in this region, managing company lands as well as providing landowner assistance for small private growers. Alluvial sites are periodically recharged with flood deposits and do not respond to fertilizer amendments. Old-field sites have been in cotton or soybean production for more than 20 years while protected from flooding by levees. Cottonwood on old-field sites responds to N, but not other nutrients. In this case, urea – ammonium nitrate solution (98 kg N ha^{-1}, 32–0–0) is injected at a depth of 45 cm during fall subsoiling. Cuttings are planted the following winter or early spring in the slit produced by the blade. This deep placement of N has proved more effective than side dressing because it is placed within reach of the tree roots, but beyond the reach of competing vegetation.

Fig. 11. Three-year-old eastern cottonwood plantation on a Commerce soil in the Lower Mississippi Alluvial Valley. Photo by US Forest Service.

Thinning

Poplar plantations offer an opportunity to produce sawlogs and veneer within 20–30 years of planting. Systematic and selective thinning regimes must be included in management for these products (Fig. 12). Timing of thinning treatments will be determined largely by initial spacing, which is affected by site quality, establishment practices, and survival.

Spacing and thinning studies on eastern cottonwood illustrate the complexity of managing plantations for sawlogs. Cottonwood is characterized by very rapid diameter and height growth in the early years, and plantations must be managed aggressively to maintain this rapid growth and avoid stagnation. Initial spacing has no affect on the rate at which diameter growth peaks, generally by the third or

Fig. 12. Thinning a hybrid poplar plantation to maximize the yield of useful products and maintain the diameter growth of residual trees. This stand will produce high quality sawtimber and veneer. Photo by Don Dickmann.

fourth year (Krinard and Johnson 1984). Because cottonwood cannot tolerate side competition, it responds poorly to release, following crowding. Wide spacing with pruning of the lower branches or closer spacing accompanied by early thinnings is necessary to maintain rapid growth of individual trees.

Anderson and Krinard (1985) summarized experience from experimental and operational plantings of five spacing intervals on two sites (medium and good), shown in Table 6. Generally dbh increases as spacing increases, from 3.7 × 3.7 m to 7.4 × 7.4 m. All spacing intervals were thinned at least once except the two widest. Sawtimber yields were greatest for stands spaced 7.4 × 7.4 m. Wider spacing, however, requires intensive pruning to maintain quality and more weed control to successfully establish plantations. A compromise adopted in the Lower Mississippi River Valley of 3.7 × 3.7 m is suitable both for pulpwood and sawlog production (Gascon and Krinard 1976). Stands established at this spacing on good sites can be systematically thinned, beginning at ages 3–5 years, removing half the trees. Stands should be selectively thinned thereafter to maintain growth.

Spacing trials of black cottonwood indicate that the 3.7 × 3.7 m spacing is best for this species as well (DeBell 1990). Black cottonwood also responds well to thinning. Hybrid poplars are grown at a variety of spacings, including a rectangular spacing of 3.05 × 2.1 m used by Fort James in western Washington. Although they use rectangular spacing for mechanical efficiency, indications are

Table 6. Total wood volume and lumber volume yields of eastern cottonwood plantations on a good site by spacing interval and thinning regime[a] (source: Anderson and Krinard 1985).

Spacing (m)	Age (years)	Residual stand stocking (stems ha^{-1})	Residual stand dbh (cm)	Total volume cut per hectare (m^3 ha^{-1})	Sawtimber volume cut per hectare (m^3 ha^{-1})
3.47 × 3.47	5	296	15.5	44.1	
	12	188	30	53.2	
	20	124	42	30.0	19.9
	30	0	55	292.3	179.6
4.88 × 5.49	8	168	24	72.8	
	18	111	43	27.4	18
	30	0	58	308.4	187.7
7.32 × 7.32	15	99	42	41.0	26.7
	30	0	63	328.2	196.3
9.75 × 9.75	30	0	63	334.4	200.0
11 × 11	30	0	64	275.1	164.5

[a]All stands were thinned once except the two widest spaced stands, 9.75 × 9.75 m and 11 × 11 m. The first thinnings in the stands spaced 3.47 × 3.47 m and 4.88 × 5.49 m were row thinnings in which every other row was removed.

that rectangular spacing may produce higher yields. This could be due to more rapid crown differentiation, and a shorter time growth is checked by competition (DeBell et al. 1997*b*).

Coppicing

The ability of poplars to sprout readily from stump or root collar provides an opportunity to regenerate and manage coppice stands in the second rotation. The coppice system of natural regeneration is an inexpensive alternative to replanting. Coppice management is currently used in eastern cottonwood plantations grown for pulpwood in the southern U.S. (Fig. 13), but it is not used elsewhere in North America. Most poplar growers continually replace old planting stock with genetically improved stock; thus, coppice is unattractive even for pulpwood production. If sawlogs or veneer logs are the product goal, replanting is the best option because of poor stem form in coppice, and stumps of larger trees sprout less vigorously.

Coppice rotations are economically attractive to non-industrial private landowners because of lower establishment costs. Clearing and site preparation following harvest of a plantation is complex and expensive. The two most important factors for coppice regeneration are age of stand and time of harvesting. Harvest should begin no later than age 10 in the rotation to insure vigorous sprouting. The time of harvest should be during the dormant season, usually between the months of

Fig. 13. Coppice regrowth of an eastern cottonwood plantation in the Lower Mississippi Alluvial Valley. Photo by Don Dickmann.

October and April, depending on region and weather patterns. Plantations harvested during the winter months are typically those that may be targeted for coppicing. Often there is a proliferation of shoots that arise from a single stump, and how these shoots are treated can potentially affect growth, yield, and average tree size through rotation. If coppice is undesirable, harvesting should begin after trees have fully flushed in the spring and can continue until trees begin dormancy.

Because of multiple sprouting, it has been customary to thin stumps back to two sprouts in the winter after the third growing season, removing up to 10 sprouts from each stump. Without this cleaning step, yields of the coppice rotation will be half or less than the first rotation because of small stem size. Recently, Crown Vantage has harvested every other row in a plantation in the winter, which encourages sprouting. After it is clear that sprouting has been successful, usually after one or two growing seasons, the residual trees are harvested in the summer to discourage sprouting. In this way, even multiple sprouts on a stump will have sufficient growing space to develop to merchantable size. For small landowners, however, this may not be cost effective, as the stand must be entered twice.

Growth and yield

Many factors influence the growth of poplars in plantations, including species or clone, site quality, climate, and spacing. After establishment, the amount of growing space available to an individual tree dominates stand yield and significantly influences the average size stem attained by harvest age. DeBell and others (1997*b*) concluded that optimal spacing and rotation length would be wider and longer, respectively, than was presented in earlier biomass research (Ranney et al. 1987). DeBell et al. (1997*a*, *b*) concluded that hybrid poplar plantations needed a minimum of 6.2 m^2 growing space per tree to yield a stand with mean tree diameter at harvest of 15 cm, regarded as the economic minimum.

Tree growth is not uniform, however, even when individuals are all from the same clone. Poplars are extremely intolerant of shading, such that crowns of eastern cottonwood do not touch even in densely spaced plantations. Belowground competition probably occurs before crown closure. Francis (1985) found that by age 8, average root length of eastern cottonwood stabilized at slightly more than half the distance between individual stems, indicating significant belowground competition. Clones of eastern cottonwood and hybrid poplar vary in their tolerance of shading; some can be planted closer together than others, a concept expressed as “stockability” (DeBell et al. 1989). Before reviewing the scant data available on growth and yield of poplar plantations, it is instructive to compare patterns of stand development in natural stands to plantations.

Switzer et al. (1976) compared patterns of biomass accumulation in eastern cottonwood natural stands to closely spaced thinned plantations and widely spaced unthinned plantations. Dry matter accumulation in the closely spaced thinned plantations was greater than natural stands early on, until age 10. The pattern of dry matter accumulation was similar in natural stands and widely spaced plantations until about age 15. Periodic and mean annual biomass increments followed similar trends. The maximum mean annual biomass increment was about the same for all three cultural regimes, between 10 and 11 tons ha^{-1} $year^{-1}$. The potential of a site to produce biomass appears to be relatively fixed, at least under a given management intensity. The time required to achieve culmination of mean annual biomass increment, however, can be influenced by manipulating growing space available to individual stems. Even more importantly, the time required to reach a minimum or average stem size can be influenced by manipulating growing space, nutrients, and water.

Growth of natural stands of eastern cottonwood (Table 7) and black cottonwood (Table 8) provide a baseline for comparing growth and yield potential of poplar plantations (Table 9). The highest values for operational plantation culture are generally for eastern cottonwood, although T×D hybrids in the Pacific Northwest rival yields from southern bottomland sites. Evidence from experiments with improved genetic stock and more intensive management practices promise significantly higher yields in the future. Directly extrapolating from small research plots to operational yield expectations, however, is dangerous. For example, Heilman and Stettler (1985) determined mean production values at age 4 for a hybrid poplar clone 11–011 (T×D) to be 28 tons ha^{-1} $year^{-1}$. DeBell et al. (1997*b*) used larger plots, attained growth equal to or better than other studies with the same clone, but estimated yield to be 18 tons ha^{-1} $year^{-1}$, similar to the

Table 7. Yields of natural cottonwood stands in the Lower Mississippi River Valley (source: Williamson 1913).

Age (years)	Volume (m^3 ha^{-1})	Stocking (stems ha^{-1})	dbh (cm)	Height (m)	Mean annual diameter increment (cm)	Mean annual height increment (m)	Mean annual volume increment (m^3 ha^{-1})
5	46		5.1	6.7	1.0	1.3	9.1
10	126	1727	14.5	17.1	0.6	1.7	12.6
15	269	682	23.4	24.7	0.6	1.6	18.0
20	343	403	31.2	29.6	0.6	1.5	17.1
25	381	282	38.1	32.9	0.6	1.3	15.3
30	408	198	44.2	35.1	0.6	1.2	13.6
35	430	146	50.0	36.9	0.6	1.1	12.3
40	450	121	55.9	38.7	0.6	1.0	11.2
45	467	104	61.5	40.2	0.5	0.9	10.4
50	483	79	67.3	41.5	0.5	0.8	9.7

Table 8. Yields of natural black cottonwood (*Populus trichocarpa*) stands by site class (source: Smith 1980, cited by DeBell 1990).

Site class	Stand age (years)	Average dbh (cm)	Stocking Stems (ha^{-1})	Height (m)	Net volume (m^3 ha^{-1})	Maximum mean annual volume increment (m^3 ha^{-1})
I	112	46	294	41	302	5.5
II	101	33	415	30	220	2.8
III	87	28	474	21	123	1.7

Table 9. Growth and yield potentials of intensively managed poplar plantations (source: Dickmann and Stuart 1983).

Parameter	Growth or yield[a]
First-year height growth	1–3.6 m
Mean annual height growth after 10–20 years	0.8–2.0 m
Mean annual diameter growth after 10–20 years[b]	1–2.5 cm
Mean annual volume increment after 10–20 years	7–25 m^3 ha^{-1}
Mean annual biomass increment after 5–20 years[c]	5–20 tons ha^{-1}

[a]Growth and yield will vary appreciably, depending upon geographic location, site quality, clone or cultivar used, and silvicultural conditions. Highest values generally are for cottonwood on southern bottomland sites.

[b]Diameter growth of individual trees depends on stocking density. Wide spacing or frequent thinnings promote rapid diameter growth.

[c]Oven-dry, leafless stems, and branches. Attainment of maximum annual increment will occur only if stands are heavily fertilized and irrigated and will occur much sooner at tree spacing of 2 m or less.

operational yields obtained at the James River Lower Columbia Fiber Farm near Camas, WA (Fig. 14). Nevertheless, clonal trials do indicate biological potential.

Diameter and height growth

Eastern cottonwood is one of the tallest hardwood species. Heights in natural stands of 53–58 m and diameters of 120–180 cm have been reported (Putnam et al. 1960; Johnson and Burkhardt 1976). Cao and Durand (1991*a*) reported mean annual height increments of 1.9–2.4 m $year^{-1}$ at age 10 for plantation stands of eastern cottonwood growing on different soil series in the Lower Mississippi Alluvial Valley, with mean annual increments on the very best sites exceeding 3 m. Heights of 13 m at age 3 and more than 30 m at age 9 have been observed for individual trees. Trees planted at wide spacing can average 29 cm dbh at age 5 (Krinard 1979).

Fig. 14. The payoff — final harvest of a short-rotation hybrid cottonwood plantation on an industrial site in western Oregon. Photo by Don Dickmann.

Annual increment of dominant and co-dominant trees in black cottonwood plantations in British Columbia and Washington can average 1.6 m in height and 1.9 cm in dbh (Silen 1947). In the lower Fraser Valley, annual increment of dbh was 2 cm and height was 1.7 m in a 10-year-old plantation (Smith and Blom 1966); growth is less on sites in the interior and at locations that are more northerly.

Hybrid poplars in the Pacific Northwest can achieve height growth of 2–3 m year^{-1} and up to 2.3 cm year^{-1} annual dbh growth (Heilman and Stettler 1985). Ceulemans et al. (1992) reported height growth up to 3.4 m year^{-1} and dbh growth of up to 2.55 cm year^{-1} for hybrids. Both reports are for 4-year old plantations including parents and hybrids.

Hybrid poplars growing under favorable conditions in the Northeast averaged 1–3 cm in annual dbh increment and height increased 1–2 m annually (Zsuffa et al. 1977). Hybrid poplars growing on silty clay loam soils in southern Ontario varied in annual height growth from 0.7 to 1.3 m and from 0.6 to 1.3 cm in annual diameter growth after 18–22 years (Marshall 1979). Clones of eastern cottonwood and its hybrid with *P. balsamifera* grew as well as natural aspen (*P. tremuloides* Michx.) root suckers on good sites in northern Ontario, well beyond the range of cottonwood (Farmer et al. 1991). Growth after 9 years for the best clones was 80 cm year^{-1} in height and over 1 cm dbh. Hybrid aspen in northern Wisconsin have grown in excess of 1 m in height and 1 cm in dbh annually (Benson 1972).

Volume growth[1]

Data on volume of managed poplar plantations are scarce, and the best data available are for pulpwood rotations of eastern cottonwood in the South. A compatible growth and yield model is available (Cao and Durand 1991*a*), which uses the individual tree volume equations developed by Krinard (1988):

[1] $\text{TVOB} = 0.06 + 0.002221D^2H$

and

[2] $\text{MVIB} = -0.86 + 0.001904D^2H$

where

TVOB = total tree volume outside bark in ft^3 from a 30-cm stump to the tree tip,

MVIB = merchantable tree volume inside bark in ft^3 from a 30-cm stump to a 7.6-cm top,

D = diameter at breast height in in., and

H = total tree height in ft.

Total and merchantable volumes can be converted from ft^3 to m^3 by multiplying by 0.02832.

The equations for total tree volume and merchantable tree volume per acre (1 acre = 0.404 ha) are:

[3] $\ln \text{TV}_i = 2.64098 + 0.00868S - 3.27063/A_i + 1.09103 \ln B_i$

and

[4] $\ln \text{MV}_i = 2.12838 + 0.01411S - 5.04889/A_i + 1.08576 \ln B_i$

where

$\ln x$ = natural logarithm of x,

TV_i = total outside-bark volume in $ft^3\ acre^{-1}$ at time i,

MV_i = total merchantable inside-bark volume in $ft^3\ acre^{-1}$ at time i,

S = site index in ft at base age 10 years,

---

[1]Volume equations presented in this section are given in English rather than metric units because the original data sets from which the equations were derived were not available to make the conversions. English–metric conversion factors for dependent variables, however, are given in each case (see Appendix for English–metric conversions).

B_i = stand basal area in ft^2 acre^{-1} at time i, and

A_i = stand age in years at time i.

Total and merchantable volumes can be converted from ft^3 acre^{-1} to m^3 ha^{-1} by multiplying by 0.06998.

Site index often is used in growth and yield models and must be estimated from stand data. Cao and Durand (1991*b*) developed polymorphic site index curves for eastern cottonwood plantations in the Lower Mississippi Alluvial Valley. Their site index curve for any base age up to 10 years is:

[5] $$\ln(H) = 5.83564 + [\ln(S) - 5.83564](I/A)^{0.41576}$$

where

$\ln(H)$ = natural logarithm of average height in ft of the dominants and co-dominants,

S = site index in ft,

I = base age in years, and

A = stand age in years. Feet can be converted to m by multiplying by 0.3048.

Stanturf and Portwood (1999) used data from three stands of different productivity classes to evaluate the economics of afforestation with eastern cottonwood. They used the Cao and Durand (1991*a*) model to estimate merchantable yield to a

Table 10. Characteristics of the stands selected to represent soil/site productivity classes and their estimated merchantable yields[a] at rotations of 10 years; stands were age 3 years when measured (source: Stanturf and Portwood 1999).

	Commerce[b]	Tunica–Bowdre[b]	Sharkey[b]
Site index (base age 10), m	24.4	22.3	20.1
Basal area, m^2 ha^{-1}	6.7	3.9	3.4
Stems ha^{-1}	682	622	642
Survival, %	91	83	86
Tons ha^{-1}, age 10	76.7	56.3	47.1
Mean annual increment, OD tons ha^{-1} at age 10	7.7	5.6	4.7
Cumulative annual increments, OD tons ha^{-1} at age 10	8.4	7.0	6.0

[a]Merchantable yields are estimated by eq. [4] (inside bark, to a 7.6-cm top). These stands are all on old field sites, protected by the river levee, with good survival. All were planted with the technology described in Table 11.

[b]Soils are Commerce (Aeric Fluvaquents), Tunica–Bowdre (Vertic Haplaquepts – Fluvaquentic Hapludolls), and Sharkey (Chromic Epiaquerts).

Table 11. Schedule of operations for cottonwood cultural practices in the southern U.S.

Dates	Activity
October year 0	Two-pass site preparation disking Row establishment and liquid nitrogen applied in trenches @ 112 kg N ha^{-1}
March year 1	Plant cottonwood
March year 1	Spray herbicide in band over dormant cuttings (oxyfluorfen @ 0.26 kg ha^{-1} + glyphosate @ 1.4 kg ha^{-1})
May year 1	One-pass disking, followed 2 weeks later by second pass at right angle to first
June and July year 1	Basal application of oxyfluorfen @ 0.7 kg ha^{-1}
August year 1	One-pass disking, followed 2 weeks later by second pass at right angle to first
Summer year 1	Insect control for cottonwood leaf beetles (carbaryl @ 0.92 kg ha^{-1})
June year 2	Insect control for cottonwood leaf beetles (carbaryl @ 0.92 kg ha^{-1})
June and July year 2	One-pass disking
Winter year 10	Cottonwood pulpwood harvest

7.6 cm top (Table 10). The silvicultural system used operationally in the Lower Mississippi Alluvial Valley is detailed in Table 11, and the system in the Lake States in Table 12.

Black cottonwood plantations in the Fraser River Valley in British Columbia yielded mean annual volume increments ranging from 10.5 to 15.4 m^3 ha^{-1} (Smith 1980), much higher than the values for natural stands shown in Table 8. Even greater mean annual volume increment was obtained in a plantation growing on deep alluvial soils in coastal Washington, 20.8 m^3 ha^{-1} over 24 years (Murray and Harrington 1983).

Hybrid poplars in Ontario have been reported to yield as much as 29 m^3 ha^{-1} $year^{-1}$ after 12 years (Zsuffa et al. 1977), and from 10 to 27 m^3 ha^{-1} $year^{-1}$ after 18–22 years (Marshall 1979). Elsewhere in the Northeast, mean annual volume increment of *P.* × *canadensis* cv. Eugenei in Indiana was 7 m^3 ha^{-1} $year^{-1}$ (Merritt and Bramble 1966). Recent work in the Lake States with several disease resistant clones of hybrid poplar (D×N crosses) showed estimated yields of up to 9.4 oven-dry (OD) tons ha^{-1} $year^{-1}$ (Hansen 1992; Netzer and Tolsted 1999), with adjacent small plot trials of the best new clones yielding up to 13.4 OD tons ha^{-1} $year^{-1}$ (Hansen et al. 1994).

Biomass growth

Leafless biomass increment is usually measured in research studies of poplar growth. Yield of eastern cottonwood in the southern U.S. and hybrid poplar in the Pacific Northwest is measured in terms of dry or green tons per ha, but

Table 12. Schedule of operations for establishment of hybrid poplar plantations in the Midwest U.S.

Dates	Activity
July year 0	On fields in sod, pasture, or hay apply glyphosate alone or in combination with 2-4-D
	Plow 10 days after herbicide application and fallow for remainder of season
October year 0	On fall-cropped fields apply glyphosate if perennial weeds are present
	Plow all planting sites to a 25-cm depth
	Seed cover crop (rye, barley, oats) if winter wind erosion is likely
Late April – early May year 1	Field cultivate or till sites as close to planting as practical
	Use tractor-mounted marking system to premark tree rows or exact planting locations
Late April – early June year 1	Plant presoaked hardwood cuttings
	Apply pre-emergent herbicides (linuron 2.25 kg ha^{-1}, imazaquin 2.8 0.2 kg ha^{-1}, or other registered herbicides (see Table 2)) directly over newly planted hardwood cuttings
July–September year 1	Shallow cultivate (<5 cm) as needed to keep site weed free
May–September year 2 – 3 or 4, depending on crown closure	Apply registered grass herbicides (such as Poast 1.2 L ha^{-1}, Fusilade 0.88 L ha^{-1} followed by 0.58 L ha^{-1} if needed) to control invading grasses
	Apply clopyralid 0.39–0.77 L ha^{-1} to control invading thistle
October year 1 – 3 or 4	Fall herbicide application when poplars are dormant but before soil freezes (sulfometuron methyl or azafenidin at label rates for soil type)
July year 2 – harvest	Fertilize to maintain leaf nitrogen content above 3%
Year 8–12	Harvest plantation when average annual growth begins to decline

extrapolation from experimental plantings to production per unit area can lead to overly optimistic expectations. Cannell and Smith (1980) noted that the annual leafless biomass increment of short-rotation poplars has not exceeded 10–12 OD tons ha^{-1} $year^{-1}$ in rotations of 4–5 years. Recent reports, however, of the range of productivities from commercial plantations in North America are higher, exceeding 12 OD tons ha^{-1} $year^{-1}$ in some regions (Table 13). Strauss and Wright (1990) estimated a benchmark yield by age 6 of 16 OD tons ha^{-1} $year^{-1}$, from data obtained in commercial poplar plantations. DeBell et al. (1997*a*) reported hybrid poplar yields from large monoclonal block plantings and operational experience, averaging 12–17 tons ha^{-1} $year^{-1}$. They noted differences between clones in tolerance to crowding or competition. In other experiments, the very best clones of

Table 13. Leafless aboveground biomass yields in commercial poplar plantations in North America.[a]

Species	Region (clone)	Site and culture	Rotation age	Spacing (m)	Mean annual biomass increment (OD tons ha^{-1} year^{-1})
Eastern cottonwood	Southern U.S.	Fair	10	3.7 × 3.7	6.7
		Good	10	3.7 × 3.7	7.7
		Very good	10	3.7 × 3.7	10.2
		Best	10	3.7 × 3.7	12.5
Hybrid poplar	Lake States	Good	10		9.4
	Pacific Northwest, eastside	Fertigated	7	3.7 × 2.3	13.0
	Pacific Northwest, westside, clone 11–011	Fertigated	7	1 × 1	18
	Pacific Northwest, westside, clone D-01	Fertigated	7	1 × 1	11
	Pacific Northwest, westside	Good	7	3.05 × 2.1	13
Black cottonwood	Pacific Northwest, westside	Fertigated	7	1 × 1	13

[a]These yields represent what is currently achieved in operational plantings. Values for narrow spacings are indicative of what is achievable on wider spacing and slightly longer rotations in the Pacific Northwest (DeBell et al. 1997*a*, *b*) and coincide with reports from industrial plantation managers. Results from experimental plantings (Strauss and Wright 1990) of an average of 16 OD tons ha^{-1} year^{-1} have been achieved operationally only in fertigated stands. Nevertheless, substantial gains are likely in the future as improved material is deployed and technological advances are implemented.

T×D hybrids were able to achieve aboveground dry biomass production of 29 tons ha^{-1} year^{-1} over 4 years, although at spacing of 1 × 1 m (Heilman et al. 1994).

Netzer and Tolsted (1999) developed a biomass regression equation for three disease resistant clones of hybrid poplar planted in western Wisconsin and Minnesota. The equation is of the form:

[6] $\text{Tree weight} = 6.16 - 2.23\text{DBH} + 0.353\text{DBH}^2$

where tree weight is oven-dry in kg and DBH is in cm.

They used this equation to estimate mean yields from plantations of improved hybrid poplar in the Lake States to be more than 6.7 OD tons ha^{-1} $year^{-1}$, with yields in better sites averaging 9.4 OD tons ha^{-1} $year^{-1}$. Peak yields at 2.4×2.4 m spacing occurred between year 7 and 10, provided weed control was adequate.

Environmental effects

Public perceptions of plantations are often negative, although poplar plantations provide many environmental benefits. The value of poplar plantations to wildlife has been documented (Twedt et al. 1999; Wesley et al. 1981). Where poplar plantations replace row cropping, additional benefits are improved water quality and greater floral diversity.

Eastern cottonwood plantations in the lower Mississippi River valley produce understory biomass exceeding 1200 kg ha^{-1} annually (Wesley et al. 1981). This luxuriant undergrowth is excellent habitat for whitetail deer and rabbits year-round. Spring nesting and brood habitat is available for wild turkey and quail. There is little difference between older cottonwood plantations and the surrounding bottomland hardwood forest for Neotropical migratory bird territory, except that there are fewer cavity nesters in the plantations (Twedt et al. 1999).

Afforestation of cropland with eastern cottonwood in conjunction with slower growing oak species is practiced in the lower Mississippi Valley (Schweitzer and Stanturf 1999). Besides early financial returns to the landowner, this interplanting system benefits forest-breeding Neotropical migratory birds (Twedt and Portwood 1997). Cottonwood also increases plant species diversity by providing perches for fruit-eating birds, thereby facilitating spread of other species into the afforestation stands (Twedt and Portwood 1997).

Despite the intensive site preparation and early weed control needed to successfully establish poplars, plantations can improve water quality as compared to continuous row cropping. The simplest effect is that soil disturbance in poplar plantations is limited to at most 3 years out of the 7–10-year rotation, while soils supporting row crops are continuously disturbed. Poplar culture can significantly improve water quality by reducing sediment loss as well as lower pesticide and nutrient movement. Surprisingly, the improvement can occur even in the first year. Thornton et al. (1998) reported on a side-by-side comparison of eastern cottonwood with cotton. Sediment loss from the cottonwood was 2.3 tons ha^{-1} the first year, significantly less than the 16 tons ha^{-1} from cotton. The tree canopy developed nearly complete cover over the 3.7 m interval between rows by late spring and probably decreased raindrop impacts on the soil surface because interception lowered runoff.

Acknowledgements

We thank the following for their assistance and suggestions for improving of this chapter: Perry Boshart, Mead Corporation; Bill Berguson, University of Minnesota, Duluth; Jake Eaton, Potlatch Corp.; Chuck Kaiser, Fort James; Peter McAuliffe; Pierre Périnet; Mike Roy, Domtar Forest Products; Brian Stanton, Fort James; Robert van den Driessche; Joanne van Oosten, SilviConsult; Tim Volk, Environmental Science and Forestry, SUNY at Syracuse; and Lisa Zabek, University of British Columbia, Vancouver.

References

Anderson, W.C., and Krinard, R.M. 1985. The investment potential of cottonwood sawtimber plantations. *In* Proceedings of the Third Biennial Southern Silvicultural Research Conference, Atlanta, Georgia, November 7–8, 1984. *Edited by* Eugene Shoulders. USDA For. Serv. Gen. Tech. Rep. SO-54. pp. 190–197.

Baker, J.B., and Broadfoot, W.M. 1979. A practical field method of site evaluation for commercially important hardwoods. USDA For. Serv. Gen. Tech. Rep. SO-36.

Benson, M.K. 1972. Breeding and establishment — and promising hybrids. *In* Proceedings of the Aspen Symposium. USDA For. Serv. Gen. Tech. Rep. NC-1. pp. 88–96.

Bernier, B. 1984. Nutrient cycling in *Populus*: a literature review with implications in intensively-managed plantations. Canadian Forestry Service, Ottawa. IEA/ENFOR Rep. 6.

Blackmon, B.G. 1976. Response of *Aigeiros* poplars to soil amelioration. *In* Proceedings of the Symposium on Eastern Cottonwood and Related Species, Greenville, Mississippi, September 28 – October 2, 1976. *Edited by* B.A. Thielges and S.B. Land, Jr. Louisiana State University, Division of Continuing Education, Baton Rouge. pp. 344–358.

Blackmon, B.G., Baker, J.B., and Cooper, D.T. 1979. Nutrient use efficiency by three geographic sources of eastern cottonwood. Can. J. For. Res. **9**: 532–534.

Brenneman, R. 1982. Electric fencing to prevent deer browsing on hardwood clearcuts. J. For. **80**: 660–661.

Buhler, D., Netzer, D.A., Riemenschneider, D.E., and Hartzler, R.G. 1998. Weed management in short-rotation poplar and herbaceous crops grown for biofuel production. Biomass Bioenergy, **14**: 385–394.

Cannell, M.G.R., and Smith, R.I. 1980. Yields of minirotation closely spaced hardwoods in temperate regions: review and appraisal. For. Sci. **26**: 415–428.

Cao, Q.V., and Durand, K.M. 1991*a*. A growth and yield model for improved eastern cottonwood plantations in the lower Mississippi Delta. So. J. Appl. For. **15**: 213–216.

Cao, Q.V., and Durand, K.M. 1991*b*. Site index curves for eastern cottonwood plantations in the lower Mississippi Delta. So. J. Appl. For. **15**: 28–30.

Ceulemans, R., Scarascia-Mugnozza, G., Wiard, B.M., Braatne, J.H., Hinckley, T.M., Stettler, R.F., Isebrands, J.G., and Heilman, P.E. 1992. Production physiology and morphology of *Populus* species and their hybrids grown under short rotation. I. Clonal comparisons of 4-year growth and phenology. Can. J. For. Res. **22**: 1937–1948.

Coleman, M.D., Dickson, R.E., and Isebrands, J.G. 1998. Growth and physiology of aspen supplied with different fertilizer addition rates. Physiol. Plant. **103**: 513–526.

DeBell, D.S. 1990. *Populus trichocarpa* Torr. & Gray. *In* Silvics of North America. Vol. 2. Hardwoods. *Edited by* R.M. Burns and B.H Honkala. USDA For. Serv. Agric. Handb. 654. pp. 570–576.

DeBell, D.S., Harms, W.R., and Whitesell, C.D. 1989. Stockability: a major factor in productivity differences between *Pinus taeda* plantations in Hawaii and the southeastern United States. For. Sci. **35**: 708–719.
DeBell, D.S., Harrington, C.A., Clendenen, G.W., and Zasada, J.C. 1997*a*. Tree growth and stand development in four *Populus* clones in large monoclonal plots. New For. **14**: 1–18.
DeBell, D.S., Harrington, C.A., Clendenen, G.W., Radwan, M.A., and Zasada, J.C. 1997*b*. Increasing the productivity of short-rotation *Populus* plantations. Final Report to the Bioenergy Feedstock Development Program, Oak Ridge National Lab, Oak Ridge. Report ORNL/M-5943. (Available at <http://bioenergy.ornl.gov/reports/debell/contents.html>).
Demeritt, M.E. 1990. Poplar hybrids. *In* Silvics of North America. Vol. 2. Hardwoods. *Edited by* R.M. Burns and B.H. Honkala. USDA For. Serv. Agric. Handb. 654. pp. 577–582.
Dickmann, D.I., and Lantagne, D. 1997. Planting oaks for timber and other uses. North Central Regional Extension Publication No. 605. Michigan State University Extension, East Lansing. 12 pp.
Dickmann, D.I., and Stuart, K.W. 1983. The culture of poplars in eastern North America. Michigan State University Press, East Lansing. 168 pp.
Ericsson, T., Rytter, L., and Linder, S. 1992. Nutritional dynamics and requirements of short rotation forests. *In* Ecophysiology of short rotation forest crops. *Edited by* C.P. Mitchell, J.B. Ford-Robertson, T. Hinckley, and L. Sennerby-Forsse. Elsevier Applied Science, London. pp. 35–65.
Farmer, R.E., Palmer, C.L., Anderson, H.W., Zsuffa, L., and O'Reilly, G. 1991. Nine-year outplanting test of cottonwood and hybrid poplar clones in northwestern Ontario. Tree Planters Notes, **42**: 49–51.
Francis, J.E. 1985. The roots of plantation cottonwood: their characteristics and properties. USDA For. Serv. Res. Note SO-314.
Gascon, R.J., Jr., and Krinard, R.M. 1976. Biological response of plantation cottonwood to spacing, pruning, thinning. *In* Proceedings of the Symposium on Eastern Cottonwood and Related Species, Greenville, Mississippi, September 28 – October 2, 1976. *Edited by* B.A. Thielges and S.B. Land, Jr. Louisiana State University, Division of Continuing Education, Baton Rouge. pp. 385–391.
Hansen, E.A. 1992. Mid-rotation yields of biomass plantations in the north central United States. USDA For. Serv. Res. Pap. NC-309.
Hansen, E.A. 1993. A guide for determining when to fertilize hybrid poplar plantations. USDA For. Serv. Res. Pap. NC-319.
Hansen, E.A., and Netzer, D.A. 1985. Weed control in short-rotation intensively cultured poplar plantations. USDA For. Serv. Res. Pap. NC-260.
Hansen, E.A., Netzer, D.A., and Tolsted, D.N. 1993. Guidelines for establishing poplar plantations in the north-central U.S. USDA For. Serv. Res. Pap. NC-363.
Hansen, E.A., Ostry, M.E., Johnson, W.D., Tolsted, D.N., Netzer, D.A., Berguson, W.E., and Hall, R.B. 1994. Field performance of *Populus* in short-rotation intensive culture plantations in north-central U.S. USDA For. Serv. Res. Pap. NC-320.
Heilman, P.E. 1985. Sampling and genetic variation of foliar nitrogen in black cottonwood and its hybrids in short rotation. Can. J. For. Res. **15**: 1137–1141.
Heilman, P.E., and Stettler, R.F. 1985. Genetic variation and productivity of *Populus trichocarpa* and its hybrids. II. Biomass production in a 4-year plantation. Can. J. For. Res. **15**: 384–388.
Heilman, P.E., and Stettler, R.F. 1986. Nutritional concerns in selection of black cottonwood and hybrid clones for short rotation. Can. J. For. Res. **16**: 860–863.
Heilman, P.E., Ekuan, G., and Fogle, D. 1994. Above- and below-ground biomass and fine roots of 4-year-old hybrids of *Populus trichocarpa* × *Populus deltoides* and parental species in short-rotation culture. Can. J. For. Res. **24**: 1186–1192.
Heilman, P.E., Stettler, R.F., Hanley, D.P., and Carkner, R.W. 1995. High yield hybrid poplar plantations in the Pacific Northwest. Pac. Northwest Coop. Ext. Bull. PNW 356. 39 pp.
Jia, H., and Ingestad, T. 1984. Nutrient requirement and stress response of *Populus simonii* and *Paulownia tomentosa*. Physiol. Plant. **62**: 117–124.

Johnson, R.L., and Burkhardt, E.C. 1976. Natural cottonwood stands — past management and implications for plantations. *In* Proceedings of the Symposium on Eastern Cottonwood and Related Species, Greenville, Mississippi, September 28 – October 2, 1976. *Edited by* B.A. Thielges and S.B. Land, Jr. Louisiana State University, Division of Continuing Education, Baton Rouge. pp. 20–30.

Krinard, R.M. 1979. Five years' growth of pruned and unpruned cottonwood planted at 40- by 40-foot spacing. USDA For. Serv. Res. Note SO-252.

Krinard, R.M. 1988. Volume equations for plantation cottonwood trees (*Populus deltoides*). USDA For. Serv. Res. Note SO-347.

Krinard, R.M., and Johnson, R.L. 1984. Cottonwood plantation growth through 20 years. USDA For. Serv. Res. Pap. SO-212.

Leech, R.H., and Kim, Y.T. 1981. Foliar analysis and DRIS as a guide to fertilizer amendment in poplar plantations. For. Chron. **57**: 17–21.

Marshall, P.L. 1979. Growth and yield of two young poplar plantations in Ontario. *In* Poplar research, management and utilization in Canada. For. Res. Info. Pap. No. 102. Ontario Ministry of Natural Resources, Toronto.

McKnight, J.S. 1970. Planting cottonwood cuttings for timber production in the south. USDA For. Serv. Res. Pap. SO-60.

Merritt, C., and Bramble, W.C. 1966. Poplar plantation research in Indiana — a five-year report. Res. Bull. No. 818, Purdue University Agricultural Experiment Station, West Lafayette.

Morris, R.C., Filer, T.H., Solomon, J.D., McCracken, F.I., Overgaard, N.A., and Weiss, M.J. 1975. Insects and diseases of cottonwood. USDA For. Serv. Res. Gen. Tech. Rep. SO-8.

Murray, M.D., and Harrington, C.A. 1983. Growth and yield of a 24-year-old black cottonwood plantation in western Washington. Tree Planters' Notes, **34**(2): 3–5.

Nelson, L.E., Switzer, G.L., and Lockaby, B.G. 1987. Nutrition of *Populus deltoides* plantations during maximum production. For. Ecol. Manage. **20**: 25–41.

Netzer, D.A. 1984. Hybrid poplar plantations outgrow deer browsing effects, USDA For. Serv. Res. Note NC-325.

Netzer, D.A., and Hansen, E.A. 1992. Seasonal variation in hybrid poplar tolerance to glyphosate. USDA For. Serv. Res. Pap. NC-311.

Netzer, D.A., and Hansen, E.A. 1994. Establishment and tending of poplar plantations in north-central U.S. *In* IEA/BA Task IX Activity 1. Short rotation intensive culture forestry. Mechanization in SRIC Forestry, March 1–3, 1994, Mobile, Alabama. *Edited by* B.J. Stokes and T.P. McDonald. USDA Forest Service, Auburn, AL. pp. 79–87.

Netzer, D.A., and Noste, N.V. 1978. Herbicide trials in intensively cultured *Populus* plantations in northern Wisconsin. USDA For. Serv. Res. Note NC-235.

Netzer, D.A., and Tolsted, D.E. 1999. Yields of ten and eleven year-old hybrid poplars in the North Central United States. Final Report to the U.S. Department of Energy, Biofuels Feedstock Development Program. USDA Forestry Services, North Central Research Station, Forestry Sciences Laboratory, Rhinelander. 9 pp.

Netzer, D.A., Riemenschneider, D.E., Bauer, E.O., and Spatcher, D. 1997. Effect of clopyralid on the growth and morphology of hybrid poplars. *In* Proceedings of the North Central Weed Science Society, Louisville, Kentucky, December 9–11, 1997. North Central Weed Science Society, Champaign, IL. Vol. 51. pp. 114–118.

OMNR. 1991. A grower's guide to hybrid poplar. Ontario Ministry of Natural Resources, Toronto.

Ostry, M.E., Wilson, L.F., McNabb, H.S., Jr., and Moore, L.M. 1989. A guide to insect, disease, and animal pests of poplars. USDA For. Serv. Agric. Handb. 677.

Putnam, J., McKnight, J.S., and Furnival, G. 1960. Inventory and management of southern hardwoods. USDA For. Serv. Agric. Handb. 181.

Ranney, J.W., Wright, L.L., and Layton, P.A. 1987. Hardwood energy crops: the technology of intensive culture. J. For. **85**: 17–28.

Schuette, B., and Kaiser, C. 1996. Weed control strategies for SRIC hybrid poplars: farmer's perspective. *In* Proceedings of the Short Rotation Woody Crops Working Group Meeting, Paducah, Kentucky, September 23–25, 1996. 9 pp.

Schweitzer, C.J., and Stanturf, J.A. 1999. A comparison of large-scale reforestation techniques commonly used on abandoned fields in the Lower Mississippi Alluvial Valley. *In* Proceedings of the Tenth Biennial Southern Silvicultural Research Conference, Shreveport, Louisiana, February 16–18, 1999. *Edited by* J.D. Haywood. USDA For. Serv. Gen. Tech. Rep. SRS-30. pp. 136–141.

Silen, R.R. 1947. Comparative growth of hybrid poplars and native northern black cottonwoods. USDA For. Serv. Res. Note PNW-35.

Smith, J.H.G. 1980. Growth and yield of poplar in British Columbia. Paper presented at the 1980 meeting of the Poplar Council of Canada. *Cited in* Debell, D.S. *Populus tricocarpa* Torr. & Gray. *In* Silvics of North America. Vol. 2. Hardwoods. *Edited by* R.M. Burns and B.H. Honkala. USDA For. Serv. Agric. Handb. 654. pp. 570–576.

Smith, J.H.G., and Blom, G. 1966. Decade of intensive cultivation of poplars in British Columbia shows need for long-term research to reduce risks. For. Chron. **42**(4): 359–376.

Solomon, J.D. 1985. Impact of insects on growth and development of young cottonwood plantations. USDA For. Serv. Res. Pap. SO-217.

Stanturf, J.A., and Portwood, C.J. 1999. Economics of afforestation with eastern cottonwood (*Populus deltoides*) on agricultural land in the Lower Mississippi Alluvial Valley. *In* Proceedings of the Tenth Biennial Southern Silvicultural Research Conference, Shreveport, Louisiana, February 16–18, 1999. *Edited by* J.D. Haywood. USDA For. Serv. Gen. Tech. Rep. SRS-30. pp. 66–72.

Strauss, C.H., and Wright, L.L. 1990. Woody biomass production costs in the United States: an economic summary of commercial *Populus* plantation systems. Sol. Energy, **45**: 105–110.

Switzer, G.L., Nelson, L.E., and Baker, J.B. 1976. Accumulation and distribution of dry matter and nutrients in *Aigeiros* poplar plantations. *In* Proceedings of the Symposium on Eastern Cottonwood and Related Species, Greenville, Mississippi, September 28 – October 2, 1976. *Edited by* B.A. Thielges and S.B. Land, Jr. Louisiana State University, Division of Continuing Education, Baton Rouge. pp. 359–369.

Thornton, F.C., Joslin, J.D., Bock, B.R., Houston, A., Green, T.H., Schoenholtz, S.H., Pettry, D., and Tyler, D.D. 1998. Environmental effects of growing woody crops on agricultural land: first year effects on erosion and water quality. Biomass Bioenergy, **15**: 57–69.

Twedt, D.J., and Portwood, C.J. 1997. Bottomland hardwood reforestation for Neotropical migratory birds: are we missing the forest for the trees? Wildl. Soc. Bull. **25**(3): 647–652.

Twedt, D.J., Wilson, R.R., Henne-Kerr, J.L., and Hamilton, R.B. 1999. Impact of forest type and management strategy on avian densities in the Mississippi Alluvial Valley, USA. For. Ecol. Manage. **123**(2–3): 261–274.

van den Driessche, R. 1999. First-year growth response of four *Populus trichocarpa* × *Populus deltoides* clones to fertilizer placement and level. Can. J. For. Res. **29**: 554–562.

Von Althen, F.W. 1981. Site preparation and postplanting weed control in hardwood afforestation: white ash, black walnut, basswood, silver maple, hybrid poplar. Rep. 0-X-325. Great Lakes Forest Research Centre, Sault Ste. Marie. 17 pp.

Wesley, D.E., Perkins, C.J., and Sullivan, A.D. 1981. Wildlife in cottonwood plantations. So. J. Appl. For. **5**: 37–42.

Williamson, A.W. 1913. Cottonwood in the Mississippi Valley. USDA Bull. 24. 62 pp.

Young, M., and Berguson, W. 2000. In-situ foliar nitrogen determination in hybrid poplar using a Minolta SPAD-502 chlorophyll meter. Research Note, Lake States Region 00-02, Champion International Corp., Norway, MI. 6 pp.

Zsuffa, L., Anderson, H.W., and Jaciw, P. 1977. Trends and prospects in Ontario's poplar plantation management. For. Chron. **53**: 195–200.

CHAPTER 6
Environmental benefits of poplar culture

J.G. Isebrands and David F. Karnosky

To plant trees is to give body and
life to one's dreams of a better world.

Russell Page

Introduction

Poplars have important values above and beyond wood or fiber production. Poplars have been planted for environmental purposes for centuries. There are reports of poplar plantings dating back to early Chinese history and biblical times in the Middle East. When immigrants came to North America in the 18th and 19th century, they often brought cuttings of their favorite poplar to plant on their newfound land or garden in the New World. Thus, the long history of planting poplars in the Old World was preserved and continued in the New World. J.E. Rogers (1906) wrote about the merits of planting cottonwood (*Populus deltoides*) — "it can be planted for shade and ornament, for windbreaks and to hold banks of streams — it endures heat and soot, and has dignity with added years." Because the early settlers were mostly agrarian, they often planted native cottonwoods for wind and snow protection of their farmsteads and animals as well as to decrease soil and wind erosion; but there were also a significant number of linear plantings of poplars in cities for protection, visual screens, and aesthetics.

In this chapter, our primary focus is on current uses of poplars for environmental purposes. The use of poplars for shelterbelts on the prairies is not new, as both the United States and Canada have had a long history of shelterbelt programs (Munns and Stoeckler 1946; Roller et al. 1972; Kort and Turnock 1996). No matter which

J.G. Isebrands. USDA Forest Service, North Central Research Station, 5985 Highway K, Rhinelander, WI 54501, U.S.A.
D.F. Karnosky. School of Forestry and Forest Products, Michigan Technological University, Houghton, MI 49931, U.S.A.
Correct citation: Isebrands, J.G., and Karnosky, D.F. 2001. Environmental benefits of poplar culture. *In* Poplar Culture in North America. Part A, Chapter 6. *Edited by* D.I. Dickmann, J.G. Isebrands, J.E. Eckenwalder, and J. Richardson. NRC Research Press, National Research Council of Canada, Ottawa, ON K1A 0R6, Canada. pp. 207–218.

environmental use one pursues with poplar planting, it is essential that one choose the appropriate clone for the intended site and use (Dickmann and Isebrands 1999). Consideration must be given to soil, microclimate, pests and diseases, and environmental ethics (i.e., exotic versus native material) in making this choice. We cannot emphasize enough the importance of this point, as in our experience there are innumerable examples of planting failures due to hasty deployment of the inappropriate poplar clone for the site.

Protection plantings: windbreaks and shelterbelts

Poplars have been used for farmstead windbreaks and field shelterbelts in North America since the days of early settlement. The early settlers used poplars because they grew rapidly under various soil, site, and climatic conditions, and provided rapid protection to farmsteads, livestock, and crops (Scholten 1988; Zsuffa et al. 1996). By definition, windbreaks are single or multiple rows of trees established for environmental purposes, usually around farmsteads. Shelterbelts are usually larger in size, surrounding fields of agricultural crops. Native poplars, (e.g., cottonwood) were primarily used in the early days (Jones and Parker 1916), but subsequently shelterbelt tree improvement programs identified hybrid poplars with low branching crowns, rapid growth, and drought resistance, which were considered more desirable for windbreaks and shelterbelts (Roller et al. 1972). Recently, the pendulum has swung back in the direction of native poplar species. For example, the Prairie Farm Rehabilitation Administration (PFRA) Shelterbelt Centre at Indian Head, Saskatchewan, Canada, currently produces only *P. deltoides* 'Walker' and three clones selected from its open-pollinated offspring (PFRA 1999).

Farmstead windbreaks of poplars protect farm homes, buildings, equipment, orchards, and livestock from cold winter winds. They moderate summer heat and winds, improve living conditions for people and animals thereby increasing farm value, beautify the landscape, decrease noise and dust, and provide wildlife habitat (Fig. 1). Field shelterbelts protect crops by decreasing soil erosion and moisture loss, filter field runoff, increase populations of wildlife and beneficial insects, serve as a site for animal manure disposal, and produce biomass for wood and energy (Andre 1960). Shelterbelts have been shown to increase crop yields by up to 20% by decreasing stress on the plants. In the prairie states and provinces, poplars are also used as living snow fences to keep blowing snow off roads and access drives. They, thereby, save energy, store snow at low cost, provide wildlife habitat, improve aesthetics, harvest water in dry regions, and provide long-term protection over the years.

A new concept called "timberbelts" recently has been introduced by agroforesters. Timberbelts are multiple-row windbreaks designed to provide all the environmental benefits of windbreaks and shelterbelts, while also producing economic benefits from salable small-diameter wood products on short rotations

Fig. 1. Farmstead shelterbelts of poplars protect orchards and vineyards along the Columbia River in eastern Oregon and Washington.

(i.e., 7–10 years) or large-diameter material on longer rotations. Poplars are ideally suited for the timberbelt concept, providing farmers additional income with potential markets for pulp, oriented strandboard, veneer, solid wood products, and carbon credits (Kuhn and Josiah 2000).

Erosion control

By integrating engineering and biological expertise, poplars and willows can be used to control the erosion of banks of streams, rivers, and reservoirs. Poplars (and willows) are combined with manmade structures (i.e., rock toes) to protect natural beauty and valuable personal and cultural properties along the rural or urban stream or reservoir banks. The poplars' roots bind to the toes and strengthen the stream banks where they are planted. This approach decreases soil erosion on valuable property and maintains a more natural stream or reservoir bank without increasing stream flow or bank erosion. The advantages of erosion control are both environmental and economic; natural plant materials are low cost, provide natural habitat, minimize flooding, and are self-sustaining (Illinois State Water Survey 1999). The approach is currently being used extensively throughout the Mississippi river watershed. It is notable that poplars and willows are also beginning to be used to preserve Native American cultural sites along its tributaries.

Riparian buffer systems

Poplars are well suited for riparian buffer systems in North America and in recent years have received increased attention from farmers and farm agencies. Riparian buffers are by definition streamside plantings of trees, shrubs, or grasses that intercept contaminants from both surface runoff water and ground water before they reach the stream (Fig. 2). Poplars (or willows) protect streambanks from soil erosion and act as filters from the adjacent fields or urban landscape before the contaminants such as excess nitrates and pesticides reach the stream. Studies show that riparian buffers of poplars can retain as much as 70–90% of nitrates and 75% of sediments compared to unbuffered streams (Haycock and Pinay 1993). The trees along the streams in many parts of North America also provide shade to keep water cool for trout and salmon habitat. Riparian poplar buffers also provide improved wildlife habitat, decreased flood damage, improved biodiversity, value from recreational hunting and fishing, and economic returns from biomass and wood products. These buffers will increase in importance as financial incentives to landowners from state and federal agencies become more popular. Such incentives are needed to make riparian buffers more economically attractive to landowners and to reward them for their environmentally based management practices.

Not all poplars are suited for planting in riparian areas. Many hybrid poplars are not flood tolerant and should not be planted on land that is subjected to frequent flooding. Those lands are more suited for plantations of native cottonwoods,

Fig. 2. Riparian buffer of poplars and willows along the Minnesota River in south central Minnesota.

balsam poplar, and cottonwood hybrids, as they are better adapted to wetter soils in riparian ecosystems.

Phytoremediation and wastewater reuse

Phytoremediation and wastewater reuse are emerging technologies that are cost effective, aesthetically pleasing, and environmentally sound approaches for cleaning up and remediating contaminated soils and water. Poplars and willows are the two most common tree genera being used for phytoremediation because they grow rapidly, are easily propagated, and adapt well to riparian sites. By definition, phytoremediation is the use of green plants for on-site risk remediation of contaminated soils, sludges, sediments, and groundwater, through contaminant removal, degradation, or confinement (U.S. EPA 1998). Poplars are being used throughout North America to clean up sites with heavy metals, salts, pesticides, solvents, explosives, radionuclides, hydrocarbons, and landfill leachates. Poplars are well suited for phytoremediation because they can remove contaminants from soil and water in several ways, including degrading them, confining them, or by acting as filters or traps. Poplars are also being used as vegetative caps on municipal and industrial landfills because they take up large quantities of water from rainfall and groundwater. This consumption of water decreases the tendency for surface and groundwater contaminates to move toward drinking water aquifers, streams, or lakes.

There is also evidence that poplars are particularly useful in degrading chlorinated solvents such as trichloroethylene (TCE). Researchers at the University of Washington are using this approach in the Pacific Northwest of the U.S., and similar approaches may be possible with poplars adapted to other regions of North America.

Poplars are also well suited for disposal of agricultural, industrial, and community wastewater. Because poplars take up large quantities of water (see Chap. 3), they are an ideal woody plant for secondary wastewater management. They are a low cost alternative to wastewater treatment systems and wastewater distribution systems. The wastewater provides the poplars with essential water and nutrients, and the trees provide an aesthetic, economic, and environmentally sound system. The trees decrease soil erosion, provide protection from wind, and produce valuable biomass and wood products while performing an important environmental benefit. Poplars have also been used in the Northeastern U.S. and Canada for strip mine reclamation (Davidson 1979).

On agricultural land, poplars can be used for taking up excess irrigation water that often contains high levels of nutrients, pesticides, and other contaminants. Poplar plantations also have potential for applications of livestock wastes from dairies, cattle, hog, and poultry confinement operations. For example, large dairy operations are expanding into the arid portions of the western U.S. Some dairy

Fig. 3. A poplar plantation in Woodburn, OR, used to take up and reuse municipal secondary treated wastewater. Note wastewater emitter to the right of the tree in the foreground.

operators are using hybrid poplars as an alternative for applying wastewater to their land. Test plantings of hybrid poplars, irrigated with dairy wastewater, will help determine their viability to protect water quality and produce wood products. The trees can also serve as a buffer between the operations and streams or public access areas. We believe these applications are likely to increase in importance as agricultural regulations become more stringent.

In communities, poplar plantations have value for managing municipal and industrial wastewater and landfill leachate (Smesrud et al. 2000). The poplars take up the nutrient-laden wastewater that might otherwise contribute to problems of nutrient loading in streams. The communities then reap the same economic and environmental benefits that are provided by poplars in agricultural settings (USDA 2000).

There are several municipalities currently using poplar plantations for their wastewater treatment. Notably, the city of Woodburn, OR, in cooperation with

CH2M Hill, Inc., of Portland, OR, is discharging secondary treated wastewater (5 million L/day) that has high levels of ammonia on a nearby 34 ha poplar plantation (Fig. 3). The results have been so successful that the project is being expanded to 120 ha.

Bioenergy

Poplars are currently being used as an environmentally acceptable source of biomass for wood and energy. Wood chips can be mixed with traditional fuels such as coal to produce electricity. This approach is cleaner, cheaper, and more suitable than coal alone. The forest products industry in the western U.S. is growing poplars on short rotations for fiber (i.e., less than 10 years), and after chipping the aboveground portion of whole trees, the residual biomass is used for hog fuel in the pulp mill to generate a renewable, low-polluting source of electricity (Fig. 4). Poplars are also used extensively as a local source of firewood throughout North America.

The U.S. Department of Energy (DOE) Biofuels Feedstock Development Program (BFBP) has been funding research on biofuels with poplars and willows

Fig. 4. Piles of residual barking chips from a Fort James Corporation whole-tree chipping operation of short-rotation hybrid poplars in western Oregon are loaded for hog fuel.

since the early 1970's. Their goal has been to develop and demonstrate environmentally acceptable crops and cropping systems for low cost, high-quality biomass for energy. At this time, there are no commercial operations that depend totally upon poplar biomass as a fuel source. However, a 50 MW power plant that utilizes whole-tree hybrid poplar biomass as a fuel is under construction in southeast Minnesota. Over the years, millions of U.S. dollars have been made available from DOE to various research institutions across North America to improve short-rotation poplar crops through genetics, physiology, pathology, silviculture, and environmental studies.

Carbon sequestration

As the world's population grew and the use of energy increased, atmospheric concentration of carbon dioxide (CO_2) has increased dramatically. Continued increases may have significant effects on the global environment and economy through climate change and global warming.

Trees provide a natural mitigation strategy by capturing and storing CO_2 in their biomass and in the soil, while in turn releasing oxygen back to the atmosphere through the process of photosynthesis. Establishing fast-growing poplar windbreaks, shelterbelts, amenity plantings, and short-rotation plantations thereby

Fig. 5. Landscape planting of hybrid poplar provides a visual screen in northern Wisconsin.

hold promise as a viable carbon sequestration strategy throughout North America. For example, Kort and Turnock (1996) reported that poplar shelterbelts on the Canadian prairies could contain between 61 and 222 tons of carbon per kilometer after 40 years, depending on the soil type. Efforts are underway in the U.S. and Canada to offer payments or carbon credits to growers for carbon sequestration and storage as part of the United Nations 1997 Kyoto Protocol for stabilizing carbon dioxide levels in the earth's atmosphere. Poplars in North America may well play an important role if this protocol is approved and implemented.

Urban amenity plantings

The genus name *Populus* is derived from the Latin "*arbor populi*" — the tree of the people. It is an appropriate name, as poplars have been used for urban amenity plantings historically throughout the world. Planting poplars provides benefits to individuals and communities throughout North America. These benefits are primarily environmental, economic, and psychological. They also provide other social benefits including aesthetics, historic preservation, and living memorials.

The environmental benefits of poplars include moderation of summer and winter climate, improvement of air quality, protection of water reserves, and creation of bird and other wildlife habitat. A single poplar tree is said to provide cooling to a house equivalent to a single room air conditioner. Poplars also help screen homes from heavy wind, and they filter the air we breathe by removing dust particles and air pollutants such as ozone and sulfur dioxide (Karnosky 1976). They intercept water, store it, and decrease runoff, thereby reducing soil erosion. Poplar trees are utilized by squirrels, songbirds, and other animals for food, shelter, and nesting sites. Psychologically, people feel more peaceful and serene in the presence of trees.

Community benefits of poplar plantings include enhanced property values and privacy, as well as visual screens of objectionable views (Fig. 5). Poplars decrease noise and decrease glare and reflection. They enhance the value of schoolyards, other public and industrial areas, serve as arboretum and landscape trees, beautify city streets and highways (Fig. 6), and make golf courses more scenic and challenging (Fig. 7). On the whole, poplar amenity plantings and other trees enhance the quality of life of community residents.

Climate change

Tropospheric carbon dioxide (CO_2) and ozone (O_3) are increasing simultaneously in the atmosphere as a result of increased human activities such as fossil fuel burning and land use change. Our research has shown that increases in these greenhouse gases will affect the future health and productivity of poplar trees and stands (Karnosky 1976; Isebrands et al. 2000). In general, poplars are very sensitive to climate change episodes; for example, poplar growth and productivity

Fig. 6. Highway planting of hybrid poplars for aesthetics and soil erosion control in northern Iowa.

Fig. 7. Hybrid poplars planted on a golf course in central Wisconsin for beauty and a challenge for players.

Fig. 8. Hybrid poplars vary in their sensitivity to tropospheric ozone exposure.

is increased by elevated CO_2 concentrations and decreased by sustained elevated O_3 concentrations (Fig. 8). Our results show that certain poplar clones are more sensitive to simultaneous interacting elevated CO_2 and O_3. With hybrid poplars, one can choose specific hybrid poplar clones for planting in high air pollution environments (Dickson et al.1998). Aspen (*Populus tremuloides*) clones also range in sensitivity from very sensitive to intermediate to very tolerant to these interacting stressors. However, our results show that on the whole, poplar growth will be less in a future climate scenario characterized by increased air pollution and elevated CO_2 than in today's climate. Future climate scenarios may also affect disease incidence as well as the size and severity of aspen leaf and wood boring insect populations. Fortunately, the large genetic diversity of poplars will likely allow us to continue our environmental use of poplars well into the future.

In summary, the environmental uses of poplars for protection, erosion control, riparian buffers, phytoremediation, bioenergy, carbon sequestration, urban plantings, and CO_2 abatement will likely increase in the future as our population and social environmental needs increase. At the same time, society will reap the important aesthetic and psychological benefits that poplars inherently provide.

Acknowledgement

The authors wish to acknowledge the support of the US Department of Energy through the Biofuels Feedstock Development Program, Oak Ridge National Laboratory.

References

Andre, F. 1960. Midwest farm handbook. Iowa State University Press, Ames, IA. pp. 404–418.

Davidson, W.H. 1979. Hybrid poplar pulpwood and lumber from a reclaimed strip mine. USDA For. Serv. Res. Note NE-232. 2 pp.

Dickmann, D.I., and Isebrands, J.G. 1999. Caveat emptor. Am. Nurseryman, **109**: 60–65.

Dickson, R.E., Coleman, M.D., Riemenschneider, D.E., Isebrands, J.G., Hogan, G.D., and Karnosky, D.F. 1998. Growth of five hybrid poplar genotypes exposed to interacting elevated CO_2 and O_3. Can. J. For. Res. **28**: 1706–1716.

Haycock, N.E., and Pinay, G. 1993. Groundwater nitrate dynamics in grass and poplar vegetated riparian buffer strips during winter. J. Environ. Qual. **22**: 273–278.

Illinois State Water Survey 1999. Streambank erosion. Misc. Publ. No. 149.

Isebrands, J.G., Dickson, R.E., Rebbeck, J., and Karnosky, D.F. 2000. Interacting effects of multiple stresses on growth and physiological processes in northern forests. *In* Responses of northern US forests to environmental change. *Edited by* R.E. Mickler, R.A. Birdsey, and J. Hom. Springer-Verlag, Berlin, Germany. pp. 149–180.

Jones, E.R., and Parker, E.C. 1916. Farm economy. Twelve courses of agriculture. H.L. Baldwin Publ., Minneapolis, MN. 416 pp.

Karnosky, D.F. 1976. Threshold levels for foliar injury to *Populus tremuloides* by sulfur dioxide and ozone. Can. J. For. Res. **6**: 166–169.

Kort, J., and Turnock, R. 1996. Biomass production and carbon fixation by prairie shelterbelts. PFRA Shelterbelt Centre Suppl. Rep. 95-6. 14 pp.

Kuhn, G.A., and Josiah, S.J. 2000. Timberbelts: windbreaks that enhance production and produce profitable wood products. Gen. Tech. Rep. NC 215. p. 101.

Munns, E.N., and Stoeckler, J.H. 1946. How are the Great Plains shelterbelts? J. For. **44**: 237–257.

PFRA Shelterbelt Centre. 1999. Description of poplar clones. <www.agr.ca/pfra/shbpub/shbpub7.htm> (accessed 14 Feb. 2001).

Rogers, J.E. 1906. The tree book. Doubleday and Page, New York. 571 pp.

Roller, K.J., Thibault, D.H., and Hindahl, V. 1972. Guide to the identification of poplar cultivars on the prairies. Environment Canada, Ottawa, ON. Catalog No. FO 47-1311. 55 pp.

Scholten, H. 1988. Farm shelterbelts, University of Minnesota, St. Paul, MN. NR-BU-0468.

Smesrud, J., Dickey, J., Asare, S., Cox, A., Lanier, A., Jordahl, J., and Madison, M. 2000. Beneficial reuse of landfill leachate with hybrid poplar. USDA For. Serv. Gen. Tech. Rep. NC-215. p. 167.

U.S. Department of Agriculture. 2000. Working trees for treating waste. USDA NPS 00-0254. 6 pp.

U.S. Environmental Protection Agency. 1998. A citizen's guide to phytoremediation. EPA 542-F-98-011. 6 pp.

Zsuffa, L., Giordano, E., Pryor, L.D., and Stettler, R.F. 1996. Trends in poplar culture: some global and regional perspectives. *In* Biology of *Populus* and its implications for management and conservation. *Edited by* R.F. Stettler, H.D. Bradshaw, Jr., P.E. Heilman, and T.M. Hinckley. NRC Research Press, Ottawa, ON. pp. 515–539.

CHAPTER 7
Insects pests of *Populus*: coping with the inevitable

William J. Mattson, Elwood A. Hart, and W. Jan A. Volney

Introduction

Trees in the genus *Populus* (the aspens, cottonwoods, poplars, and their hybrids) are highly regarded for their phenomenal potential for producing wood. This reputation derives from their high physiological capacity for exploiting light- and nutrient-rich environments. They are classic examples of "growth-dominated" plant species, i.e., ones that consistently allocate a high proportion of their gathered resources to several key growth-enhancing processes, such as continuous foliage canopy enlargement, during a prolonged growing season.

Tradeoffs: high growth, low resistance to pests

As desirable as fast growth traits are, they may often come at an expense, i.e., a tradeoff with other desirable traits. For example, some high-growth-adapted plants may exhibit poor stress resilience, and high susceptibility to pathogens, insects, and vertebrate herbivores (Chapin et al. 1993; Herms and Mattson 1997). For example, in North America, the number of insects and mites commonly found on the 12 species of *Populus*, at least 300 species, ranks among the highest on any native tree genus (Drooz 1985; Ives and Wong 1988; Peterson and Peterson 1992). In Europe, the recorded total number is almost twice as large, about 525 species of insects and mites (Delplanque et al. 1998). The pathogens are just as numerous, there being more than 250 species of decay fungi on just *Populus*

W.J. Mattson. USDA Forest Service, North Central Research Station, Forestry Sciences Laboratory, 5985 Highway K, Rhinelander, WI 54501, U.S.A.
E.A. Hart. Department of Entomology and Department of Forestry, 403 Science II, Ames, IA 500011, U.S.A.
W.J.A. Volney. Northern Forestry Centre, Canadian Forest Service, 5320-122 St., Canadian Forestry Service, Edmonton, AB T6H 3S5, Canada.
Correct citation: Mattson, W.J., Hart, E.A., and Volney, W.J.A. 2001. Insects pests of *Populus*: coping with the inevitable. *In* Poplar Culture in North America. Part A, Chapter 7. *Edited by* D.I. Dickmann, J.G. Isebrands, J.E. Eckenwalder, and J. Richardson. NRC Research Press, National Research Council of Canada, Ottawa, ON K1A 0R6, Canada. pp. 219–248.

tremuloides (Newcombe 1996). Some very excellent, color handbooks are now available to permit rapid identification of the most common and most important pests (for North America, see Ostry et al. 1989 and Ives and Wong 1988; for Europe, see Delplanque et al. 1998).

What this means is that *Populus* trees are prone to have large numbers of insect and pathogen species that attack them. Moreover, this typically translates into many kinds of injuries that can often be quite substantial and detrimental. For example, a midsummer survey of foliage damage in a 3-year-old trembling aspen sucker stand, regenerated after a logging clearcut in western Upper Michigan, revealed that of nearly 8000 sample leaves examined from several hundred plants, only a paltry 5% completely escaped insect injury.

Practically every single leaf in the young stand was injured to some degree by insect feeding. Averaging the amount of leaf surface area removed or damaged across all plants showed that defoliation was about 20%. Leaf area losses of this magnitude are quite common in *Populus*, and ought to be expected as the norm. This particular level of injury, although seemingly substantial, may not be anything to worry about because *Populus* is quite defoliation-tolerant, i.e., capable of compensating for low-to-moderate reductions in leaf surface area (Robison and Raffa 1994; Reichenbacker et al. 1996).

Thus, although high-growth traits may often be correlated with low resistance to insects and pathogens, the good news is that some plants, such as *Populus*, may likewise be well equipped for compensating for most leaf damage until it exceeds a moderately high threshold (Herms and Mattson 1992; Reichenbacker et al. 1996). Therefore, growers should think about management strategies for holding pest damage below the limits of the plant's compensation threshold. The thresholds will vary with cultivar and with soil/site, and weather conditions.

Not all insects are equally important

Although folivores, those 200 or so species eating whole leaves or parts of leaves are the most common insects; they as a group are not necessarily the ones of greatest long-term concern because the sum total of their injuries seldom goes beyond normative levels (ca. 30%) of defoliation. Likewise, among the other species that attack other parts of the plant, only a handful are seriously threatening. The poster insects, the ones that need to be kept front and center in our vigilance, are those few species that most seriously impair the optimal functioning of the apical meristems, and the lateral or cambial meristems (Mattson et. al. 1988). The former generates new shoots and buds, and the latter, new phloem and xylem tissues. This is not to dismiss the defoliator class entirely. Growers need to be concerned about just a few folivore species, those that for some reasons have the capacity to generate prodigious outbreaks. These insects are very important because, through their incredible abundance, they not only greatly diminish photosynthetic area, they also can substantially diminish the generative capacities of the apical and cambial meristem, and hence overall growth. Among the other insects, growers need to be concerned about a half dozen or so that directly damage the young extending shoots, and the main stems.

Selected insect problems

Therefore, rather than present a general overview of the many insects of *Populus*, we will instead address a few species more thoroughly, those perceived to be among the most important insects affecting *Populus* culture (aspens, cottonwoods, hybrid poplars) in North America. Among them are three outbreak defoliators, two shoot feeders, and four stem borers. For each insect species, we will concisely outline their life history and damage, and spell out reasonable management suggestions for them. Other common insects on *Populus* that may sometimes become serious pests are listed in Table 1 along with references to obtain more information about them.

Insects feeding on leaves

Cottonwood leaf beetle

The cottonwood leaf beetle (CLB), *Chrysomela scripta* F. (Coleoptera: Chrysomelidae), is considered to be a major defoliating insect of *Populus* throughout most of the United States and southern Canada, with the exception of the coastal regions of the Pacific Northwest. This native insect is especially damaging to

Table 1. Other insects that may be commonly observed in *Populus* plantations, sometimes as significant pests, and an appropriate reference for more information.

Insect common name	Insect Latin name	Plant part attacked	Reference
Poplar tent maker	*Clostera inclusa*	Leaves	Ostry et al. 1989
Spiny-elm caterpillar	*Nymphallis antiopa*	Leaves	Ostry et al. 1989
Imported willow leaf beetle	*Plagiodera versicolora*	Leaves	Ostry et al. 1989
Gypsy moth	*Lymantria dispar*	Leaves	Delplanque 1998
Aspen blotch leaf miner	*Phyllonorycter salicifoliella*	Leaves (mining)	Ostry et al. 1989
Pale green weevil	*Polydrusus* spp.	Leaves, roots	Delplanque 1998
Cottonwood borer	*Plectrodera scalator*	Roots, base of trunk	Solomon 1995
Agrilus beetles	*Agrilus* spp.	Roots, trunk	Ostry et al. 1989
Tarnished plant bug	*Lygus lineolaris*	Shoots	Ostry et al. 1989

species and hybrids of sections *Aigeiros* and *Tacamahaca* (Table1). For the most part, the CLB is not considered to be a major pest of material with section *Populus* parentage, but recent observations in west-central Minnesota indicate that some hybrids from this section definitely are susceptible under heavy outbreak conditions. Both larvae and adults feed on the foliage (Fig. 1).

The number of generations each year depends upon local climate and weather conditions. In the northern part of its range, the CLB may have only one generation each growing season; in the southern United States, as many as seven generations have been recorded. In central Iowa, three generations per year were noted from 1989 through 1996, but because of warm, extended growing seasons in 1997–1999, four full generations occurred each year. The implication is that if warming trends continue on this continent, additional generations and additional damage also are likely to occur in many areas. The thermal requirements for development from egg to adult are reported to be 230–280 degree days (Burkot and Benjamin 1979; Jarrard 1997).

Fig. 1. Cottonwood leaf beetle damage.

The CLB overwinters as an adult in duff or ground vegetation, and emerges at the same time as native *Populus* buds begin to break in the spring. Following aggregation of both sexes on terminals for feeding and mating, adults disperse and lay eggs in masses of 30–80, preferentially on the underside of immature foliage on a growing terminal (Bingaman and Hart 1992). The younger leaves are preferred for feeding by both larvae and adults; fully expanded, mature leaves usually are non-preferred on most clones. In outbreak populations, all acceptable leaf tissue may be completely devoured and the more succulent stem tissue girdled, causing multiple leaders (Fang and Hart 2000).

The newly-hatched larvae are dark brown to black, feed gregariously, and because of their small size (ca. 1 mm) only graze on the leaf surface. Second and third larval stages are somewhat lighter colored but have large, paired defensive glands on the dorsal surface that when the insect is disturbed emit a defensive chemical. These two stages become progressively less gregarious, and feed on the entire leaf blade, leaving only the midrib and larger veins on the older leaves (Fig. 1). Late third stage larvae wander to various parts of the tree or move to undergrowth, fasten themselves with a posterior adhesive pad, and pupate. Adults are 5.4–9.0 mm long, with longitudinal, ivory-to-gold stripes intermixed with brownish-black stripes on the wing covers (Fig. 2). New emerged adults disperse from the pupation sites, both sexes aggregating, feeding, mating, and then dispersing for egg-laying as did the overwintering adults.

The impact of defoliation is most severe when a high percentage of the foliage on a tree is in a susceptible state of development. There is evidence that trees during the first 1–3 years of growth have the highest percentage of preferred foliage for the first two generations of the CLB (Fang 1997). Studies using artificial defoliation during the first 2 years of growth indicate that growth and biomass losses may surpass one third of the potential when defoliation on LPI 0-8 reaches 75% (Reichenbacker et al. 1996). A recent field study shows that heavy natural defoliation, often approaching 100%, during the first 2 years leads to greater than 50%

Fig. 2. Cottonwood leaf beetle adult.

production loss. Additional studies need to be performed through harvest to determine the economic implications of such natural defoliation over a complete rotation and to determine whether or not it is economically justifiable to consider CLB management after the plantation establishment phase.

Naturally-occurring resistance mechanisms may hold some promise in breeding and selection programs. The role of phenolic glycosides, although important in discouraging feeding by insects that are not closely co-evolved with *Populus*, seems to have limited impact on the preference by the cottonwood leaf beetle (Bingaman & Hart 1993). A combination of chemicals occurring on the leaf surface, however, has been found to affect egg-laying behavior of the insect, and may be useful in breeding and selection for resistance (Lin et al. 1998*a*, 1998*b*). Currently, however, no selections with natural resistance to the CLB are commercially available.

Populus selections were among the first trees to be genetically modified for enhanced insect resistance (McCown et al. 1991). These selections included the *Bacillus thuringiensis* (*B.t.*) endotoxin genes and proved to be successful in increasing mortality in Lepidoptera. Transformations that include a Coleoptera-active *B.t.* gene are receiving attention in other programs. Another approach, the inclusion of a novel gene that interferes with the digestive functions of the CLB, has been only marginally successful in affecting CLB biology (Kang et al. 1998), and probably holds little promise as an effective management tool. Other *Populus* selections, transformed with yet a different digestion inhibitor, show promise (Leple et al. 1995) against a related European species, and may warrant additional research in North America.

To date, most CLB management programs in the United States and Canada have depended upon broad-spectrum insecticides applied as ground or aerial sprays. Although they are currently effective, the development of resistance to these chemicals with continued use should be of concern. Recent management efforts have successfully incorporated several commercial *B.t.* formulations (Coyle et al. 1999, 2000). These formulations are effective only against larvae, and particularly the first two larval stages. There are two important considerations for effective use of *B.t.* sprays: (1) applications are effective only during the first one or two generations each year when the development cycle is relatively synchronized and when nearly all of the life stages present are larvae; (2) monitoring activities must be conducted to insure that applications are applied at a time when most of the eggs have hatched or will hatch within a few days. A strict, narrow reliance on *B.t.* as a means of suppression should be avoided because resistance to the *B.t.* toxin can develop (Bauer 1995; James et al. 1999). With a lack of other effective options, as CLB populations become less synchronized through the growing season, broad-spectrum insecticides become the management tool of choice.

The role of natural enemies in CLB suppression has been evaluated in several areas (Head et al. 1977; Burkot and Benjamin 1979; Jarrard 1997). Although there is evidence of their impact on CLB populations, insufficient information is available at this time to make recommendations beyond using biorational management materials, such as *B.t.*, in a management program to conserve natural enemy populations.

Another leaf-feeding beetle, identified as *Phratora californica*, also native to North America, has emerged as an important problem in the coastal regions of the Pacific Northwest. Limited information is available on the biology and impact of this insect. Host resistance has been observed and is currently under study.

Forest tent caterpillar

The forest tent caterpillar (FTC), *Malacosoma disstria* Hübner (Lepidoptera: Lasiocampidae), is one of the most widely distributed defoliators in North America. This insect also has a wide host range, severely defoliating water tupelo (*Nyssa aquatica* L.), blackgum (*N. sylvatica* Marsh.), sweetgum (*Liquidambar styraciflua* L.), and oaks (*Quercus spp.*) in the southern U.S. (Fitzgerald 1995). The insect has been recorded as feeding on 29 native forest tree species in the north but seems only to sustain outbreaks in stands where trembling aspen (*P. tremuloides* Michx.) is a principal component (Prentice 1963). Although it also feeds on large tooth aspen (*P. grandidentata* Michx.) and balsam poplar (*P. balsamifera* L.), the FTC does not seem to thrive on these species of *Populus*. Within aspen populations, clones show a variation in susceptibility to feeding by FTC, and this may be indicative of constitutive defenses mediated by levels of proteins and phenolic glycosides in aspen foliage (Lindroth and Bloomer 1991). The basis for these differences in susceptibility might be worth investigating because of the insect's importance in northern areas of poplar culture and the opportunities this research might suggest in producing resistant stock.

Bands that consist of as many as 200 eggs are deposited, encircling twigs by mated FTC female moths in July. The neonate larvae overwinter within the egg band and emerge at about budbreak of the earliest flushing clones of trembling aspen. Delays in hatching relative to budbreak decreases the survival of larvae (Parry et al. 1998). Larvae are able to mine developing trembling aspen buds (Ives and Wong 1988) and are thus able to survive on ramets of later flushing clones. Larvae feed gregariously for the first instars, returning to a silken mat between feeding bouts. They eventually become solitary and often wander off the host tree in the final instar, especially if most of the foliage on the host trees has been consumed (Fig. 3). Cocoons are spun between aspen leaves if defoliation is not severe. In high populations, cocoons can be found in any available crevice.

Reasons for population changes in FTC have not been determined definitively (Fitzgerald 1995). There is a suggestion that cold winters and warm springs favor population increases and that unfavourable weather during the larval stage can cause populations to collapse. Although diseases and starvation of larvae may play a role in the collapse of some outbreaks, it would appear, based on a large

Fig. 3. Forest tent caterpillar late instar larvae.

number of reports, that pupal parasitoids are typically correlated with the decline of populations (Fitzgerald 1995). At endemic levels, experiments suggest that predation of pupae by the northern oriole (*Icterus galbula* L.) may be the factor responsible for maintaining low densities (Parry et al. 1997*a*). The mechanisms of release of endemic populations from control by predation remain unexplained.

Large areas of northern aspen-dominated forests are completely defoliated by FTC when its populations erupt. Records dating from the early years of the last century suggest that outbreaks have been a constant problem somewhere in the range. In the last series of Canadian outbreaks, the size of the areas involved peaked at 10.0, 12.8, and 13.0 million ha in Manitoba, Saskatchewan, and Alberta, respectively (Cerezke and Volney 1995). Outbreaks seem to recur at 10–12 year intervals and may last from 2 to 5 years in individual stands. These particular outbreaks were located on the southern margin of the boreal forest and the grassland/forest ecotone. Thus, their ubiquity in the region likely to support intensive poplar culture and the ability of the moths to disperse over several kilometers make the risk of damage from FTC a constant threat to investments in intensively managed stands.

The damage caused by severe defoliation results in an immediate reduction in growth and is reflected in a suppression of radial increment throughout the stem. There is a corresponding increase in mortality of stems in subsequent years that can be attributed directly to FTC defoliation. However, secondary pests further contribute to the reduction in stand yield in the years following outbreaks. In simulating the impacts of FTC defoliation on young aspen stands, Mattson and Addy (1975) suggest that in addition to a yield loss of 25% at age 40 years, one severe FTC outbreak lasting 3 years affected the stand's productive capacity for a decade following population decline. The decline of stands following defoliation may be further exacerbated by diseases such as hypoxylon canker, and climatic conditions that include late frosts and drought (Witter et al. 1975).

Management of FTC populations has included attempts at direct control through the use of insecticides and microorganisms (Fitzgerald 1995). This approach is a reactive strategy that may be viewed as a stopgap method to protect valuable stands with established populations that erupt. Maintainance of a healthy natural enemy complex that includes the preservation and enhancement of oriole populations may be a significant element in the protection of intensively managed *Populus* stands that risk being damaged by FTC. The development and management of resistance to FTC in genetically modified planting stock warrants serious examination. If the benefits from these technologies are not to be squandered, given the evolutionary potential of most pest species to swamp resistance mechanism bred into planting stock, then the spatial and temporal arrangement of resistant, susceptible and native stands must be designed to mitigate this contingency. The concerns regarding the use of resistant stock range from the evolution of resistant target insect populations, the extirpation of non-target populations that harbor important natural enemies, the depletion of genetic diversity within intensively managed stands, reducing future opportunities for genetic gain, and the susceptibility of the stock to other insect species that are presently not considered pests.

Large aspen tortrix

The large aspen tortrix (LAT), *Choristoneura conflictana* (Walker) (Lepidoptera: Tortricidae), feeds primarily on aspen, with only 3, 2, and 1% of the collections made by the Forest Insect Survey of Canada being recovered from willow (*Salix* spp.), balsam poplar, and bigtooth aspen, respectively (Prentice 1965). There is one generation per year. Females lay eggs in masses containing up to 450 eggs on the upper surfaces of leaves (Prentice 1955). Eggs hatch in mid to late July, and first instars feed on the leaf epidermis before dispersing and finding suitable overwintering sites to spin hibernacula. Larvae moult to the second instar and overwinter in this condition. In spring, second instars emerge to feed on developing aspen buds. There are five instars; the later instars (Fig. 4) web developing aspen leaves together and feed within the shelter thus created. Pupation also occurs within these shelters. Moths emerge in late June to early July, often leaving pupal cases protruding from the shelters. There is some variability in many of these traits, depending on stand conditions and geographical location. Eggs may be laid on a variety of available surfaces in severely defoliated stands and overwintering larvae may be found throughout the crown in mild climates. Prentice (1955) reports that eggs are found on leaves and overwintering larvae are restricted to lower tree boles. Presumably overwintering larvae are thus protected from extreme winter conditions or predation below the snow line.

The LAT is an occasional defoliator of aspen without any apparent defined periodicity to their outbreaks. The reliance on defoliation surveys and the confusion with FTC defoliation may have resulted in under-reporting of LAT outbreaks (Volney and Cerezke 1995). Outbreaks that have been observed last for 2–3 years. Declines in population densities of the LAT have been associated with starvation

Fig. 4. Large aspen tortrix late instar larva.

and subsequent reduced fecundity. Numerous parasitoid species have been reared from LAT, with the tachinid *Omotoma fumiferanae* Toth., accounting for as much as 64% mortality in one case. Disease caused by *Beauvaria bassianna* (Bals.) Vuill. seemed to be an important contributor to overwintering larval mortality (Prentice 1955). Ants (*Formica fusca* L. and *F. sanguinea subnuda* Emery) also are known to prey on larvae emerging from hibernacula. Ives (1981) associated an increased chickadee population in decreasing populations of LAT.

The locations most severely affected by LAT tend to be north of those affected by FTC. The areas affected can be as large as the 1.3 million ha outbreak near Fairbanks, Alaska (Beckwith 1968). Because of the sporadic nature of outbreaks and their short duration, damage to trees seems to be restricted to losses of radial increment and twig mortality rather than outright tree mortality (Cerezke and Volney 1995). No studies have reported secondary pest effects following severe defoliation by LAT. Temperatures as low as –40°C do not seem to hinder survival of overwintering larvae (Beckwith 1968), and Prentice's (1955) observation that the overwintering sites are on the lower bole may reflect the effect of chickadees on these populations. In analyzing long-term Forest Insect Survey data, Ives (1981) concluded that cold weather with heavy precipitation early in the winter followed by mild weather with light precipitation enhanced survival and was associated with periods of LAT population increases. Late spring frosts also adversely affect populations if the developing foliage is severely damaged. Extremely hot weather, by elevating temperatures within the larval feeding shelter, has been considered a factor in larval mortality (Criddle quoted by Ives 1981).

Management of LAT has not been necessary in the extensive northern aspen forests. The increasing importance of the resource and a commitment to intensive culture may make intervention increasingly necessary, however. The likelihood that world climate change will alter the risk of outbreaks and thus increase the need for management intervention is uncertain. Indications are that winter and spring temperatures have increased more than the summer and fall temperatures over the past century in Canada. Although this risk is mitigated somewhat by slightly warmer summer temperatures decreasing survival, this is not completely compensated for by the positive effects that the fewer late spring frosts and the increasing winter and spring temperatures will have on elevating survival of LAT larvae.

Direct control agents such as *B.t.* may be problematical for use against this insect in the spring because the shelter makes it difficult to deliver a sufficiently toxic deposit on the feeding surfaces efficiently. Alternatives such as applications of *Beauvaria bassiana* or *B.t.* could be developed for treating first instars when they feed on the upper surfaces of fully developed leaves. Such a strategy may also augment the mortality of overwintering larvae from the diseases caused by these organisms. Other natural enemy populations, such as ants, downy woodpeckers, red-eyed vireos (Prentice 1955), and chickadees (Ives 1981), should also be investigated for their potential in reducing or maintaining low LAT populations through cultural practices. Options here include retention of residual forest structure to provide shelter, nesting opportunities, and alternate food sources in stumps, snags, and standing live and dead trees on harvested sites. Late flushing clones of aspen pose some problems for newly emerged LAT larvae establishing feeding sites in spring (Parry et al. 1997*b*). Although larvae are able to compensate by spinning an additional hibernaculum on developing buds, their survival suffers because of the increased exposure to the elements, dispersal losses, and natural enemies.

Insects feeding on elongating shoots

Spotted poplar aphid

The spotted poplar aphid (SPA), *Aphis maculatae* Oestlund (Homoptera: Aphididae), a dark aphid having powdery patches along the sides of its body (Fig. 5), is primarily a pest of very young plantings, having greatest impact during the first three summers of development. During the summer months, they feed on the tips of the young, long shoots and the expanding leaves of *Populus*. In the fall, they transfer to dogwoods, where they overwinter. Populations of the SPA can enlarge substantially over the summer because the aphid is incredibly prolific, capable of producing many generations, weather permitting. Consequently, one may see dense clusters containing hundreds to thousands of aphids, infesting all of the expanding long shoots and immature leaves of susceptible trees. When aphid populations attain such high levels, they can significantly reduce *Populus* canopy enlargement because their feeding diminishes long shoot geometric growth, which in turn determines how much photosynthetic area the plant produces during a growing season. Overall growth of the trees is directly correlated with canopy architecture and size.

Management of the SPA can be done by, first of all, planting less susceptible *Populus* clones (Table 1). Next, growers should encourage populations of aphid natural enemies (ladybird beetles, parasitic wasps, lacewings, hover flies, etc.) by avoiding the application of any broad spectrum pesticides. Coupled with plant resistance, natural enemies are the main line of natural defense against aphid outbreaks. When all else fails, aphicides would be the first choice.

Cottonwood twig borer

The cottonwood twig borer (CTB), *Gypsonoma haimbachiana* (Kearfott) (Lepidoptera: Tortricidae), a small caterpillar in the bellmoth family, is native to the eastern United States and infests *Populus* species and hybrids throughout the range of its hosts. In southern U.S., where there is higher probability of high populations, the CTB may be a limiting factor on the success of commercial *Populus* plantings (Payne et al. 1972). Overwintering is passed in an early larval stage under bark scales or leaf scars; in spring these larvae move to the growing shoots,

Fig. 5. White spotted poplar aphids.

bore in, and tunnel and feed to complete development. The pupal stage is completed in bark crevices or in litter beneath trees. Eggs of the next generation are deposited on leaves, and the first larval stage mines into the leaf veins. The second larval stage moves to the tender shoot, and larval development is completed inside the shoot. Infested tips often die back, resulting in multiple leaders, which may in turn be attacked by the next generation of CTB, leading to stunted growth and malformed stems. Multiple infestations in each shoot are quite common in the southern United States, and may lead to heavy shoot mortality (Stewart and Payne 1975).

There are two generations each year in central Iowa (McMillin et al. 1998) and as many as five in Mississippi. Generations per year and population levels seem to be related to climate and to weather conditions.

Some indication of host plant resistance has been noted in Texas (Woessner and Payne 1971), but the role that it may play in management of the insect is uncertain. Management with broad-spectrum insecticides can be accomplished (Morris

1986), but whether or not it is economically or ecologically feasible is uncertain as well.

Insects feeding within woody stems

Poplar borer

The poplar borer, *Saperda calcarata* Say (Coleoptera: Cerambycidae), a large beetle, roughly 30 mm long, is a pest of both young and older stands, living primarily at and below the root collar zone in young trees and throughout the bole on larger trees (Solomon 1995). Damage is invariably more common in open than in dense stands, and often along the edges of stands, owing to the beetle's apparent preference for higher light conditions for oviposition. Egg-laying begins typically in late June to early July when the females cut crescent-shaped slits in the bark and insert their eggs into the phloem. About three weeks later, the eggs hatch and young larvae begin tunneling at the inner bark – sapwood interface. The next season, the larvae leave this interface area and tunnel into the sapwood and heartwood where they eventually weaken the stems and predispose them to storm breakage (Fig. 6). After feeding for 2–3 years, the larvae attain a length of 40–50 mm, and pupate inside the stem behind a plug of wood chips. Adults emerge in late June. Besides the damage done by the borer itself, woodpeckers exacerbate the problem as they create even larger wounds while trying to find the poplar borer larvae. These wounds may become important infection courts for fungi.

Management of the poplar borer is best done by maintaining dense, thrifty stands. Nothing is presently known about resistant cultivars, but such selections could eventually be very important in minimizing damage by this species. Highly infested trees should be rogued from the stand and cut into small pieces, chipped, or burned to cause rapid desiccation and death of the larvae.

Poplar gall saperda and the poplar branch borer

The poplar gall saperda (PGS), *Saperda inornata* Say (Coleoptera: Cerambycidae), in contrast to the poplar borer, is a smaller beetle, about 12 mm long (Fig. 7), and is mainly a problem in young stands, less than 5 years old (Nord et al. 1972*a*). As does its larger relative, it bores into the stems of young trees; in older trees, it bores mainly into the branches where its injuries are mostly insignificant, unless they facilitate fungal invasion of the tree. Adults seem to prefer higher light conditions and thus their damage is more common in low-density stands and on edges. Also, there is a correlation of high PGS incidence on poor sites (Nord et al. 1972*a*). In mid–late June, females deposit their eggs under horseshoe-shaped egg niches cut into the bark of small stems or branches. Often a female will cut 2–3 egg niches at about the same relative height but spaced around the stem. The larvae, when they hatch, begin feeding at the inner bark – wood

Fig. 6. Poplar borer adult, larvae, and damage.

interface, and stimulate the growth of a globose or spindle-shaped gall (Fig. 8). As they grow for the next 1–2 years, larvae bore into the wood, creating winding tunnels that weaken the stem and often predispose them to storm breakage. Not only the beetles galleries but also woodpeckers can seriously damage the stem when they hammer into the saperda galls in search of larvae. Both beetles and woodpeckers also create infection courts for the highly damaging and typically lethal fungus, hypoxylon canker (Nord and Knight 1972; Ostry and Anderson 1998), especially in the aspens, and hybrid poplars in section *Populus* (Ostry et al. 1989).

The poplar branch borer (PBB), *Oberia schaumii* LeConte (Coleoptera: Cerambycidae), is also a small beetle, similar in size to the PGS (Solomon 1995). Just as for the PGS, PBB preferentially attacks stems of young saplings and branches on older trees, and is most prevalent in low-density stands (Myers et al. 1968; Nord et al. 1972*b*). In mid–late June, the female gnaws an elongate, rectangular egg niche in the outer bark and inserts her eggs into the inner bark. Larvae bore downward (15–30 cm) from this point and eventually tunnel into the wood. They do not trigger an obvious swelling of the wood, and thus no gall develops as it does for the PGS. Their hidden tunneling and feeding is often revealed by either bleeding sap or golden sawdust-like frass emanating from the egg niche or 1–3 small shot

Fig. 7. Poplar gall saperda adult.

Fig. 8. Poplar gall saperda larva in a stem.

holes in the stem 10–30 cm below the oviposition scar. It typically takes 3 years to complete their life cycle. In young trees, their wounds can likewise predispose trees to storm breakage. Similarly, the wounds may also enhance hypoxylon infections. This could be much more important than their direct damage by tunneling in the wood.

Management of both the PGS and PBB is best done by maintaining dense plantings on good sites. Stocking levels of less than 20 000 stems/ha are highly suitable for beetle infestation (Myers et al. 1968). Sanitation is recommended along with employing resistant clones when they are known. Slow-growing trees on poor soils may be more susceptible to these beetles because the trees' induced defenses, such as rapid callus formation and strong hypersensitive reactions to eggs and young larvae, may be debilitated. Fast-growing individuals have more potent rapid inducible defenses, which are effective against poorly mobile, invasive herbivores such as the small, young larvae of these beetles (Herms and Mattson 1992).

Poplar–willow borer

The poplar–willow borer (PWB), *Cryptorhychus lapathi* (L.) (Coleoptera: Curculionidae), introduced from Europe, is a robust weevil about 10 mm long that attacks cottonwoods, poplars, willows, and alders (Solomon 1995; Schoene 1907). They seem to prefer stems that are more than 2 years old, and greater than 25 mm basal diameter. They may also attack branches as do the PGS and PBB. Adults emerge from infested stems during the late summer and early fall. After a week or so of feeding and mating, gravid females chew slits in the corky bark, often in lenticels, scar tissue, branch bases, and injured areas, typically within 40 mm from the root collar, and insert eggs therein. The developing larvae feed at the inner bark – sapwood interface and only later bore into the wood itself as they mature. The tree is thus weakened by their excavations and may break during wind, snow, or ice storms, or die from stem girdling (Fig. 9). Development from egg to sexually mature adult takes 1–2 years, but adults may overwinter (in the duff) and live up to 2 years. These carryover adults will emerge as warm weather arrives the next spring and quickly begin egg-laying (Furniss 1972).

Management of the PWB is best done by maintaining well-stocked, thrifty stands. Planting young trees near older trees that may be infested is not recommended. The use of resistant clones would be desirable if they were known. Finally, sanitation, i.e., complete removal and chipping or burning of infested trees parts, is recommended.

What to plant? Choosing low-susceptibility clones

With such a large number of insect and mite species that are capable of feeding upon *Populus*, and the dangers inherent in ramping up to large acreage, it

Fig. 9. Poplar–willow borer damage.

behooves growers to make careful selections of the cultivars. It is especially important to match the clones to the climate and soils (Dickmann and Isebrands 1998). Unfortunately, there has not yet been a comprehensive, in-depth study of the insect-resistance traits of most *Populus* clones. As a result, the available data on insect resistance remains quite spotty, and none has been fully and unequivocally confirmed by repeated trials over many different environments. And to make matters worse, there are only a few cases where insect resistance has been linked to pathogen resistance or to other desirable traits such as high plant growth rates and high wood quality. Therefore, we are walking on thin ice with respect to cultivar selection and the challenges that insects and diseases are likely to throw at us. But, we cannot wait to obtain all this desired information. Instead, we need to proceed with what knowledge we have and make creative adjustments, i.e., adaptive management, as we encounter problems. Anything that is learned about resistance/susceptibility to insects from first-hand experience in the field should be duly and carefully recorded and brought to the attention of specialists in the genetics and pests of *Populus*. In other words, learning by doing is one of the best options.

Table 2 compiles what has been learned so far about hybrid poplar insect resistance to many different insects, each species usually studied at only one locality. Because growers need to consider the simultaneous impacts of many different insect pests, it is not yet obvious if any clone will have the needed traits to render them of moderate to high resistance to all of the major insect problems. However, at least one hybrid poplar clone looks broadly interesting, NC5339. Several others, NC4872, 5270, 5271, and 5272, may exhibit low susceptibility to defoliators. Growers should be advised, however, that few of the clones listed in Table 2 are commercially available, and none has received extensive evaluation for field performance under a wide range of conditions. Nor are any of the mechanisms of resistance or tolerance sufficiently well understood to use in breeding and selection programs. Other traits such as plant resistance to pathogens, adaptations to prevailing climatic and site conditions as well as growth characteristics must also be considered.

Landscape considerations: how to plant, knowing that more plants means more insects

It may be a law of nature that as crop acreage increases so does the likelihood of more serious insect and pathogen damage. Ecological literature has long reported on the positive correlations that exist between the numbers of species of insects occurring on a plant and the geographic area covered by the plant. It might be called the target hypothesis; the bigger the target area, the more insect species from the surrounding environment are eventually capable of finding and thus "hitting" the target plant. An important corollary of the target hypothesis is that as the number of close relatives of said target plant increases in its surrounding environment, the more insects (coming from nearby relatives) will find and colonize the target plant.

Foresters, farmers, and others have long known that large monocultures of any kind of plant are somehow inevitably linked to outbreaks of pests. In fact, Mattson et al. (1991) analyzed the character of natural North American forests

Table 2. Ranking of tested hybrid poplar clones according to their relative susceptibility (high, medium, low) to different insect pests, and tolerance to defoliation in the Great Lakes region, N.A.

Clone numbers	Clone parentage[a]	Spotted aphid rank[b]	Tarnished plant bug rank[c]	Forest tent caterpillar rank[d]	Cottonwood leaf beetle rank[e]	Defoliation tolerance rank[d]
D38	*deltoides*	Low				
DBJ21	× *jackii*	High				
DBJ22	× *jackii*	High				
DN1	× *canadensis*		Med.			
DN17	× *canadensis*	Low				
DN18	× *canadensis*	Low	Low			
DN19	× *canadensis*	Low				
DN21	× *canadensis*		High			
DN22	× *canadensis*	Med.	Low			
DN31	× *canadensis*		High			
DN55	× *canadensis*	Low	High			
DN9	× *canadensis*		Low			
DN96	× *canadensis*	Low				
DTAC2	*deltoides* var. *angulata* × *berolinensis*			Med.		Med.
ELJ14	× *jackii*	High				
FRS1	Unidentified (Fry nursery)		Low			
FRS2	Unidentified (Fry nursery)		High			
GRJ6	× *jackii*	Med.				
I4551	× *canadensis*	Low	Med.			
LJ14	× *jackii*	High				
LUJ7	× *jackii*	High				
NC11004	*deltoides*			High		Med.
NC11382	*nigra* var. *charkowiensis* × *berolinensis*			Med.		High
NC11396	*maximowiczii* × *berolinensis*			High		Med.
NC11432	*deltoides* var. *angulata* × *trichocarpa*			Med.		High

Table 2 (*continued*).

Clone numbers	Clone parentage[a]	Spotted aphid rank[b]	Tarnished plant bug rank[c]	Forest tent caterpillar rank[d]	Cottonwood leaf beetle rank[e]	Defoliation tolerance rank[d]
NC11445	*nigra* × *laurifolia*			High		High
NC11505	*maximowiczii* × *trichocarpa*	Med.	Low	Med.		High
NC238	*deltoides* × *nigra* Volga		Low			
NC4877	*alba*				Low	
NC4878	× *canadensis*				High	
NC4879	× *canadensis*		Low		Med.	
NC5258	*Populus* sp.	Low	Low		Med.	
NC5260	*tristis* × *balsamifera*		Low	Med.	Med.	
NC5261	*deltoides* × *balsamifera*				Med.	
NC5262	*balsamifera* var. *candicans* × *berolinensis*	Med.	High	Med.	Med.	High
NC5263	*balsamifera* var. *candicans* × *berolinensis*	High	High		High	
NC5264	*deltoides* var. *angulata* × *nigra* var. *plantierensis*	Low	Low		High	
NC5265	*deltoides* var. *angulata* × *trichocarpa*				Med.	
NC5266	*deltoides* var. *angulata* × *trichocarpa*		Low		Med.	
NC5267	*deltoides* × *nigra* var. *caudina*				Med.	
NC5268	*deltoides* × *trichocarpa*				Med.	
NC5270	*deltoides* × *trichocarpa*				Low	

Table 2 (*continued*).

Clone numbers	Clone parentage[a]	Spotted aphid rank[b]	Tarnished plant bug rank[c]	Forest tent caterpillar rank[d]	Cottonwood leaf beetle rank[e]	Defoliation tolerance rank[d]
NC5271	*nigra* var. *charkowiensis* × *nigra* var. *caudina*			Med.	Low	High
NC5272	*nigra* × *laurifolia* Strathglass	Med.	Med.		Low	
NC5273	*deltoides*				High	
NC5277	× *canadensis*		Med.			
NC5318	*deltoides*				Med.	
NC5319	*deltoides*				High	
NC5321	× *canadensis*				Med.	
NC5322	× *canadensis*				Med.	
NC5323	× *canadensis*	Low	Low		High	
NC5324	× *canadensis*				Med.	
NC5325	× *canadensis*	Med.	Low		High	
NC5326	× *canadensis*	Low	High		Med.	
NC5327	× *canadensis*				High	
NC5328	× *canadensis*				Med.	
NC5331	*nigra* var. *betulifolia* × *trichocarpa*	Low	Low	Med.	Med.	Med.
NC5332	*nigra* var. *betulifolia* × *trichocarpa*	High	High		Med.	
NC5334	*deltoides* var. *angulata* × *trichocarpa*	Med.	Low		High	
NC5335	*deltoides* × *trichocarpa*				High	
NC5339	*alba* × *grandidentata*		Low	Low	Low	
NC5351	*Populus* sp.		Med.			
NC5377	× *canadensis*			Med.	High	Med.
NC9921	*Populus* sp.		High			
NC9922	*Populus* sp.		Low			
NE10	*nigra* × *trichocarpa*	Med.				

Table 2. (*continued*).

Clone numbers	Clone parentage[a]	Spotted aphid rank[b]	Tarnished plant bug rank[c]	Forest tent caterpillar rank[d]	Cottonwood leaf beetle rank[e]	Defoliation tolerance rank[d]
NE19	*nigra* var. *charkowiensis* × *nigra* var. *caudina*	Low	Low			
NE20	*nigra* var. *charkowiensis* × *nigra* var. *caudina*	Low	Low			
NE206	*deltoides* × *trichocarpa*	Low				
NE207	*deltoides* × *trichocarpa*	Low				
NE209	*deltoides* × *trichocarpa*	High				
NE214	*deltoides* × *trichocarpa*	Low				
NE224	*deltoides* × *nigra* var. *caudina*	Low				
NE225	*deltoides* × *nigra* var. *caudina*	Med.				
NE238	*deltoides* × *nigra* Volga	Low				
NE255	*deltoides* var. *angulata* × *trichocarpa*	Low				
NE264	*deltoides* var. *angulata* × *nigra* Volga	Med.				
NE265	*deltoides* var. *angulata* × *nigra* Volga	Med.				
NE300	*nigra* var. *betulifolia* × *trichocarpa*	Low				
NE308	*nigra* var. *charkowiensis* × *nigra* var. *incrassata*	Low	Low			

Table 2 (*concluded*).

Clone numbers	Clone parentage[a]	Spotted aphid rank[b]	Tarnished plant bug rank[c]	Forest tent caterpillar rank[d]	Cottonwood leaf beetle rank[e]	Defoliation tolerance rank[d]
NE318	*deltoides* var. *charkowiensis* × *deltoides*	Low				
NE332	*simonii* × *berolinensis*			Med.		
NE346	*deltoides* × *trichocarpa*	Low				
NE351	*deltoides* × *nigra* var. *caudina*	Low				
NE359	*deltoides* × *nigra* var. *caudina*	Med.	Low			
NE360	*deltoides* × *nigra* var. *caudina*	High				
NE373	*deltoides* var. *angulata* × *trichocarpa*	Low				
NE374	*deltoides* var. *angulata* × *trichocarpa*	Low				
NE41	*maximowiczii* × *trichocarpa* Androscoggin	Med.				
NM6	*nigra* × *maximowiczii*			Med.		High
RAV	× *canadensis*	Low	Med.			

[a]Several Latin names in this column do not reflect current taxonomic priority. See Chap. 1, especially Tables 1–3, for correct synonyms.
[b]Source: Wilson and Moore 1986.
[c]Source: Wilson and Moore 1985; Sapio et al. 1982.
[d]Source: Robison and Raffa 1994.
[e]Source: Harrell et al. 1981; Caldbeck et al. 1978.

that are notorious for expansive, severe outbreaks by insects, and concluded that virtually all such forests are typically dominated (≥50% composition) by one or a few tree species, the commonest ones being the primary hosts of the outbreak insects. Outbreaks refer to insect or pathogen populations that are so abundant that they cause plant injuries to vastly overshoot the plant's natural compensation threshold. In the case of trembling aspen, for example, during outbreak peaks of the forest tent caterpillar and the large aspen tortrix, the caterpillar populations can reach millions per hectare and typically remove all of the foliage from the tree canopies with the result that wood growth is nearly negligible. This commonly continues for 2–3 consecutive years, causing substantial losses in wood yield.

Polycultures are in; monocultures are out

The particular cultivars used and their spatial deployment are obviously important, if not crucial, considerations in trying to minimize the development of future pest problems. Yet, there are few hard rules to live by. Because forest crops are likely to be in the ground for 10+ years, it makes sense to select several of the most resistant lines known. However, they should not be planted in monocultures, but in polycultures. Polyculture stands ought to be constructed of several clones, varying in their susceptibility to the major pests anticipated (Gould 1991). Three clones is probably the minimum. Using polycultures is desirable because there will be heightened within-stand heterogeneity, causing a multiplicity of plant selection factors to influence the growth and survival of the insect populations. The goal is to prevent insects from responding uniformly to the resistance traits of the most resistant plants and in so doing developing counter-adaptations (Gould 1991). By employing some clones with only low–moderate resistance, there also will likely be enough insects around to sustain the highly valuable populations of natural enemies (e.g., predators, and parasitic wasps and flies) in the stands to enable them to take part in keeping pests in check, i.e., below the damage thresholds.

Checkerboarding: keeping "islands" small and difficult to find increases pest extinction

Based on the theories of island biogeography and metapopulation dynamics, stand size and patterns across the landscape are another important level of consideration in pest management. When possible, small stands, relatively isolated from one another, will work to minimize pest issues. Small remains to be defined, but perhaps keeping stands in the 10–20 ha range is a reasonable consideration. Likewise, keeping the small stands distant or separated (by non-host crops, forests) from one another will help to minimize outbreaks and to increase the pests' extinction rates within each individual stand. All insect populations are prone to fluctuate, but small populations are more prone to fluctuate to extinction or near extinction, caused by random mortality factors such as unfavorable weather. When and if this happens, the stand is at least temporarily free from the impact of the pest species until an immigrant female from a surrounding similar stand happens to discover the empty island. This is why it is important that each island be difficult to find, owing either to its long distance from the pest inoculum, or its concealing surroundings of non-host trees. As yet, this is not an exact science, and hard knowledge will only come from trial and error.

Managing natural enemies to encourage presence, persistence, and efficacy

Although it is desirable to make the pest populations prone to extinction and unlikely to discover the crop stands, the opposite is true for their natural enemies. To promote natural enemy abundance and their efficacy in finding the pests, one needs special, detailed information about which natural enemies are important for each and every significant pest (for example, the nine listed above) and what factors limit natural enemy abundance and searching capacity. This detail is beyond the scope of this article, but nevertheless the principles will be addressed here in at least a cursory fashion. Planting poplars next to another crop that will not share its pests but will share its natural enemies is one approach. For example, some generalist parasitoids coming from defoliators in a spruce–fir forest might very likely search for pest defoliators in a neighboring poplar stand. The same might be true for natural enemies derived from an adjacent pine stand, maple stand, or even a marsh. Mattson et al. (1968) reported, for example, that blackbirds that nested in marshes adjacent to a several hundred hectare jack pine forest flocked in hundreds to prey for several weeks on jack pine budworms, *Choristoneura pinus* Freeman, which were abundant in the jack pines. In the same vein, fostering birds by conserving patches of their habitat or certain limiting resources that they need for nesting may provide a measure of resistance to stands at risk to attack by defoliators such as FTC and LAT.

The nutrition of adult parasitoids is often limiting, and hence their capacity for searching for and parasitizing pests is likewise limited (Cappuccino et al. 1999). Because many parasitic flies and wasps require plant nectar and (or) honeydew from aphids and scales to bolster their energy demands while egg-laying, enhancing the abundance of nectar and honeydew sources could pay dividends in pest management. For example, fresh honeydew and dried honeydew on the leaves of a nominal number of aphid-susceptible trees, purposely planted within poplar stands, might significantly aid parasites. Likewise, border trees, and plants that are honeydew and (or) nectar producers might substantially enhance parasitism rates.

Concluding remarks

Insect and disease problems are inevitable and can be severe when growing *Populus*. Therefore, growers must be prepared for their appearance. First and foremost, selecting several clones that have some evidence of resistance to the main insect and disease problems must be the first line of preparation. Next, at the stand level, growing polycultures (mixtures) of many clones rather than monocultures or near monocultures of few clones is strongly advised. Hopefully, both disease and insect resistance will be incorporated into the best clones. Mixtures of several carefully selected clones may actually have higher yields per hectare than equivalent stands of monocultures. At the landscape level, whenever feasible, arranging stands so that there will be minimal movement of pests among them and minimal immigration of pests into them from natural stands is recommended. Encouraging populations of natural enemies of insect pests is also highly advised. This may be accomplished through many means: providing nesting sites for important birds, encouraging wild flowers, and weeds that offer nutrition for parasitic flies and wasps, and establishing poplar plantations close to plant communities that are natural sources of predators and parasites, but not pests. There are no simple, guaranteed recipes for success. Instead, employing common sense and adaptive management are the key principles for achieving success.

References

Bauer, L.S. 1995. Resistance: a threat to the insecticidal crystal proteins of *Bacillus thuringiensis*. Fla. Entomol. **87**: 414–443.

Beckwith, R.C. 1968. Large aspen tortrix, *Choristoneura conflictana* (Wlkr.), in interior Alaska. USDA For. Serv. Res. Note PNW-81.

Bingaman, B.R., and Hart, E.R. 1992. Feeding and oviposition preferences of adult cottonwood leaf beetles (Coleoptera: Chrysomelidae) among *Populus* clones and leaf age classes. Environ. Entomol. **21**: 508–517.

Bingaman, B.R., and Hart, E.R. 1993. Phenolic glycosides and host selection behavior of *Chrysomela scripta* (Coleoptera: Chrysomelidae). Environ. Entomol. **22**: 397–403.

Burkot, T.R., and Benjamin, D.M.. 1979. The biology and ecology of the cottonwood leaf beetle *Chrysomela scripta* (Coleoptera: Chrysomelidae) on tissue cultured hybrid Aigeiros (*Populus euramericana*) subclones in Wisconsin. Can. Entomol. **111**: 551–556.

Caldbeck, E.S., McNabb, H.S., and Hart, E.R. 1978. Poplar clonal preferences of the cottonwood leaf beetle. J. Econ. Entomol. **71**: 518–520.

Cappuccino, N., Houle, M.-J., and Stein, J. 1999. The influence of understory nectar resources on the parasitism of the spruce budworm, *Choristoneura fumiferana*, in the field. Agric. For. Entomol. **1**: 33–36.

Cerezke, H.F., and Volney, W.J.A. 1995. Status of forest pest insects in the western-northern region. *In* Forest insect pests in Canada. *Edited by* J.A. Armstrong and W.G.H. Ives. Forestry Canada, Ottawa. pp. 59–72.

Chapin, F.S., Autumn, K, and Pugnaire, F. 1993. Evolution of suites of traits in response to environmental stress. Am. Nat. **142**: S78–S92.

Coyle, D.R., McMillin, J.D., and Hart, E.R. 1999. Pupal or adult parameters as potential indicators of cottonwood leaf beetle, (Coleoptera: Chrysomelidae) fecundity and longevity? Great Lakes Entomol. In press.

Coyle, D.R., McMillin, J.D., Krause, S.C., and Hart, E.R. 2000. Laboratory and field evaluations of two *Bacillus thuringiensis* formulations, Novodor® and Raven®, for control of the cottonwood leaf beetle. J. Econ. Entomol. In press.

Delplanque, A. (*Editor*). 1998. Les insects associes aux peupliers. Editions Memor, Rue G. Biot, 23-25 B-1050 Bruxelles. 350 pp.

Dickmann, D.I., and Isebrands, J.G. 1999. Caveat emptor. Am. Nurseryman, **189**(5): 60–65.

Drooz, A.T. 1985. Insects of eastern forests. USDA For. Serv. Misc. Publ. 1426. 608 pp.

Fang, Y. 1997. Effects of cottonwood leaf beetle (Coleoptera: Chrysomelidae) larval population levels on *Populus* terminal damage. M.S. thesis. Iowa State University, Ames, IA.

Fang, Y., and Hart, E.R. 2000. Effect of cottonwood leaf beetle (Coleoptera: Chrysomelidae) larval population levels on *Populus* terminal damage. Environ. Entomol. **29**: 43–48.

Fitzgerald, T.D. 1995. The tent caterpillars. Cornell University Press, Ithaca, NY.

Harrell, M.O., Benjamin, D.M., Berbee, J.G., and Burkot, T.R. 1981. Evaluation of adult cottonwood leaf beetle, *Chrysomela scripta*, feeding preference for hybrid poplars. Great Lakes Entomol. **14**: 181–184.

Head, R.B., Neel, W.W., and Morris, R.C. 1977. Seasonal occurrence of the cottonwood leaf beetle, *Chrysomela scripta* (Fab.), and its principal insect predators in Mississippi and notes on parasites. J. Georgia Entomol. Soc. **12**: 157–163.

Herms, D.A., and Mattson, W.J. 1992. The dilemma of plants: to grow or defend. Quart. Rev. Biol: **67**: 283–335.

Herms, D.A., and Mattson, W.J. 1997. Trees, stress, and pests. *In* Plant health care for woody ornamentals. University of Illinois Cooperative Extension Service, Urbana-Champaign, IL. pp. 13–25.

Ives, W.G.H. 1981. Environmental factors affecting 21 forest insect defoliators in Manitoba and Saskatchewan, 1945–69. Can. For. Serv. Inf. Rep. NOR-X-233.

Ives, W.G.H., and Wong, H.R. 1988. Tree and shrub insects of the prairie provinces. Can. For. Serv. NOR-X-292. 327 pp.

Jarrard, J.A. 1997. Natural enemies of the cottonwood leaf beetle in central Iowa. M.S. thesis. Iowa State University, Ames, IA.

Kang, H., Hall, R.B., Heuchelin, S.A., McNabb, H.S., Jr., Mize, C.W., and Hart, E.R. 1997. Transgenic *Populus*: in vitro screening for resistance to cottonwood leaf beetle (Coleoptera: Chrysomelidae). Can. J. For. Res. **27**: 943–944.

Leplé, J.C., Bonadé-Bottino, M., Augustin, S., Pilate, G., Dumanois Lê Tân, V., Delplanque, A., Cornu, D., and Jouanin, L. 1995. Toxicity to *Chrysomela tremulae* (Coleoptera: Chrysomelidae) of transgenic poplars expressing a cystein proteinase inhibitor. Mol. Breed. **1**: 319–328.

Lin, S., Binder, B.F., and Hart, E.R. 1998*a*. Insect feeding stimulants from the leaf surface of *Populus*. J. Chem. Ecol. **24**: 1781–1790.

Lin, S., Binder, B.F., and Hart, E.R. 1998*b*. Chemical ecology of cottonwood leaf beetle adult feeding preferences on *Populus*. J. Chem. Ecol. **24**: 1791–1802.

Lindroth, R.L., and Bloomer, M.S. 1991. Biochemical ecology of forest tent caterpillar; responses to dietary protein and phenolic glycosides. Oecologia, **86**: 408–413.

Mattson, W.J., and Addy, N.D. 1975. Phytophagous insects as regulators of forest primary production. Science, **190**: 515–522.

Mattson, W.J., Knight, F.B., Allen, D.C., and Foltz, J.L. 1968. Vertebrate predation on the jack pine budworm in Michigan. J. Econ. Entomol. **61**: 229–234.

Mattson, W.J., Lawrence, R.K., Haack, R.A., Herms, D.A., and Charles, P.-J. 1988. Defensive strategies of woody plants against different insect feeding guilds in relation to plant ecological strategies and intimacy of association with insects. *In* Mechanisms of woody plant defenses against insects: search for pattern. *Edited by* W.J. Mattson, J. Levieux, and C. Bernard-Dagan. Springer-Verlag, NY. pp. 3–38.

McCown, B.H., McCabe, D.E., Russell, D.R., Robison, D.J., Barton, K.A., and Raffa, K.F. 1991. Stable transformation of *Populus* and incorporation of pest resistance by electric particle discharge. Plant Cell Rep. **9**: 590–594.

McMillin, J.D., Anderson, M.A., Butin, E.E., and Hart, E.R. 1998. Phenology and infestation patterns of the cottonwood twig borer (Lepidoptera: Tortricidae) in Iowa. Great Lakes Entomol. **31**: 181–190.

Myers, W.L., Knight, F.B., and Grimble, D.G. 1968. Frequency of borer attacks as related to character of aspen sucker stands: A comparative study of *Oberea schaumii* and *Saperda inornata*. Ann. Entomol. Soc. Am. **61**: 1418–1423.

Nord, J.C., and Knight, F.B. 1972. The relationship of the abundance of *Saperda inornata* and *Obera schaumii* (Coleoptera: Cerambydicae) in large trembling aspen, *Populus tremuloides*, to site quality. Great Lakes Entomol. **5**: 93–97.

Nord, J.C., and Knight, F.B. 1972. The importance of *Saperda inornata* and *Oberea schaumii* (Coleoptera: Cermabycidae) galleries as infection courts of *Hypoxylon pruinatum* in trembling aspen, *Populus tremuloides*. Great Lakes Entomol. **5**: 87–92.

Nord, J.C., Grimble, D.G., and Knight, F.B. 1972*a*. Biology of *Saperda inornata* (Coleoptera: Cerambycidae) in trembling aspen, *Populus tremuloides*. Ann. Entomol. Soc. Am. **65**: 127–135.

Nord, J.C., Grimble, D.G., and Knight, F.B. 1972*b*. Biology of *Oberea schaumii* (Coleoptera: Cerambycidae) in trembling aspen, *Populus tremuloides*. Ann. Entomol. Soc. Am. **65**: 114–119.

Ostry, M.E., and Anderson, N.A. 1998. Interactions of insects, woodpeckers, and hypoxylon canker on aspen. USDA For. Serv. Res. Pap. NC-331. 15 pp.

Ostry, M.E., Wilson, L.F., McNabb, H.S., and Moore, L.M. 1989. A guide to insect, disease, and animal pests of poplar. USDA Agric. Handbk. 677. 118 pp.

Parry, D., Spence, J.R., and Volney, W.J.A. 1997*a*. Responses of natural enemies to experimentally increased populations of the forest tent caterpillar, *Malacosoma disstria*. Ecol. Entomol. **22**: 97–108.

Parry, D., Volney, W.J.A., and Currie, C.R. 1997*b*. The relationship between trembling aspen phenology and larval development of the large aspen tortrix. Can. For. Serv. Inf. Rep. NOR-X-350.

Parry, D., Spence, J.R., and Volney, W.J.A. 1998. Budbreak phenology and natural enemies mediate survival of first-instar forest tent caterpillar (Lepidoptera: Lasiocampidae). Environ. Entomol. **27**, 1368–1374.

Payne, T.L., Woessner, R.A., and Mastro, V.C. 1972. Evaluation of cottonwood clonal selections for resistance to cottonwood twig borer attack. J. Econ. Entomol. **65**: 1178–1179.

Peterson, E.B., and Peterson, N.M. 1992. Ecology, management, and use of aspen and balsam poplar in the prairie provinces, Canada. For. Can. Northwest Reg. North. For. Ctr. Spec. Rep. 1.

Prentice, R.M. 1955. The lifehistory and some aspects of the ecology of the large aspen tortrix, *Choristoneura conflictana* (Wlkr.) (N. Comb.) (Lepidoptera: Tortricidae). Can. Entomol. **87**: 461–473.

Prentice, R.M. 1963. Forest Lepidoptera of Canada recorded by the Forest Insect Survey. Vol. 3. Lasiocampidae, Thyatridae, Drepanidae, Geometridae. For. Entomol. Path. Br. Can. Dept. For. Publ. 1013.

Prentice, R.M. 1965. Forest Lepidoptera of Canada recorded by the Forest Insect Survey. Vol. 4. Microlepidoptera. For. Entomol. Path. Br. Can. Dept. For. Publ. 1142.

Reichenbacker, R.R., Schultz, R.C., and Hart, E.R. 1996. Impact of artificial defoliation on *Populus* growth, biomass production, and total non-structural carbohydrate concentration. Environ. Entomol. **25**: 632–642.

Robison, D.J., and Raffa, K.F. 1994. Characterization of hybrid poplar clones for resistance to the forest tent caterpillar. For. Sci. **40**: 686–714.

Sapio, F.J., Wilson, L.F., and Ostry, M.E. 1982. A split-stem lesion on young hybrid P*opulus* trees caused by the tarnished plant bug, *Lygus lineolaris* (Hemiptera (Heteroptera): Miridae). Great Lakes Entomol. **15**: 237–246.

Schoene, W.J. 1907. The poplar and willow borer (*Cryptorhynchus lapathi* L.). NY Agric. Exp. Stn. Bull. No. 286. pp 83–104.

Solomon, J.D. 1995. Guide to insect borers of North American broadleaf trees and shrubs. USDA For. Serv. Agric. Handbk. 706. 735 pp.

Stewart, J.W., and Payne, T.L. 1975. Seasonal abundance and impact of the cottonwood twig borer on cottonwood trees. J. Econ. Entomol. **68**: 599–602.

Wilson, L.F., and Moore, L.M. 1985. Vulnerability of hybrid *Populus* nursery stock to injury by the tarnished plant bug, *Lygus lineolaris* (Hemiptera: Miridae). Great Lakes Entomol. **18**: 19–23.

Wilson, L.F., and Moore, L.M. 1986. Preference for some nursery-grown hybrid *Populus* trees by the spotted poplar aphid and its suppression by insecticidal soaps (Homoptera: Aphididae). Great Lakes Entomol. **19**: 21–26.

Witter, J.A., Mattson, W.J., and Kulman, H.M. 1975. Numerical analysis of a forest tent caterpillar (Lepidoptera: Lasiocampidae) outbreak in northern Minnesota. Can. Entomol. **107**: 837–854.

Woessner, R.A., and Payne, T.L. 1971. An assessment of the cottonwood twig-borer attacks. South. Conf. For. Tree Improv. Proc. **11**: 98–107.

CHAPTER 8
Poplar diseases

George Newcombe, Mike Ostry, Martin Hubbes, Pierre Périnet, and Marie-Josée Mottet

Introduction

Wherever poplars grow, whether in natural or planted stands, diseases will be present, some causing catastrophic damage, others hardly noticeable. Because diseases are so pervasive and have caused so much damage in the past, the avoidance of disease problems preoccupies most poplar growers. There still is no effective and economical way to treat existing disease problems; a dependence on the strategy of "better living through chemistry" using pesticide application simply will not be effective if trees already are diseased. The only way to fight diseases is to plant poplar clones inherently resistant to them. The focus of this chapter, therefore, is on plant breeding and biotechnological methods as a means to this end. This chapter follows close on the heels of other reviews of diseases of *Populus* in North America (Callan 1998; Newcombe 1996; Newcombe 1998). To avoid redundancy, we have tried to be forward-looking and prognostic, and to present new information. Fortunately, there are untouched topics and others that need to be updated.

Conventional breeding efforts have produced thousands of new clones in recent years in North America, but serendipitous selection for disease resistance remains the *modus operandi*. Productive, disease-resistant hybrid poplar clones may be selected in one region or another, but the disease problems of these clones when grown elsewhere are not widely appreciated. Nor is the dynamism of pathogen populations generally understood. The generation of new pathogenic

G. Newcombe. Department of Forest Resources, University of Idaho, Moscow, ID 83844-1133, U.S.A.
M. Ostry. USDA Forest Service, North Central Forest Experiment Station, 1992 Folwell Avenue, St. Paul, MN 55108, U.S.A.
M. Hubbes. Faculty of Forestry, University of Toronto, 33 Willcocks Street, Toronto, ON M5S 3B3, Canada.
P. Périnet. Direction de la Recherche Forestière, Ministère des Ressources Naturelles du Québec, 2700, rue Einstein, Sainte Foy, QC G1P 3W8, Canada.
M.-J. Mottet. Direction de la Recherche Forestière, Ministère des Ressources Naturelles du Québec, 2700, rue Einstein, Sainte Foy, QC G1P 3W8, Canada.
Correct citation: Newcombe, G., Ostry, M., Hubbes, M., Périnet, P., and Mottet, M.-J. 2001. Poplar diseases. *In* Poplar Culture in North America. Part A, Chapter 8. *Edited by* D.I. Dickmann, J.G. Isebrands, J.E. Eckenwalder, and J. Richardson. NRC Research Press, National Research Council of Canada, Ottawa, ON K1A 0R6, Canada. pp. 249–276.

variation can force poplar breeders to start again with new sources of resistance. This is especially true for leaf rust. Although this chapter emphasizes North America, it is in western Europe that this need may be especially poignant due to new pathogenic variation in the population of *Melampsora larici-populina*.

Conventional breeding may also run up against agents of disease that are rarely problematic for species or first-generation hybrids, but may be especially damaging to advanced-generation hybrids. An example of this was seen in the effects of anthracnose (*Glomerella cingulata*) on a hybrid poplar F_2 progeny (Newcombe 2000*a*). The mortality of the triploid offspring of a particular *P. tremula* male (discussed later in this chapter under "Influence of disease on current and future aspen management in the Lake States") caused by an equally obscure fungus, *Lahmia kunzei*, appears to be a related phenomenon. Cryptic pathogens may complicate interspecific, back-cross breeding in which one attempts to incorporate a gene for rust or canker resistance while selecting against all other traits from the donor in successive generations. It is possible that similar phenomena would compromise transgenic disease control developed through genetic engineering.

There are many problems in breeding for disease resistance in *Populus* that are currently unsolved, but new tools are emerging. Certainly, back-cross breeding to capture desirable traits from *P. trichocarpa* (T) while retaining the Septoria canker resistance of *P. deltoides* (D) is not straightforward. It is complicated both by recessiveness of canker resistance (Newcombe and Ostry, unpublished), and by the susceptibility to leaf rust and to *Marssonina brunnea* of the first back-cross generation (i.e., TD × D). Fortunately, a good canker assay (Mottet et al. 1991) makes it possible to test other genetic hypotheses relating to resistance to canker. Fortunate also is the advent of genetic linkage maps that have improved our understanding of disease resistance in recent years (Bradshaw 1996).

Domestication of hybrid poplar is somewhat advanced when contrasted with aspen. However, in the Lake States, intensive management of aspen is beginning to result in new disease problems that could not have been anticipated. The pathological rotation (i.e., "the age when decay losses exceed annual increment" (Edmonds et al. 2000)) may be reduced by intensive management in a way that may become clearer in aspen than in any other forest tree.

No approach to disease control is viewed with more optimism than genetic engineering, but the state of the art for *Populus* is not widely known. Are there currently genes that can tranform poplar clones to a state of multiple disease resistance? If not, are such transgenic solutions just around the corner?

No disease of hybrid poplar is more important than Septoria canker, and yet there is a general lack of awareness of the pattern of presence and absence of this disease in North America. If productive, canker-susceptible clones can be grown in certain situations without risk; growers will likely want to have a method to recognize such canker-suppressive sites. The most detailed information is from

Quebec, but some general, and as-yet-unpublished, information is available for the U.S. as well.

In this chapter, we focus on the following topics:

- The major poplar diseases in North America;
- Regional variation in diseases of hybrid poplar (Newcombe);
- Influence of disease on current and future aspen management in the Lake States (Ostry);
- The transgenic approach to disease resistance in poplars (Hubbes);
- Patterns of presence and absence of Septoria canker in the U.S. (Newcombe);
- Distribution of Septoria canker in Quebec (Périnet and Mottet);
- Breeding for resistance to Septoria canker in Quebec (Périnet and Mottet).

The major diseases of *Populus* in North America

1. Leaf rust caused by species of Melampsora. Poplar leaf rust is seen as yellow or orange pustules, termed "uredinia" (Fig. 1*a*). The rust fungi parasitize living cells in leaves of their hosts. However, in the case of the poplar cultivar "Crandon," rust may also attack stems (Fig. 1*b*). Rust damages poplar by reducing growth and by predisposing trees to secondary pathogens. In extreme cases, young, rust-susceptible trees may be killed (Fig. 1*c*). It is important to remember that rust fungi are competing with other pathogens for the poplar leaf resource (Fig. 1*d*); not all of the damage that one sees on rusted plants should necessarily be attributed to rust.

Rust is generally controlled by planting resistant cultivars or clones of poplar. Genes for resistance can fail, however, due to selection for virulence (i.e., virulent individuals of the fungal pathogen, even if rare at the time of field deployment of the resistant cultivar, will reproduce asexually and become common). This so-called pathogenic variation has greatly complicated efforts to control rust with resistant hybrid cultivars in the Pacific Northwest and in Europe. In eastern North America, attempts to improve eastern cottonwood (*P. deltoides*) are also affected by pathogenic variation in the rust population.

2. Stem canker caused by *Septoria musiva*. This is a very damaging disease of branches and main stems of many hybrid poplar clones (Figs. 2*a* and 2*b*). The fungus also produces leaf spotting or lesions (Fig. 2*c*). Fortunately, there are many areas in North America where susceptible clones can be grown without canker (i.e., "disease escape" occurs; see discussion below). In canker-conducive areas, resistant cultivars or clones must be planted to avoid the disastrous consequences of infection by *S. musiva* — cankered stems breaking in mid-rotation, top dieback, or outright death of infected trees. Fungal inoculum that can cause

Fig. 1*a*. Poplar leaf rust following inoculation of greenhouse-grown plants with a virulent isolate of the rust fungus (*Melampsora* spp.). Rust can appear within a week following inoculation, and in the field, repeated waves of infection and sporulation of the rust fungus may result in an epidemic.

Fig. 1*b*. Rust of both preformed leaves and young stem of poplar cultivar Crandon (*Populus alba* × *P. grandidentata*), as seen west of the Cascade Mountains in the Pacific Northwest. Crandon escapes this rust when it is grown in the Midwest.

Fig. 1*c*. Rust can kill young trees in the first year of growth. The trees, photographed in the spring following first-year rust attack, are all full sibs of a progeny, family 545, produced by the Poplar Molecular Genetics Cooperative, University of Washington, Seattle. One half of the progeny was resistant to the prevailing rust at that time in Puyallup, WA, and these displayed no dieback. Interestingly, the susceptible half of the progeny segregated further; some died whereas others suffered dieback of lower branches only.

Fig. 1*d*. Other pathogens may cohabit damaged leaves with rust fungi. In this case, rust is limited to the periphery of a leaf that is also affected by bronzing caused by a northwestern eriophyid mite.

cankered stems is not only produced from the cankers themselves but also from leaf lesions (Fig. 2*d*). Spores overwintering in leaf litter, in fact, may be a major source of inoculum the following year.

3. Leaf and shoot blight caused by *Venturia* spp. Leaf (Fig. 3*a*) and shoot blight (Fig. 3*b*) of aspen caused by *Venturia* is common across North America in the spring (Fig. 3*c*). In contrast, Venturia blight of species of poplars and cottonwoods (sections *Aigeiros* and *Tacamahaca*) is relatively uncommon. However, some interspecific hybrid poplar clones can be affected, and even severely damaged (Fig. 3*d*) in some parts of North America.

4. Leaf spot caused by *Marssonina brunnea.* Symptoms of *Marssonina* are evident in Fig. 4. Although found on hybrid poplar across North America,

Fig. 2*a*. New stem canker caused by *Septoria musiva* in Minnesota. Susceptible, 1-year-old trees such as this one sometimes escape infection for a couple of years only to succumb in the end. This temporary disease escape complicates early selection, as susceptible trees may be wrongly classified as resistant.

Fig. 2*b*. Old stem cankers so weaken a tree that it may break.

Fig. 2*c*. *Septoria musiva* also causes leaf spots, or lesions, in addition to stem cankers.

Fig. 2*d*. Leaf lesions are sources of inoculum, and thus increase the risk of canker. This is a magnified view of a single leaf lesion in which the fungus, in this case *Septoria populicola*, is producing thousands of potentially infective, microscopic spores in the central, buff-colored portion of the lesion.

Fig. 3*a*. Leaf and shoot blight appears in the spring and is caused by fungal species in the genus *Venturia*. Aspen (pictured here) is commonly affected across North America.

Fig. 3*b*. Young shoots of susceptible hybrid poplar clones can also be killed by Venturia and form characteristic "shepherd's crooks."

M. brunnea is most damaging to susceptible clones on the Coastal Plain in the southeastern U.S. The southeastern Coastal Plain is odd in other respects as well, as leaf rust damage is limited by hyperparasitism, and Septoria stem canker is absent. Other species of *Marssonina* cause disease to *Populus* species and hybrids, but they are generally not as damaging as *M. brunnea* is to *P. trichocarpa* × *P. deltoides* and *P. deltoides* × *P. nigra* hybrid poplar clones.

5. Hypoxylon canker of aspen. Hypoxylon canker is a canker disease that damages or kills stems and branches (Figs. 5*a*–5*c*). It is to aspen what Septoria stem canker is to hybrid poplar, although there are differences. For instance, the fungus which causes Hypoxylon canker, *Entoleuca mammata*, does not cause leaf lesions as *Septoria musiva* does.

6. Leaf bronzing of aspen. The premature death of leaves in late summer associated with leaf bronzing is a symptom that can be caused by different pathogens. Aspen bronzing of the Lake States (Figs. 6*a* and 6*b*) is caused by something other than the eriophyid mite that causes poplar/cottonwood bronzing in the Pacific

Fig. 3*c*. Spores of fungal pathogens of trees are microscopic. Eight, two-celled spores of *Venturia* spp. are held together in a tiny, transparent sac in this photo. Parts of other sacs, and an empty sac, are also seen. The spores are shot off in the spring as the leaves of their hosts emerge from buds. Some land on susceptible leaves and infect if wet weather permits.

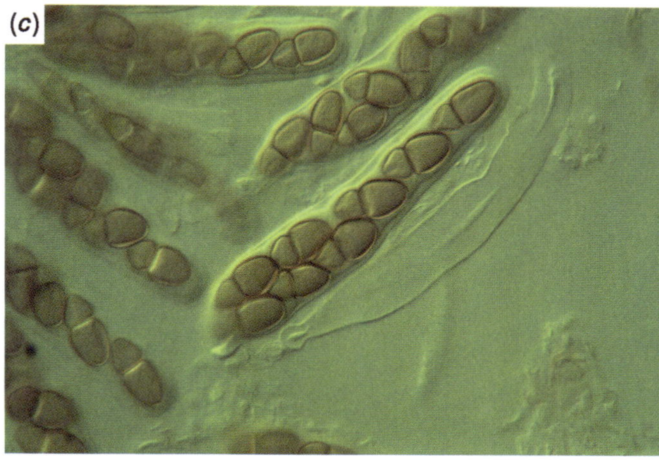

Fig. 3*d*. Repeated spring defoliation due to Venturia is associated with dieback and mortality. A Venturia-susceptible tree with a very thin crown is seen here in a plantation on Vancouver Island.

Fig 4. *Marssonina brunnea* is another fungus that causes lesions on stems and leaves.

Northwest (Fig. 1*d*). Ongoing research is clarifying the etiology or causes of aspen leaf bronzing.

7. White trunk rot of aspen. In general, conks that indicate trunk rot (Fig. 7*a*) are usually seen on older trees in natural stands. In short rotations of hybrid poplar, conks are never observed. However, *Phellinus tremulae* is an unusally aggressive rot fungus (Fig. 7*b*) that can damage aspen specifically.

Regional variation in diseases of hybrid poplar

Given the absence of Septoria stem canker in the Pacific Northwest, the most serious disease problem of hybrid poplar in the region is leaf rust. Dramatic changes in the past decade stem from the introduction of *Melampsora medusae*, and its hybridization with *M. occidentalis*. *Populus trichocarpa* × *P. deltoides* clones were free of rust until 1991 at which time *M. medusae* was introduced into the region

Fig. 5*a*. Hypoxylon canker, caused by the fungus *Entoleuca mammata*, is a damaging disease in aspen stands in the Great Lakes Region in particular. This canker on the basal portion of the stem, just above the grass, could be the site of stem breakage.

Fig. 5*b*. Hypoxylon canker affects branches as well as stems.

Fig. 5*c*. A closer view of a main stem affected by Hypoxylon canker.

Fig. 6*a*. Leaf bronzing of most of the crown of an aspen tree in Minnesota.

Fig. 6*b*. A closer view of leaf bronzing in the highly susceptible *P. alba* × *P. grandidentata* cultivar Crandon.

Fig. 7*a*. Conks of *Phellinus tremulae*, a major cause of trunk rot of aspen.

Fig. 7*b*. White heart rot is evident in this cross section of a decayed trunk. Black zone lines delimit the decay column.

(Newcombe 1996). The native rust, *M. occidentalis*, had up until that time remained confined to native *Populus trichocarpa*. Between 1991 and 1994, no pathogenic variation was detected in the *M. medusae* population affecting *P. trichocarpa* × *P. deltoides* hybrid clones. Only some clones were susceptible to *M. medusae*, and this was a predictable phenomenon throughout the region. However, everything changed in 1995 as clones that had been resistant were found to be susceptible. Hybridization between *M. medusae* and *M. occidentalis* is now known to account for the new morphological and pathogenic variation that characterizes the current leaf rust population on hybrid poplar in the Northwest (Newcombe et al. 2000). This new hybrid rust, *M.* × *columbiana*, also led to the discovery of an older hybrid population (Fig. 8) that is geographically intermediate between the type locality of *M. medusae* (the southeastern Coastal Plain of the U.S.) and that of *M. occidentalis* (the Willamette Valley of Oregon). However, the practical implications of rust hybridization for poplar growers mainly involve new pathogenic variation.

Pathogenic variation presents a new challenge to breeders. Clones that are selected for resistance in one location may prove to be susceptible elsewhere due to differences in local pathotypes. The resistance of all *P. trichocarpa* × *P. deltoides* hybrid clones grown commercially in the Pacific Northwest is now in question. These clones have been the mainstay of the Northwestern hybrid poplar industry, and all are now susceptible to at least one pathotype of *M.* × *columbiana*. A running total of 15 pathotypes of *M.* × *columbiana* have been characterized to date. This situation is potentially as problematic as that in Western Europe.

Even when pathogenic variation per se is not an issue, disease may still limit the geographic range in which a poplar clone or cultivar can be successfully grown. The most important and striking example of this is stem canker of *P. trichocarpa* × *P. deltoides* F_1 hybrids. Known to be genetically susceptible as a hybrid class to canker caused by *Septoria musiva*, *P. trichocarpa* × *P. deltoides* clones are grown without canker in the Pacific Northwest, and on the Coastal Plain in the Southeast. In contrast, attempts to grow the best Northwestern *P. trichocarpa* × *P. deltoides* clones in bottomlands of the Mississippi and St. Lawrence River drainages, or in the Lake states, are likely to fail due to canker. The "commercial range" of these Northwestern hybrids is thus restricted due to disease.

There are many other examples of this limitation due to disease. For example, the poplar cultivar Crandon, a natural hybrid of *P. alba* × *P. grandidentata*, is a productive, rust-free clone in some parts of North America, although its susceptibility to leaf bronzing is increasingly problematic in parts of the Midwest. However, when planted in the maritime Pacific Northwest, its performance has been poor due to systemic rust infections by *Melampsora populnea*. Uredinia, or rust pustules, have been observed in early spring not only on leaf laminae, but also on newly flushed petioles and shoot apices. Densely clustered uredinia were observed in late April each year from 1997 to 1999 inclusive in a manner that left little doubt that the fungus overwinters on this poplar clone. The result is branch

Fig. 8. Distribution of the old and new populations of the hybrid rust, *Melampsora* × *columbiana*. The new, pathogenically variable population in Washington and Oregon on *P. trichocarpa* × *P. deltoides* hybrids is shown in black, arbitrarily set at a density of 10 samples per state. The old population is also imperfectly known and is shown in gray, arbitrarily set at a density of 1 per state. Some adaptive radiation is apparent in the old population; the following *Populus* taxa are implicated: CA, *P. fremontii*; ID, *P. trichocarpa*; WY, *P.* × *acuminata* and *P. angustifolia*; CO, *P.* × *acuminata*; MT, *P. angustifolia* and *P. deltoides*; SD, NE, MN, and IL, *P. deltoides*; WI, *P. balsamifera*.

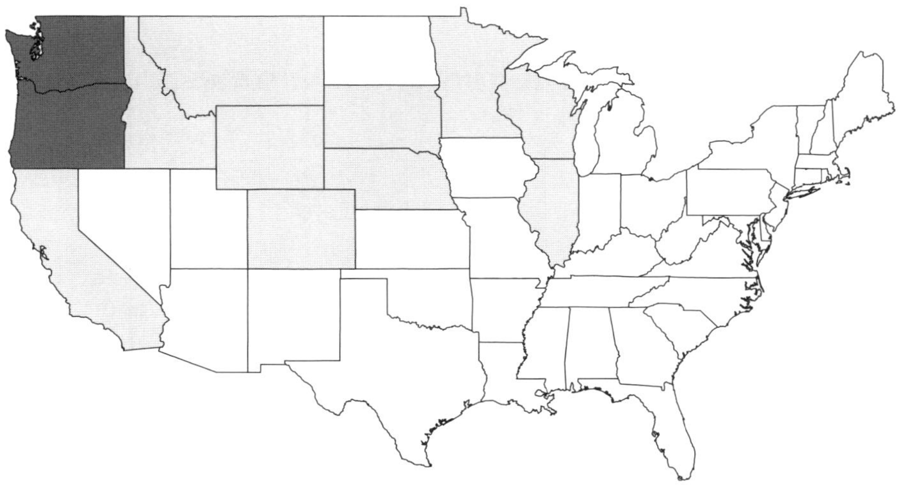

dieback. Clones in the same block planting, with resistance to *M. populnea*, have not been affected by dieback. Fortunately, the Eurasian *M. populnea* has never expanded in geographic range in North America; it has always been known as an exotic rust that occurs on *P. alba* on the Pacific and Atlantic coasts. Thus, Crandon may be grown as a productive disease-escape only in places where this rust and leaf bronzing do not occur.

Complicating the issue of commercial-range limitations due to disease, there are pathogens that are currently expanding in geographic range. *Marssonina brunnea*, native to eastern North America, was initially found in the Pacific Northwest only on the coast. It has since spread inland from the mouth of the Columbia River, so that it is now present in the commercial poplar-growing Clatskanie Valley of Oregon. New or "borderline" pathogens, such as *Pestalotiopsis populi-nigrae* (Newcombe 2000*b*), continue to be introduced into North America. This fungus is reputed to cause poplar shoot blight in Japan, but its incidence and ecological role in North America are unclear.

Pathogens also may be affected by their own parasites in different ways in different regions. For example, hyperparasitism of the species of *Melampsora* that cause poplar leaf rust varies by region in North America. In the Southeast on the coastal plain, rust epidemics caused by *M. medusae* have started as early as the first week of June. Yet the expected premature defoliation does not materialize. A hyperparasitic fungus is present on the coastal plain and appears to be responsible for the aborted epidemics. Rust pustules turn white, or are replaced by the hyperparasite. However, the same hyperparasitic fungus is not present elsewhere in North America on poplar leaf rust. In the presence of effective hyperparasites, it is possible that even rust-susceptible clones might be successfully grown. But strategies to introduce the Southeastern hyperparasite into other regions, and to enhance its biological control of rust, remain to be developed.

Influence of disease on current and future aspen management in the Lake States

Aspens — trembling (*Populus tremuloides*) and bigtooth (*P. grandidentata*) — are the most abundant and commercially important of the native poplar species in the Lake States of Minnesota, Wisconsin, and Michigan. Considering their value as pioneer species on disturbed sites, the contribution of their brilliant gold foliage in the fall to the aesthetics of the "north woods," and the array of critical wildlife habitat attributes they provide, they also are among the most ecologically important species in this region.

Past aspen management was necessarily on an extensive basis, primarily because of economic factors associated with supply and demand. The unprecedented increase in the demand for aspen has encouraged, if not required, managers to examine intensive management strategies similar to those used in hybrid poplar plantations to increase supplies. Under this new scenario, management of pests and diseases will be essential.

Although aspen are host to a large number of endemic insect pests and pathogens that may reduce the productivity and quality of affected trees (Ostry et al. 1989), relatively few of them have caused concern among managers. Many diseases have not been studied in detail, and no effective control strategies have been developed for even the most serious damaging agents. Potential impacts from exotic invasive insect pests such as the gypsy moth (*Lymantria dispar*) and Asian long horn beetle (*Anoplophora glabripennis*) and their interactions with pathogens and climate change are relatively unknown at this time. Aspen are shallow-rooted, and large changes in soil moisture and temperature could have significant effects on their growth and disease tolerance. With a growing dependence on this species and the need to shorten rotations and increase fiber and wood quality, managers are becoming much more aware of the impacts of insects and diseases on this crop. Aspen is now frequently managed for multiple rotations rather than harvested and allowed to convert to more tolerant hardwoods and conifers or planted to red pine as was so frequently done in the past.

Numerous foliar and shoot pathogens affect aspen, some causing highly conspicuous blights and in some years premature defoliation. However, no studies have been undertaken to determine the long-term impact of foliage diseases on aspen growth in native stands. In studies of the impact of insects and diseases on developing aspen stands, shoot blight of aspen suckers caused by *Venturia* (Fig. 3*a*) had a profound effect on height growth and crown position of infected individuals (Ostry and Ward, unpublished data; Perala 1984). What is not known, however, is what effect yearly incidence of this and other foliage diseases has on growth and productivity of affected stands. While periodic losses in growth potential may be subtle and of little consequence on aspen managed on 50–70 year rotations, they become far more important in the future when expensive management practices are applied and projected rotations are shortened to 20 years.

So why has this short-lived tree been so successful in becoming the dominant cover type in the Lake States forests when a myriad of biotic and abiotic damaging agents attack, infect, and affect virtually all parts the tree throughout their lives? Aspen is a pioneer species and among the first tree species to become established on sites disturbed by logging, wind, and particularly by fire that exposes mineral soil. Aspen produces abundant seed, and seedlings can quickly become established under favorable conditions. After harvest or fire, existing root systems develop dense sucker reproduction that over time produces clones of varying sizes that vary in a multitude of traits, including disease resistance.

Two silvical traits of aspen may be the key to this species success and should be considered when decisions to intensify management of not only native aspens but also plantings of hybrid aspen and poplars: (1) aspen is a clonal species that reproduces from root suckers, creating stands made up of a mosaic of genetically different clones, and (2) after clearcutting, from 25 000 to 75 000 suckers per ha can develop within 2 years.

Tree monocultures can be much more susceptible to damaging agents than mixed species stands, however, the relatively pure aspen stands in the Lake States have not met with devastating failures. Certainly this in part may be because of the rich genetic diversity among the clones that have developed within the stands. Over time, superior clones expand and replace clones that cannot compete or that are susceptible to damage. Past aspen research in growth and yield, management, and insect and disease biology, for example, has not adequately addressed the genetic variability within aspen stands and how this diversity affects biological response variables over a range of sites.

Aspen is very intolerant of shade throughout its life, and competition among ramets, especially for light, within a stand reduces the number of stems over time. In addition, aspen naturally self-prunes, producing clean stems in well-stocked stands. Numerous examples exist that illustrate the importance of tree density and poplar diseases. For example, significantly fewer stem cankers developed on pruned aspen than unpruned stems in a paired test. Various insect wounds on branches provide an entry for the fungus *E. mammata* that can result in a high incidence of stem cankers in under-stocked stands where natural pruning is not effective. "Spatial resistance" is the term that describes the interactions of pathogens, disease incidence and severity, and tree density in native aspen and hybrid poplar plantations. The large numbers of aspen suckers that regenerate after harvest or other disturbances are gradually reduced throughout the life of the stand. In large part, this mortality is governed by available light, moisture, nutrients, and the activities of insects and diseases. Eventually adapted, disease-resistant clones enlarge, while maladapted, disease-susceptible clones slowly die and fail to regenerate.

The most important disease that often results in aspen mortality in the Lake States is Hypoxylon canker. Genetic resistance among clones and aspen families (Bucciarelli et al. 1998, 1999; Enebak et al. 1997, 1999) and stand density has a major influence on the incidence and severity of this disease. Managers applying strategies to increase growth and yield and shorten aspen rotations — planting selected clones and families and thinning stands — may need to consider how complex potential biological interactions (Ostry and Anderson 1998) can alter disease relationships within a stand or plantation. Reducing stem density within a stand may have the unexpected consequence of increasing Hypoxylon incidence and severity.

White trunk rot of aspen results in more wood volume loss than any other disease of aspen. The affected volume in individual trees and within stands is often underestimated. In Minnesota, current inventory methods underestimated aspen decay volume by 38% (Jones and Ostry 1998). This disease, more than any other stem disease, imposes a pathological rotation on aspen stands. Attempts to correlate site, tree age and size, and clones to the incidence and extent of decay have been largely unsuccessful. Generally thought to be an indicator of decadent, over-mature stands, there is evidence that the incidence of white trunk rot can be significantly increased in young, vigorously-growing stands when stems are wounded during mechanical thinning operations (Ostry and Ward, unpublished data). The incidence of stem damage by a number of wood borers has also increased in thinned aspen plots. Subsequent invasion and damage by stain, decay, and canker fungi is known to be more severe on aspen stems attacked by wood boring insects.

Although many of the same pathogens are found in native aspen stands as in planted hybrid poplar plantations, severe epidemics of diseases caused by these pathogens are rare in native stands, and when they do occur, they certainly do not have the same impact as in plantations. In plantations, an increase in disease severity may be related to maladapted clones not having disease resistance, stem densities that are too high or too low for a particular site or clone, and inputs such as fertilization and mechanical cultivation that may favor pathogen development, spread, and infection.

Similar to modern agriculture, we are attempting to domesticate a wild plant system to meet our needs. Insects and diseases are essential components in native aspen stands, regulating stand density and structure, and influencing the genetic makeup of the final stand. As we manipulate these stands to increase their productivity, however, we need to guard against making unintentional changes to the natural resiliency and resistance to environmental stresses and damaging agents inherent in natural, native stands. Although great potential exists for the genetic improvement of aspen, long-term field-testing is required prior to the release of new selections. Two examples of disease problems that impacted aspen selections early in an aspen improvement program illustrate the importance of field tests and underscore the damage that new diseases can cause on non-native aspen species and new hybrids.

Hybrid triploid aspen (*P. tremuloides* × *P. tremula*) test plantings in Wisconsin and Michigan experienced over 90% mortality associated with the fungus *Lahmia kunzei* (syn. *Parkerella populi*) while neighboring native diploid and triploid aspen were unaffected (Enebak et al. 1996). This fungus was identified for the first time in the United States in the mid-1980's (Ostry 1986). A common male *P. tremula* parent was implicated in the failure of all the hybrid triploid families in all eight field trials.

Leaf bronzing of aspen is a disease with a poorly understood biology but thought to be caused by the fungus *Apioplagistoma populi*. Although it has been found occasionally on native aspens, the disease has been most severe on *P. grandidentata* hybrids with *P. alba* and *P.* × *canescens*, resulting in the decline and eventual death of affected trees. Several selections have been discontinued, and the disease has had a major impact in the utilization of these hybrids. Recently, this disease has been found affecting several other cultivars commonly used in landscape plantings and may result in their failure and removal from commercial trade.

Diseases that are now of moderate importance will most certainly become more of a concern, and many diseases that are now considered only of interest to mycologists or forest pathologists may become more widely noticed when management objectives are negatively impacted. Multiple entries into aspen stands and associated mechanical wounding of trees will increase the frequency and severity of root and stem injuries that can lead to increases in stain and decay. In addition, aspen harvested on shorter rotations potentially can have a higher incidence of root and butt rot caused by *Armillaria* spp., a disease that is now usually important only in highly stressed stands such as those that have been repeatedly defoliated by forest tent caterpillar (*Malacosoma disstria*).

Finally, some consideration must be given to the potential effects that may develop as we increase the number and size of intensively managed aspen stands and hybrid poplar plantations across the landscape. Could these intensively managed stands become reservoirs of pathogens or create "bridges" allowing pathogens or insect pests to move into new locations and become damaging to not only the managed stands but also to the native stands that serve many important ecological functions? Genetic diversity in forest trees such as aspen provides us with one of the greatest advantages we have in managing trees as a crop. We must exercise caution that we do not modify the natural system to the point where this diversity is no longer functional.

The transgenic approach to disease resistance in poplars

The discovery of the double helix structure of DNA signaled the advent of a new era in biology. With the introduction of recombinant DNA and genetic engineering techniques, biotechnology boomed. New tools became available that allowed the insertion of genes controlling desired traits not readily available in sexually accessible gene pools. The belief arose among foresters that greater disease resistance could be achieved in a much shorter time using biotechnology than with traditional tree-breeding methods. Transformation with genes mediating disease resistance is possible, but this feat has not yet been achieved in forestry.

Populus was quickly identified as more amenable to this new technology than any other forest tree. In fact, *Populus* is an ideal model for the study of biological and molecular mechanisms in trees that cannot be studied in annual plants. Past problems encountered in poplar transformation systems required for the expression of foreign genes have been largely overcome. However, some problems such as gene silencing, transformation frequency, and collateral genetic damage may still be encountered (Han et al. 1996; Bent and Yu 1999).

The remaining problem in developing disease resistant poplars is the lack of availability of appropriate genes that mediate resistance. This is somewhat surprising, since diseases such as *Dothichiza populea*, *Marssonina brunnea*, *Septoria musiva*, *Venturia* spp., *Xanthomonas populi* pv. *populi*, and *Melampsora* rusts have often threatened the success of poplar culture. Occasionally, the replacement of old susceptible clones by new selections solved a few of the disease problems — for example, replacement of Dothichiza canker-susceptible *P. nigra* with resistant *P. deltoides* × *P. nigra* hybrid clones — but for others the problem persisted (Frey and Pinon 1997). Thus, there is an urgent need for the development of poplar clones with durable disease resistance.

The development of resistant clones and the isolation of corresponding genes can only be achieved by a thorough understanding of the molecular and biochemical basis of host resistance and pathogen virulence. This would ensure that host tree resistance would not be of only short duration.

Trees cannot run from pathogenic microorganisms, insect pests, fire, and unfavorable weather conditions. How do they then survive? During their evolution trees, have developed a subtle but very effective protection system against stress factors that threaten their survival. It is a very sophisticated defense system, although many aspects are not yet fully understood. Some mechanisms may only become evident during or after an encounter with a pathogen, while others are already in place prior to infection. As a consequence, resistance to potential pathogens is the rule whereas susceptibility is the exception. In host–pathogen interactions, specific signaling compounds released by the pathogen trigger a cascade of defense responses in the attacked host. To activate these responses, the host cells must possess a genetic system that perceives, transduces, and translates potential pathogen signals into biochemical and physical responses. These in turn check the pathogen invasion. Many of these defense reactions, which also are seen in poplars, can be encountered as clonal resistance (Laurans and Pilate 1999), organ-specific resistance (Newcombe 1996), age-related resistance (Enebak et al. 1997), and induced resistance (Flores and Hubbes 1979). Although the defense mechanisms are present in some clones, they are often not sufficient to fully protect the clones against specific pathogens. Traditional tree breeding coupled with disease screening has been successful in producing new clones resistant against specific diseases. The drawback of this approach, as with all traditional tree breeding programs, is that it is very lengthy, and the strength and duration of the underlying mechanisms of resistance may not be known. Resistance to one pathogen may still leave a clone susceptible to other pathogens. Single-gene resistance may also be overcome quickly by the emergence of new virulent isolates (Pinon and Frey 1997).

One approach to identify genes of disease resistance is to conduct simultaneous inoculations of naturally resistant and susceptible hosts with a given pathogen. Then, the temporal physical reactions of the two hosts are compared by light, scanning, or transmission electron microscopy. A biochemical analysis of the tissues implicated in the host–pathogen interaction measures the physiological reactions of the resistant or susceptible host tissue. What should follow is the identification of proteins tightly linked to these defense reactions. Their amino acid sequence and DNA probes derived from their sequence would lead to the isolation of the respective genes found in the corresponding genomic library. A shorter way to detect genes involved in the defense reactions of poplar tissue would be cDNA sequencing, i.e., reverse transcription from mRNA expressed during the defense reaction.

It is evident that there are a number of genes in poplars that are multifunctional. They control normal tissue development as well as defense reactions against invading pathogens and other stress factors. Pathogen-specific compounds or host-specific products released during the host–pathogen interaction trigger induction of the defense system. For example, a key factor in the defense reaction of trees is the rapid lignification of cell walls to form a physical barrier zone that restricts and blocks the rapid spread of the pathogen. In tree species, lignin can constitute 20 or 30% of the dry weight of the wood and presents an economic and environmental problem in paper making. Therefore, efforts are made to produce transgenic poplars in which the lignification process is down-regulated. However, this may interfere in the defense mechanisms of the transformed clones in rendering them more vulnerable to pathogen attack, unless other defense mechanisms (e.g., phytoalexins or foreign transgenes) can fill the gap.

The introduction of foreign genes into trees may also be problematic, given their long breeding cycle. If resistance is only based on a single, race-specific gene, pathogens will evolve to overcome the defense mechanism. Control of spatial and temporal gene expression may need to be engineered as a supplementary measure.

Recombinant DNA technology or genetic engineering, has become a dominant tool in biology. The potential problem with the release of an engineered organism is the loss of strict control over the transgene. With long-lived, wind-pollinated species like poplars, measures to restrict pollen flow to non-engineered plants must be considered. Furthermore, genetically engineered parts of trees may disperse from leaves and roots; e.g., via insect or bacterial vectors. To counteract these potential risks, guidelines regarding the release of genetically engineered plant material have been developed and are discussed in numerous publications (e.g., Hubbes 1990, 1993; James, 1997). Finally, the rise of ecoterrorist groups that are strongly opposed to genetic biotechnology has already led to the sabotage of the work of several poplar geneticists. This mischief will continue and must be recognized as an occupational hazard.

The transgenic approach to producing disease-resistant poplars is still in its infancy. Genetic engineering has progressed much faster in developing herbicide-resistant or insect-resistant poplar clones. The rapid progress in the identification of poplar genes that control tissue development as well as those genes implicated in general defense mechanisms will stimulate further molecular work on clonal resistance to diseases.

Patterns of presence and absence of Septoria canker in the U.S.

Stem cankers of hybrid poplars are typically caused by *Septoria musiva*. The related *Septoria populicola* may also cause cankers (Zalasky 1978), but this disease may be exceptional or atypical. Author Newcombe has not obtained cankers in an inoculation with an isolate of *S. populicola* whereas author Ostry, using a different isolate, has obtained cankers. *Septoria populicola* certainly does not cause cankers in *P. trichocarpa* and *P. balsamifera* in their native ranges. However, the genetic susceptibility to *S. musiva* of *P. trichocarpa*, and its hybrids with *P. deltoides* is beyond dispute. This susceptibility is in evidence whenever trees are planted in canker-conducive areas in eastern North America.

However, there are many gaps in our knowledge of stem canker of poplar. People have assumed the worst about the "Septoria limitation" of the commercial range of *P. trichocarpa* and its hybrids. Some have assumed that *Septoria musiva*, and the killing stem canker that it produces, will inevitably spread to wherever hybrid poplar is grown. Some feel that every attempt needs to be made to keep inoculum of *S. musiva* out of the Pacific Northwest in particular. The situation may not be quite that bleak. *Septoria musiva* has shown its potential for spread; e.g., it was reported long ago in Argentina (Sarasola 1944). Surely, if it could spread to Argentina, there would have been ample anthropogenic and natural opportunity for it to establish itself in the Pacific Northwest. Not only is there currently a hybrid poplar resource in the Northwest that is uniformly susceptible, but *P. trichocarpa*, as a widely distributed species in the region, is also susceptible. However, the fact is that *S. musiva* and the canker it produces do not occur in the Pacific Northwest (Newcombe 1996), nor does *S. musiva* appear to be common on the coastal plain in the Southeastern U.S.

In fact, it may be absent on the Southeastern coastal plain. *Septoria musiva* and *S. populicola* have been recorded in North Carolina, but these records are based on only one specimen of each species. Moreover, they were collected by the same individual, F.L. Stevens, in 1908 and 1909, and from a host that is only identified as "*Populus* sp." Author Newcombe did not find *Septoria* on hybrid poplar and *P. deltoides* in surveys conducted in Virginia, North and South Carolina, and Georgia in recent years. Genetically susceptible poplar clones are being grown on the Southeastern coastal plain without canker. Thus, it is tempting to think that the absence of the causal pathogen *Septoria musiva* in both the Northwest and the southeastern Coastal Plain is the only factor necessary to explain the absence of canker.

Again, the situation does not appear to be that simple. There is evidence that genetically susceptible *P. trichocarpa* × *P. deltoides* hybrid clones can be grown without canker in some sites within regions where *S. musiva* does occur. In western Kentucky and adjacent areas, it appears that bottomland sites are canker-conducive whereas upland sites are less so. To remove lingering doubts about disease escape due to lack of inoculum, susceptible clones in putative canker-suppressive sites should be inoculated with *S. musiva*. Until this is done, growers will not be able to grow productive but canker-susceptible clones with any confidence.

Distribution of Septoria canker in Quebec

Since 1969, more than 250 experimental plantations of hybrid poplar have been established all over Quebec by the Poplar Improvement Group of the Ministere des Ressources naturelles. Regular surveys of Septoria canker have been conducted in those plantations, each one containing susceptible as well as resistant clones. So far, its distribution appears to be limited to southern Quebec in bioclimatic domain 1 (sugar maple – bitternut hickory) and 2 (sugar maple – basswood), where severe damage was observed on susceptible clones in all plantations (Fig. 9).

Septoria canker has never been observed in natural stands in Quebec, but leaf spots are found on two native poplars, *P. balsamifera* and *P. deltoides*. Hence,

Fig. 9. Distribution of Septoria canker in hybrid poplar plantations in Quebec in relation to bioclimatic domains. See text for detailed explanation.

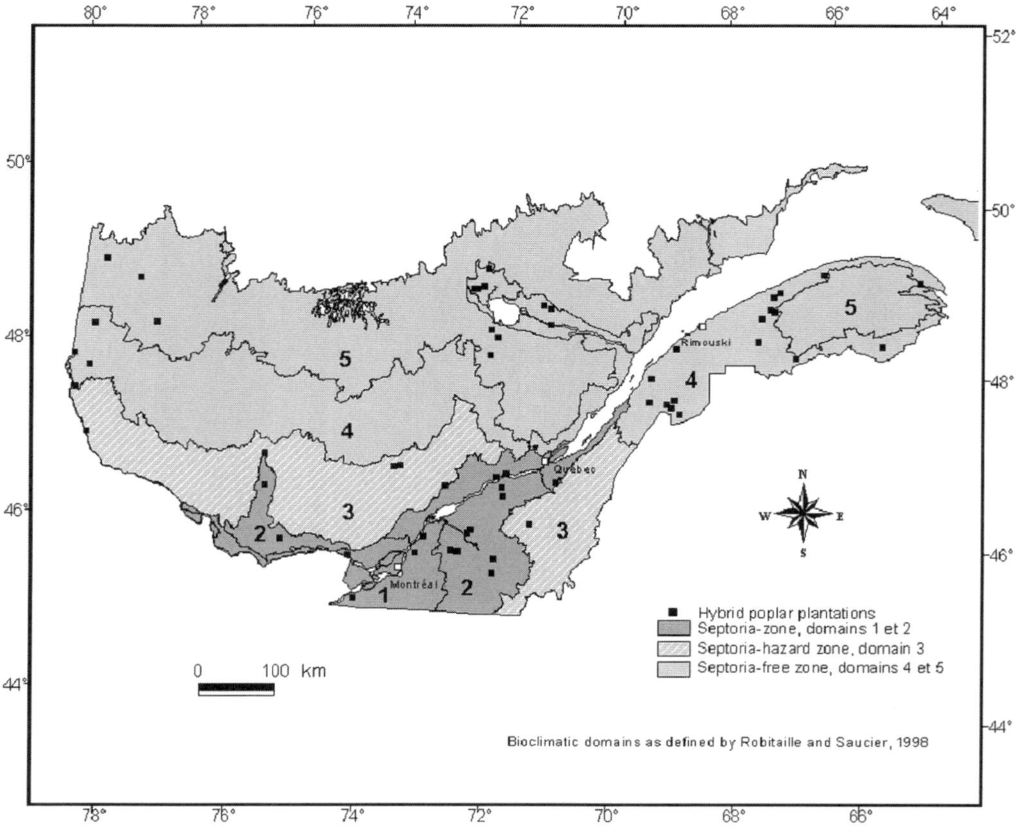

these two species are often considered as the initial source of inoculum for new hybrid poplar plantings. However *S. musiva* is uncommon on *P. balsamifera* according to regional herbarium specimens. In field surveys made by the Ministère des Ressources Naturelles in Quebec, only *Septoria populicola* was associated with *P. balsamifera* whereas *S. musiva* was found on *P. deltoides*, which must be considered as the initial source of inoculum. Furthermore, *P. balsamifera* is distributed throughout Quebec whereas *P. deltoides* is mainly confined to the southern part of the province where it inhabits the riparian zones along the tributaries of the St. Lawrence River. *Populus deltoides* also colonizes disturbed environments and can expand its distribution to the whole area of domains 1 and 2, and some parts of the domain 3 (sugar maple – yellow birch). As *P. deltoides* is actually the main source of inoculum, this could explain the close relationship between Septoria canker distribution and *P. deltoides* range in Quebec. As in the southeastern U.S., it is tempting to think that the absence of *Septoria musiva* is the only factor necessary to explain the absence of canker.

Outside domains 1 and 2, Septoria canker was detected in only one plantation located on an upland site near Thetford Mines. In this 5-year-old clonal test, 16 cankered trees were detected in 1999 on 10 susceptible clones among 1623 trees representing 243 clones. This area is located south of the St. Lawrence River and surrounded to the west and north sides by the canker-conducive zone. In addition to the presence of inoculum from *P. deltoides*, the trees could have been infected in the nursery, which is located in the conducive zone. Within the area close to the conducive zone, only three other plantations are located in domain 3. One of them (near the boundary of domain 2) is only 3 years old, while the two others are 7 and 8 years old, but these are relatively farther north. In the Témiscamingue region at the far west of domain 3, no Septoria canker was reported in a plantation established in 1986. Although little damage has been observed in domain 3, the ability of the disease to develop and form cankers on susceptible seedlings in this domain was demonstrated by an artificial inoculation assay in the Quebec City region. Hence, this domain is expected to become a Septoria-hazard zone with the extension of intensive poplar cultivation, particularly near natural cottonwood populations. Thus, domain 3 in Quebec appears to be similar to areas in the U.S. Midwest such as that near Cloquet, Minnesota, where cankers have been slow to develop, even on genetically susceptible clones.

Even if *Septoria musiva* inoculum were found throughout domain 3, it is unlikely that the canker disease would develop in domains 4 (balsam fir – yellow birch) and 5 (balsam fir – white birch). There is evidence that the environment is unfavorable in these domains. Since 1969 and even before, infected cuttings have been introduced throughout Quebec. Clones that were overly susceptible to Septoria canker in the south have often been sent to northern regions (mainly domains 4 and 5) for testing. Surveys have been performed in two plantations where cankered trees originating from the Septoria zone were accidentally introduced in the boreal zone. In all cases, the cankers stopped developing, and no new infections accrued. Although conditions favoring development of Septoria canker are not fully understood, it appears that Septoria-free (canker-suppressive) zones do exist in Quebec. Bioclimatic or edaphic conditions encountered in domains 4 and 5 seem to limit the extension of the disease. What is needed are more investigations of the disease triangle comprising *Septoria musiva*, genetically susceptible poplars, and the environment.

Breeding for resistance to Septoria canker in Quebec

In Quebec, an artificial inoculation screening procedure has been used for the past decade for the evaluation of Septoria canker resistance (Mottet et al. 1991). In July, stump sprouts are typically inoculated. For each stem, after removal of the sixth leaf, mycelium plugs are placed on the resulting fresh leaf scars. The clones are then evaluated 3 months later by calculating the mean percentage of stem circumference girdled by the pathogen. For the least and most susceptible clones, responses have been comparable to ratings of canker damage observed in field tests. In one study, when inoculating 252 clones with four isolates (three originating from southern Quebec and one from Ontario), differences in aggressiveness were found. Similar results were obtained by Krupinski (1989).

The inoculation responses of 221 clones from section *Tacamahaca* demonstrated that most of them were highly susceptible to *S. musiva*. Some clones (10% of tested clones) of *P. balsamifera*, however, proved to be more resistant. This native local species seems more resistant than either *P. maximowiczii* (now *P. suaveolens*) or *P. trichocarpa*. In field trials with high incidence of *S. musiva* inoculum, balsam poplar clones showed varying degree of resistance to canker. In any event, this species is canker-resistant in natural stands. In contrast, many resistant clones have been found in the section *Aigeiros* (35% of tested clones) and also among the hybrid *P. balsamifera* × *P. nigra* (17% of tested clones) in response to inoculation.

Since 1986, the artificial inoculation method has been regularly used for preliminary screening in addition or prior to other field tests. New isolates are collected in field trials and systematically evaluated on clones representing a range in Septoria resistance. Since the first collections in 1986, no increase in *S. musiva* aggressiveness has been observed even though all plantations included highly susceptible clones. Monitoring Septoria canker in the long term will be necessary, particularly with the extension of intensive poplar cultivation with more resistant clones.

Conclusions

In North America, the totality of pathogens affecting *Populus* is large (Callan 1998; Newcombe 1996; Ostry et al. 1989), although we have discussed only a few of the most important among them. This totality is partitioned in a more or less patchy manner across the continent. The result with respect to any one disease is the phenomenon known as disease escape, which is common and locally and regionally significant for diseases as serious as Septoria canker. Proponents of quarantine would argue that this is a precarious situation. They would oppose free exchange of poplar cuttings on the grounds that the present patchiness and disease escape will be replaced by pathogenic homogenization. In particular, they would argue that the "canker-suppressive" sites discussed in this paper should be shielded from inoculum of *Septoria musiva*. After all, introduced diseases have run rampant in poplar plantations in the Southern Hemisphere.

The fact is that in North America where natural populations of *Populus* are still robust, there have been no devastating epidemics resulting directly from exotic introductions. The Eurasian poplar rust fungus, *Melampsora larici-populina*, came and went quietly on the North American scene (Newcombe 1996), and the brief flurry of quarantine activity in the early 1990s appears to have been much ado about nothing. *Melampsora medusae*, introduced from the southeastern corner of the U.S. to the northwestern corner, quickly lost its identity through absorption into a new hybrid population (Fig. 8). The herbarium evidence of *M.* × *columbiana* in middle America demonstrates that hybridization is an old mechanism for blunting a new parasite and potentially forcing local adaptation back to host–parasite equilibrium. In general, poplar growers will always be tempted to capitalize on disease escape when genetically susceptible clones are the best available. This "gamble" has continued to pay off in the Pacific Northwest where Septoria canker is absent.

Poplar breeding for disease resistance remains experimental. Fortunately, there are always new experiments to try, and biotechnology offers an exciting new avenue for such experiments. But there are no guarantees, and little likelihood of a "silver bullet" cure for serious poplar diseases. Nonetheless, the future looks bright for the wide availability of a variety of disease-resistant clones for different regions of North America.

References

Bent, A.F., and Yu, I.C. 1999. Applications of molecular biology to plant disease and insect resistance. Adv. Agron. **66**: 251–298.

Bradshaw, H.D., Jr., 1996: Molecular genetics of *Populus*. *In* Biology of *Populus* and its implications for management and conservation. *Edited by* R.F. Stettler, H.D. Bradshaw, Jr., P.E. Heilman, and T.M. Hinckley. NRC Research Press, Ottawa, ON. pp. 183–199.

Bucciarelli, B., Jung, H.G., Ostry, M.E., Anderson, N.A., and Vance, C.P. 1998. Wound response characteristics as related to phenylpropanoid enzyme activity and lignin deposition in resistant and susceptible *Populus tremuloides* inoculated with *Entoleuca mammata* (*Hypoxylon mammatum*) Can. J. Bot. **76**: 1282–1289.

Bucciarelli, B., Ostry, M.E., Fulcher, R.G., Anderson, N.A., and Vance, C.P. 1999. Histochemical and microspectrophotometric analyses of early wound responses of resistant and susceptible *Populus tremuloides* inoculated with *Entoleuca mammata* (= *Hypoxylon mammatum*). Can J. Bot. **77**: 548–555.

Callan, B.E. 1998. Diseases of *Populus* in British Columbia: A diagnostic manual. Canadian Forest Service, Pacific Forestry Centre, Victoria, BC <http://www.pfc.cfs.nrcan.gc.ca>.

Edmonds, R.L., Agee, J.K., and Gara, R.I. 2000. Forest health and protection. McGraw–Hill. 630 pp.

Enebak, S.A., Ostry, M.E., Wyckoff, G.W., and Li, B. 1996. Mortality of hybrid triploid aspen in Wisconsin and Upper Michigan. Can. J. For. Res. **26**: 1304–1307.

Enebak, S.A., Bucciarelli, B., Ostry, M.E., and Li, B. 1997. Histological analyses of the host response of two aspen genotypes to wounding and inoculation with *Hypoxylon mammatum*. Eur. J. For. Path. **27**: 337–345.

Enebak, S.A., Ostry, M.E., and Anderson, N.A. 1999. Inoculation methods for selecting *Populus tremuloides* resistant to Hypoxylon canker. Can. J. For. Res. **29**: 1192–1196.

Flores, G., and Hubbes, M. 1979. Phytoalexin production by aspen (*Populus tremuloides* Michx.) in response to infection by *Hypoxylon mammatum* (Wahl.) Mill., and *Alternaria* sp. Eur. J. For. Pathol. **9**: 288-298.

Frey, P., and Pinon, J. 1997. Variability in pathogenicity of *Melampsora allii-populina* expressed on poplar cultivars. Eur. J. Forest Pathol. **27**(6): 397–407.

Han, K.H., Gordon, M.P., and Strauss, S.H. 1996. Cellular and molecular biology of *Agrobacterium*-mediated transformation of plants and its application to genetic transformation of poplar. *In* Biology of *Populus* and its implications for management and conservation. *Edited by* R.F. Stettler, H.D. Bradshaw, Jr., P.E. Heilman, and T.M. Hinckley. NRC Research Press, Ottawa, ON. pp. 201–222.

Hubbes, M. 1990. Development of biotechnology programmes for energy forestry. Biomass, **22**: 75–90.

Hubbes, M. 1993. Impact of molecular biology on forest pathology: A literature review. Eur J. For. Pathol. **23**: 201–217.

James, R.R. 1997. Utilizing a social ethic toward the environment in assessing genetically engineered insect-resistance in trees. Agr. Human Values, **14**: 237–249.

Jones, A.C., and Ostry, M.E. 1998. Estimating white trunk rot in aspen stands. North. J. Appl. For. **15**: 33–36.

Krupinski, J.M. 1989. Variability in *Septoria musiva* in aggressiveness. Phytopathology, **79**: 413–416.

Laurans, F., and Pilate, G. 1999. Histological aspects of a hypersensitive response in poplar to *Melampsora larici-populina*. Phytopathology, **89**: 233–238.

Mottet, M.-J., Bussieres, G., and Vallee, G. 1991. Test précoce pour l'évaluation de la sensibilité de peupliers hybrides au chancre septorien. For. Chron. **67**: 411–416.

Newcombe, G. 1996. The specificity of fungal pathogens of *Populus*. *In* Biology of *Populus* and its implications for management and conservation. *Edited by* R.F. Stettler, H.D. Bradshaw, Jr., P.E. Heilman, and T.M. Hinckley. NRC Research Press, Ottawa, ON. pp. 223–246.

Newcombe, G. 1998. A review of exapted resistance to diseases of *Populus*. Eur. J. For. Pathol. **28**: 209–216.

Newcombe, G. 2000*a*. Inheritance of resistance to *Glomerella cingulata* in *Populus*. Can. J. For. Res. **30**: 639–644.

Newcombe, G. 2000*b*. First report of *Pestalotiopsis populi-nigrae* on poplar in North America. Plant Dis. **84**(5): 595.

Newcombe, G., Stirling, B., McDonald, S.K., and Bradshaw, H.D., Jr. 2000. *Melampsora* × *columbiana*, a natural hybrid of *M. medusae* and *M. occidentalis*. Mycol. Res. **104**: 261–274.

Ostry, M.E. 1986. Association of *Parkerella populi* with declining hybrid aspen in Wisconsin. Can. J. Bot. **64**: 1834–1835.

Ostry, M.E., and Anderson, N.A. 1998. Interactions of insects, woodpeckers, and hypoxylon canker on aspen. USDA For. Serv. Res. Pap. NC-331. 15 pp.

Ostry, M.E., Wilson, L.F., McNabb, H.S., Jr., and Moore, L.M. 1989. A guide to insect, disease, and animal pests of poplars. USDA For. Serv. Agric. Hndbk. 677. 118 pp.

Perala, D.A. 1984. How endemic injuries affect early growth of aspen suckers. Can. J. For. Res. **14**: 755–762.

Pinon, J., and Frey, P. 1997. Structure of *Melampsora larici-populina* populations on wild and cultivated poplar. Eur. J. Plant Pathol. **103**(2): 159–173.

Sarasola, A.A. 1944. Dos septoriosis de las alamedas Argentinas. Rev. Argent. Agron. **11**: 20–43 (abstract in Rev. Appl. Mycol. **23**: 365–366).

Zalasky, H. 1978. Stem and leaf infections caused by *Septoria musiva* and *S. populicola* on poplar seedlings. Phytoprotection, **59**: 43–50.

CHAPTER 9
Properties and utilization of poplar wood

John J. Balatinecz and David E. Kretschmann

Introduction

To satisfy the increasing demand for forest products, fast-growing trees such as poplar grown on managed plantations are being seriously considered for future supply needs. This chapter discusses the properties of poplar that make it appealing for a number of different utilization options and the characteristics that cause it to have some unique drawbacks. Utilization options for hybrid poplar will also be evaluated.

The genus *Populus* includes trees that are commonly called "poplar." The genus has a very wide distribution in North America, with various species inhabiting a triangular region stretching from Louisiana to Newfoundland to Alaska. *Populus* includes the species trembling aspen *(P. tremuloides)*, bigtooth aspen *(P. grandidentata)*, balsam poplar *(P. balsamifera)*, eastern cottonwood *(P. deltoides)*, and black cottonwood *(P. trichocarpa)*. All of these species can be characterized as fast-growing, moisture-loving, and shade-intolerant medium to large trees with a short life span. The most frequently used and widely distributed commercial species are trembling aspen, bigtooth aspen, and the hybrid poplars.

In the past, poplar trees were regarded as weed trees that needed to be removed from timber stands. Most harvested aspen was used for pulp, lumber, hardboard, and insulation board. With the introduction of waferboard and oriented strandboard (OSB), aspen utilization exploded, increasing threefold from 1975 to 1989 (Fig. 1, Youngquist and Spelter 1990), with a utilization level today almost four times greater than in 1975. The utilization level has increased so much that there is a concern that the aspen cut will exceed growth and the aspen supply will not be adequate to support the growing solid wood, composite, and paper industries

J.J. Balatinecz. Ph.D. Emeritus Professor, Wood Science, University of Toronto, 33 Willcocks Street, Toronto, ON M5S 3B3, Canada.
D.E. Kretschmann. P.E. Research Engineer, USDA Forest Service, Forest Products Laboratory, One Gifford Pinchot Drive, Madison, WI 53705, U.S.A.
Correct citation: Balatinecz, J.J., and Kretschmann, D.E. 2001. Properties and utilization of poplar wood. *In* Poplar Culture in North America. Part A, Chapter 9. *Edited by* D.I. Dickmann, J.G. Isebrands, J.E. Eckenwalder, and J. Richardson. NRC Research Press, National Research Council of Canada, Ottawa, ON K1A 0R6, Canada. pp. 277–291.

Note: The use of trade or firm names is for information only and does not imply endorsement by the U.S. Department of Agriculture of any product or service.

Fig. 1. Utilization of aspen solid wood products (converted to cords) from 1975 to 1997.

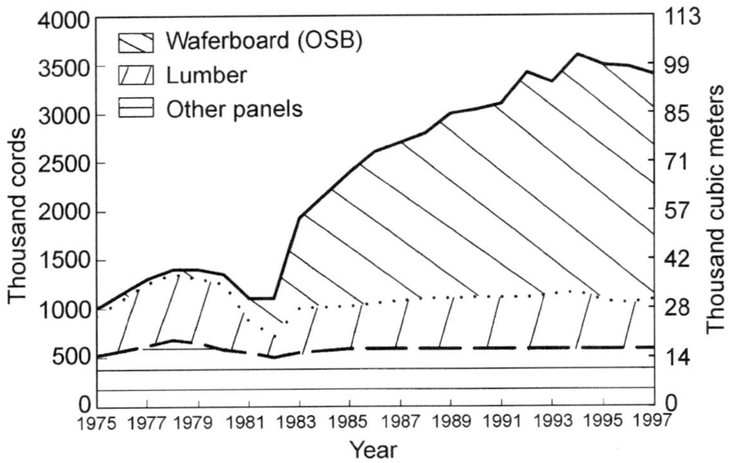

in the Lake States region (Youngquist and Spelter 1990). Poplar trees are no longer regarded as a weed species.

Hybrid poplars offer a possible solution to the potential shortage of native poplars. By hybridizing poplars, the positive characteristics of two fast-growing species can be combined to make an even more hardy and faster growing variety. Poplars are one of the more easily cloned woody species, which allows for greater availability of promising crosses.

Hybrid poplar plantations in Europe have for centuries been based on crosses of European black poplar *(P. nigra)* and the North American species eastern cottonwood *(P. deltoides)*. It was not until the 1930's that North Americans began discussing the potential of hybrid poplar plantations.

During the past 30 years, much work has focused on genetics of *Populus* species (Riemenschneider et al. 1996*a*, *b*) to develop improved hybrids. Numerous experimental studies have been planted across the Lake States, the Pacific Northwest, and Canada to investigate improved *Populus* clones. In most improvement programs, the focus has been on growth rate, form, adaptability, and disease resistance. Chemical properties have also been investigated (Dickson et al. 1974). They found large clonal differences among hybrid poplar clones in wood chemical composition. Additional emphasis has recently been placed on improving utilization properties of the material.

Wood from hybrids that have superior growth, improved form, greater adaptability, and improved fiber characteristics for paper may be less suited to solid wood processing than wood from either parent tree. The mechanical properties of particular hybrid poplar clones for structural lumber have been investigated (Holt and Murphey 1978; Bendtsen et al. 1981; Hall et al. 1982; Brashaw 1995; Kretschmann et al. 1999). This research has shown that the mechanical properties of these trees are comparable with similar native poplar species. Fast-growing clones, however, reach harvestable size more rapidly and therefore contain greater proportions of juvenile wood (juvenile wood is the first few years growth near the pith) compared with current aspen harvests. There is still a need for a better means of predicting the final mechanical properties of a hybrid based on its parent types.

Properties

The wood of all poplar species has relatively low density and diffuse porous structure. The average relative density (that is, specific gravity) of species grown in natural forests in North America is in the range of 0.30–0.39. The strength properties of poplars are relatively low. However, in bending strength and stiffness they compare favorably with common construction species such as spruce, pine, and fir (Table 1). Thus, poplar-based wood products (such as lumber, composite panels, and structural composite lumber products) can compete successfully with softwood-based products in the huge construction markets of North America. This is especially true for OSB, laminated veneer lumber (LVL), and structural composite lumber (such as parallel strand lumber (PSL), and laminated strand lumber (LSL)). These products are discussed in greater detail later in this chapter.

Standing poplar trees have high moisture content, typically about 100%, with only minor differences between sapwood and heartwood. Seasonal fluctuations exist, with summer values being somewhat lower than winter values. These high moisture levels make the wood suitable for cutting strands or wafers without

Table 1. Specific gravity and flexural properties of indigenous North American poplar species and a few selected softwoods.[a]

Species		Specific gravity	Modulus of rupture (MPa)	Modulus of elasticity (GPa)
Aspen	Trembling	0.35	35.0	5.9
	Bigtooth	0.36	37.0	7.7
Cottonwood	Eastern	0.37	37.0	7.0
	Black	0.31	34.0	7.0
Balsam poplar		0.31	27.0	5.2
Black spruce		0.38	42.0	9.5
Jack pine		0.40	41.0	7.4
Balsam fir		0.33	38.0	8.6

[a]The properties are for green wood. Source: Forest Products Laboratory (1999).

steaming. Considering their low density, poplar species have high volumetric shrinkage (11–12%). Poplars also have a high ratio of tangential to radial shrinkage, which is the main cause of form defects during drying (such as cupping and diamonding).

The volumetric composition of poplar wood is dominated by the relatively high proportion of fibers (53–60%), followed by vessel elements (28–34%), ray cells (11–14%), and a negligible proportion of axial parenchyma (0.1–0.3%) (Panshin and de Zeeuw 1980). The relative density of wood is most strongly influenced by the vessel-to-fiber ratio, as well as the diameter and wall thickness of fibers and vessel elements. Notably, the aspens (such as trembling and largetooth aspen), which have higher relative density than the cottonwoods (such as eastern and black cottonwood), are characterized by vessels with smaller diameters and fibers with slightly thicker walls.

Key characteristics of the cellular structure of poplars include short fibers with small cells compared with many other hardwoods. Poplars also have considerable shrinkage (Koubaa et al. 1998*a*, *b*). The fiber length, important for papermaking, can vary considerably by clone type and height in the tree (Fig. 2, Koubaa et al. 1998*b*).

The average length of vessel elements in mature poplar wood is in the range of 0.58–0.67 mm, whereas average fiber length ranges from 1.32 to 1.38 mm (Panshin and de Zeeuw 1980). The "paper-making fibers" (that is, tracheids) of softwoods are considerably longer (3–4 mm long). These fundamental differences in fiber length explain why softwood pulps have better properties, especially in tear, burst, and breaking length. On the other hand, the vessel elements of poplar significantly enhance the smoothness and opacity of sheets, making poplars well suited for printing papers.

The chemical composition of poplar wood is characterized by high polysaccharide content (approximately 80% holocellulose, made up of 50% cellulose and 30% hemicelluloses) and low lignin content (about 20% or less). Consequently, sulfate pulp yields are in the range of 52–56%, which is considerably higher than the 44–46% yields achieved for most softwoods. The extractives content of poplar toxic to fungi is low, which makes the wood susceptible to decay.

A number of factors make poplar appealing for growing as a forest crop. Poplar is a fast-growing, moisture-loving, full-sun-loving, large tree with a short life span. In addition, poplars can be cloned, so heritable traits can be improved more rapidly than in trees species that cannot be cloned. Hybrid poplars are specifically bred to improve disease resistance and improve the volume production and length of wood fibers for a particular site condition. Once established, poplars do not require replanting because they will reestablish themselves by their coppicing root system. After trees are harvested, the new crop emerges from suckers originating from roots or stumps. An increase in woody biomass produced after coppicing has also been reported (Phelps et al. 1987). Many plantations, however, will

Fig. 2. (*a*) Average trend line for fiber length as a function of annual ring breast height; (*b*) trend line for fiber length as a function of stem height for various annual rings (plots created from data of Koubaa et al. 1998*b*).

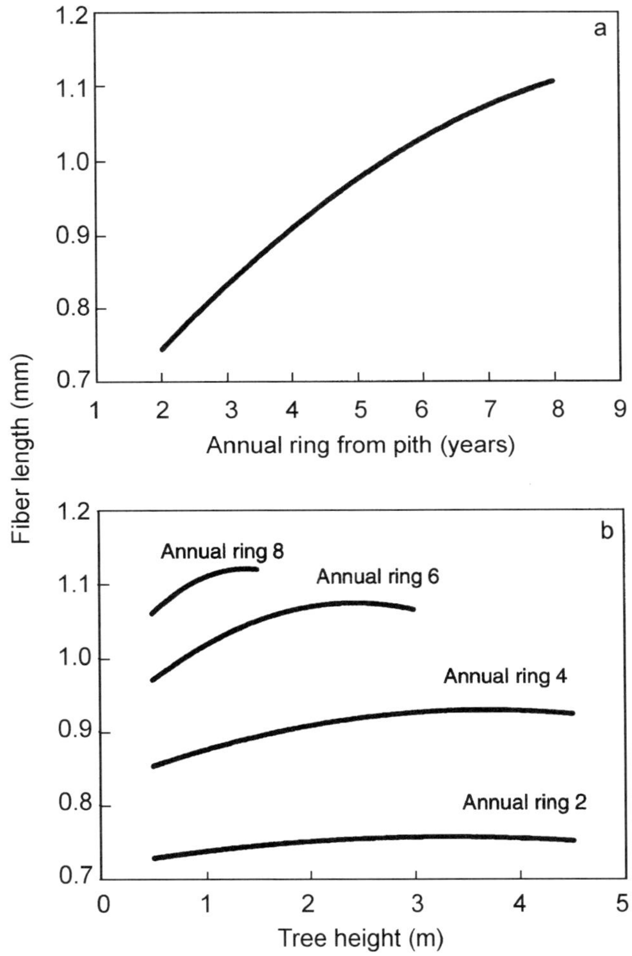

likely be replanted in spite of coppicing potential in order to capitalize on the availability of new and better clones.

Poplars also possess a number of characteristics that present challenges to utilization. Poplars in general are known to have stems with wet wood pockets, which makes uniform drying difficult. Poplar stems are susceptible to discoloration and decay. Discolored and decayed wood can be a major defect that limits the value of wood for certain finished solid-wood products such as cabinetry or moldings. Results suggest that the compartmentalizing capacity (the ability of a stem to restrict spread of discoloration or decay) of hybrid poplar trees is under genetic control (Eckstein et al. 1979). Eckstein and others found that the major difference between a strong compartmentalizing capacity, type A, and a weak compartmentalizing capacity, type C, was the vessel system (Fig. 3). Their results show that anatomical differences of the xylem existed between groups of several hybrid poplar clones. They further suggested that selecting clones based on conducting tissue may be an effective strategy for breeding trees for use in secondary lumber manufacture (molding and millwork, cabinetry).

Poplars also tend to be prone to considerable warping when dried. Special drying methods like "Saw–Dry–Rip" significantly decrease stability problems in final lumber products (Kretschmann et al. 1999).

Finally, poplars develop tension wood quite readily (Isebrands and Parham 1974; Parham et al. 1977; Holt and Murphey 1978). Tension wood is reaction wood that is formed on the upper sides of branches and the upper, usually concave, side of leaning or crooked stems. It is characterized anatomically by the lack of cell wall lignification and often by the presence of a gelatinous layer in the fibers. Holt and Murphey's work also showed that planting hybrid poplar trees at different spacings does not affect the physical, chemical, or anatomical properties of one hybrid poplar clone. Tension wood is also higher than normal in cellulose and ash content but lower in lignin and hemicellulose.

The machining, bonding, and finishing properties of poplars are quite good, making the wood well suited for a variety of conversion technologies, from sawing to veneer peeling and flaking. The relative density of a wood species determines the ideal peeling and flaking temperatures. For poplars, because of their low density, the projected "ideal" peeling and flaking temperature is in the range of 16–20°C, but acceptable results can be achieved between 7 and 30°C. Consequently, poplar requires little or no preconditioning because of both low density and high green moisture content. The low wood density is also an advantage in bonding of flakes and particles during the manufacture of composite panels because moderate pressure will bring the individual flakes and particles into intimate contact, thus ensuring a medium-density board with good strength. The pores in poplar wood are generally small enough to allow surface finishing without filler treatment.

Utilization options

Poplar wood is used for the manufacture of a large number and variety of primary and secondary forest products in North America. These products include pulp and paper, lumber, veneer and plywood, composite panels, structural composite lumber, pallets, furniture components, fruit baskets, containers, and chopsticks. The wood-using industries have turned more to indigenous poplar resources, both in the United States and Canada, during the past 30 years because of the rapidly escalating costs of softwood fiber and the broad availability of poplars at a much lower cost. The same utilization possibilities exist for hybrid poplars. Hybrid poplars will have the added advantage that they can be produced closer to markets and user industries and perhaps be genetically engineered with quality traits for

Fig. 3. Apparent relationship of the compartmentalization of discolored wood cambium (A, strong; B, weak; C, very weak) to mean vessel size for the number of growth rings (from Eckstein et al. 1979).

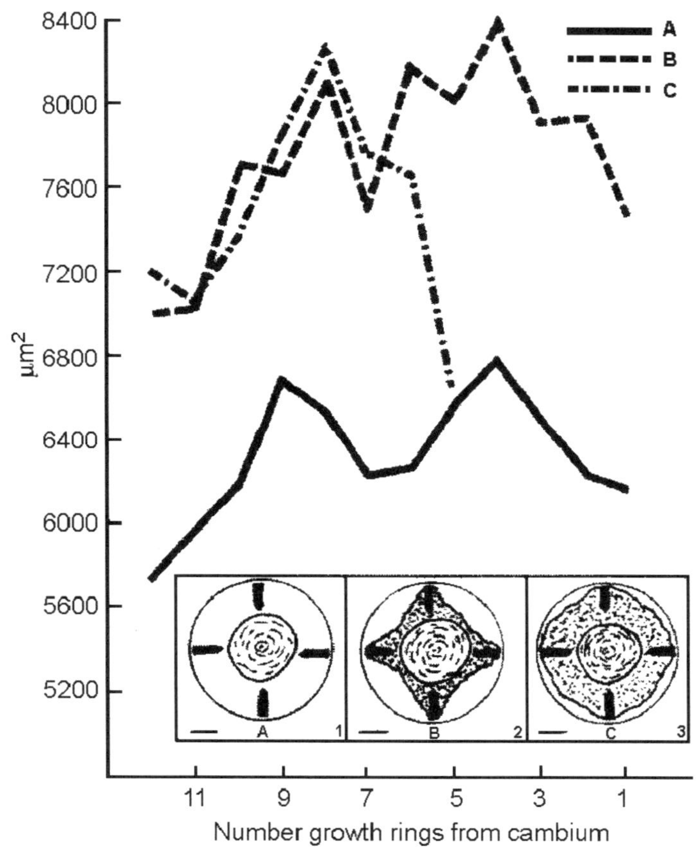

specific products. Hybrid poplars have the potential to be a major source of wood fiber in the next century.

Pulp and paper

One of the major uses of the indigenous poplar resource is for pulp and paper products. Poplar wood can be pulped by all commercial pulping methods. Mechanical, semi-chemical, kraft (or sulfate), and sulfite processes are now being used. Pulp mills designed for hardwood pulping can use up to 100% poplar.

The major uses of poplar pulp fall into three categories:

(*a*) Specialty paper products, such as napkins, tissues, towels, fine paper, paper board for packaging, and roofing felt;

(*b*) Building boards, such as insulation board, ceiling tiles, and hardboard;

(*c*) General purpose pulp, including groundwood, kraft, semi-chemical, and bleached sulfite.

Poplar kraft pulps are particularly well suited to fine paper manufacture because of inherently desirable properties such as excellent sheet formation, high opacity, good bulk, and good printability. Poplar kraft pulps can, however, have large amounts of fines and debris. Hybrid poplars have been shown to have properties similar to those of native poplar (Zarges and Neuman 1980).

Poplar kraft pulps are not made directly into paper; rather they are first blended with long-fibered softwood pulp to facilitate the development of wet web strength on fast-running paper machines. Poplar sulfite pulps (mainly from aspen) have been produced in North America for nearly 40 years. These pulps are used mainly in admixtures with bleached softwood kraft for the manufacture of high-quality printing and writing papers. Relatively recent technical advances in anthraquinone-catalyzed sulfite pulping are helping to increase pulp yield and strength properties of paper (Wong 1987).

The suitability of different poplar clones for paper making has been investigated by Labosky et al. (1983) and Law and Rioux (1997). Their work suggests that in general, hybrid poplars have a high proportion of very short cells (<0.2 mm) and high lignin content compared with trembling aspen. The chemithermomechanical pulps that were produced from this material were of acceptable quality but may require more energy for refining than aspen. Other research has shown that selection of a faster growing hybrid does not affect the fiber length (DeBell et al. 1998). Growth rate of short-rotation poplar can be increased without concern that fiber length may be negatively affected, which has led to investigation of the quality of a second rotation crop. No significant differences in total kraft pulp yields were found between first and second rotations (Labosky et al. 1983). Hybrid poplars, however, are not immune to the difficulties caused by the large amounts of fines and debris found in *Populus* tree pulps.

Lumber

Sawmills in the aspen belt of Canada and the United States have been manufacturing poplar lumber for the past several decades. Production volumes, however, remained quite low mainly because of economic factors. Due to small log diameters and the high incidence of decay, the average cost of sawing aspen is generally higher than for other hardwoods and is much higher than for softwoods. Markets for poplar lumber, while diverse, are also limited by strength properties and grade. The huge residential construction market in North America has always been dominated by softwood.

Stress-graded poplar dimension lumber (including native aspens, cottonwoods, and balsam poplar) is acceptable for framing applications. It is classified under the Northern Species group in Canada and as cottonwood or aspen in the United States and has lower allowable design stresses than the predominant S–P–F (Spruce–Pine–Fir) species group, thus putting poplar at a disadvantage. Non-stress-graded poplar lumber is used in a broad range of applications and products (pallets, crates, boxes, furniture components, lumber-core panels, and interior trim).

Davidson (1979), Hall et al. (1982), Hernandez et al. (1998), and Kretschmann et al. (1999) investigated the use of hybrid poplar for lumber. Hybrid poplar clones have mechanical properties similar to native cottonwood but slightly lower than those of aspen. Poplar clones have been shown to have a distinctive juvenile wood period. Strength-related properties increase with distance from the core. Therefore, the longer a poplar stand is allowed to mature, the more high-strength material will be available. Bendtsen et al. (1981) indicated that if hybrid poplar were harvested within the first few years after planting, it would have lower properties than cottonwood because of the high juvenile wood content. There is a weak but significant correlation between growth rate, density, and mechanical properties at breast height. Significant differences in mechanical properties have been found between various clones. This finding suggests the potential for genetically selecting higher strength clones specifically for structural lumber use.

The most significant drawback to the use of hybrid poplar for lumber is its tendency to warp. Special drying is required to improve the yield of material cut. Also, the relatively low mechanical properties of poplar species in general indicate that it is unlikely that poplars could ever compete with higher value commercial species like Douglas-fir and the Southern pine.

Composite products

A composite is any combination of two or more materials in any form and for any use. Composites take advantage of the beneficial characteristics of each component material and often have more useful properties than any of the constituents on their own. This broad definition includes a wide range of wood products, from composite panels (particleboard, fiberboard, waferboard, OSB, and even plywood) to composite lumber (LVL, PSL, LSL (Fig. 4), and composite wood I beams). Depending on the type of adhesive system used in the manufacture of these products, they may be destined for "interior" (generally decorative) or "exterior" (generally structural) applications. One of the many advantages of composites is that they use wood fiber more efficiently than sawn lumber. Typical conversion efficiencies are in the range of 52% for LVL, 64% for PSL, 76% for LSL (Nelson 1997), 80–90% for OSB, and 85–95% for particleboard and fiberboard, whereas that of sawn lumber is around 40%.

Fig. 4. Example of laminated strand lumber, TimberStrand® (Trus Joist, a Weyerhaeuser Business, Boise, ID), made from aspen.

Because OSB and other structural composites will probably play a dominant role in the future production and utilization of hybrid poplars, following is a brief of these products and their process technologies. The intent is to provide poplar breeders and growers with some relevant information about wood property and wood quality traits that will be desirable for composites.

The manufacture of all wood-based composite products requires the initial conversion of logs into smaller elements (veneers, wafers, strands, flakes, particles, and fibers) that are subsequently reassembled and bonded into efficient structural shapes and sizes (panels and lumber-like profiles) with appropriate adhesive systems. The end result is increased product yield, improved product uniformity, and enhanced fiber value.

Composites are more flexible and tolerant of wood property variation than sawn lumber, but their manufacture and properties are nevertheless influenced by wood quality. For example, the high incidence of tension wood has a negative impact on all wood machining operations involved in the manufacture of composites (veneer peeling, flaking, stranding, and sanding). Similarly, if the proportion of weak juvenile wood in the log furnish is high, the mechanical properties of the end product will be reduced. Thus, good communication between breeders, growers, and users of poplar wood will continue to be important.

The historical development of waferboard and OSB and their manufacturing processes was reviewed recently by Lowood (1997). The first OSB plant was built in Edson, Alberta, in 1982, and since then it has become a global industry. OSB is made from long, narrow strands (flakes) of wood. It is made of several layers of these strands, with the strands in each layer aligned parallel to one another. Adjacent layers of strands are perpendicular to one another, like the cross-laminated veneers of plywood. This unique mat construction gives OSB strength and stiffness properties equivalent to plywood.

The manufacture of OSB includes the following main process steps: debarking of logs, cutting and drying of strands, blending of strands with synthetic resin and wax, mat forming, hot pressing, and finishing. Markets and applications of OSB are broad and diverse. The product can be used in virtually any structural or nonstructural application where a large, thin, uniform, and dimensionally stable panel is needed. While its principal markets are in residential construction (floor, roof, and wall sheathing, siding), OSB is used in many other industries for numerous applications (including concrete forms, packaging and crating, chair seats and backs, hardwood floor core, stress skin panels, structural insulated panels, I-joist webs, pallets, shelving and display racks, and furniture frames). The continuing growth of the OSB industry will provide a major market opportunity for hybrid poplar wood from actively managed plantations.

From the family of structural composite lumber products, LVL and LSL deserve special mention because of the suitability of poplar wood for the manufacture of these products. Nelson (1997) provided an excellent overview of the manufacturing processes for LVL and LSL.

In a typical LVL process, rotary-peeled dried veneers coated with a waterproof structural adhesive are laid up into a thick sandwich with parallel grain orientation between all layers of veneer. The veneer sandwich is consolidated into a solid billet under heat and pressure. LVL is manufactured to either a fixed length using a stationary or staging press or to an indefinite length in a continuous press. The solid billets exiting the press are cut into desired cross sections and lengths. The process facilitates the placing of lower-grade veneers into the core and higher-grade veneers on the faces. Trials and tests undertaken in Canada have shown that LVL made from industrial-grade poplar veneer was nearly as strong and stiff as LVL made from Douglas-fir veneer. Veneer from poplar has shallower lathe checks than those that occur when dense softwoods are peeled. This reduces adhesive penetration and produces a more uniform bond. The advantages of LVL over sawn lumber include greater product uniformity, predictability of performance, broader range of available sizes, dimensional stability, and treatability.

In the manufacture of LSL, long and slender dry strands (more than 300 mm long) of wood are coated with waterproof adhesive and formed into a thick mat with parallel grain orientation between strands. The mat is consolidated into a thick (up to 140 mm thick), wide (2.4 m), and long (up to 15 m) billet, which is subsequently sawn into desired structural sizes. The principal advantage of LSL is the ability to convert small-diameter (even crooked) logs. Other advantages are similar to those of LVL.

Hybrid poplar has been successfully used as a substitute for aspen in a number of structural panel products in the United States and Canada (Geimer and Crist 1980; Zhou 1990; Roos and Brashaw 1993). Poplar clones have been shown to have similar properties to products made completely of aspen. However, a noticeable difference between specific gravities of aspen and hybrid poplar may require adjustments during processing. Spacing and rotation age can also have a significant impact on the quality of the resulting structural panel product.

Biomass for energy

During the energy crisis of the 1970's, hybrid poplar was seen as a savior for power companies. During this time, a shortage of petroleum and increased petroleum prices were responsible for a significant move towards establishing intensively cultivated and managed short-term rotation plantations of hybrid poplars. These were to be used as short-rotation crops that would feed co-generation power plants (Hansen et al. 1983). "Poplar farming" became a common term meant to describe plantations that would be harvested within 2–5 years.

The amount of energy that can be produced from short-rotation intensively cultured poplar trees is substantial. Caloric values for poplar biomass components have been reported to be between 4.3 and 4.8 kcal/g, which is equivalent to approximately 27 barrels of oil per hectare per year (Isebrands et al. 1979). The interest in biomass has been greatly reduced as energy costs have declined during the 1990's. However, recent volatility in energy prices has reinvigorated the discussion of poplars as an energy source. Poplars also offer a supplemental source of fuel that can be mixed with coal to reduce unwanted power plant emissions.

Other uses

Other utilization options have been investigated for hybrid poplars. For example, phytoremediation is an emerging use of poplars (see Chap. 6). Also, the protein-rich foliage of the poplar trees has been considered as a source of forage (Nelson et al. 1984; Morley and Balatinecz 1993). In addition, hybrid poplars have been used in secondary manufactured products like pallets, boxes, millwork, and hockey sticks.

Summary

Hybrid poplars are fast-growing, moisture-loving, full-sun-loving large trees that can be a rapid source of wood fiber. With the introduction of waferboard, oriented strandboard (OSB), and laminated strand lumber (LSL), aspen utilization has dramatically increased. Indigenous and hybrid poplars, however, present their own challenges, such as high discoloration potential, difficulty in drying, and high tension wood content. Further research is needed for improved selection of clones to ensure that desirable physical and mechanical qualities of poplar wood are produced for the anticipated site location and final utilization.

To date, the most promising utilization possibilities for hybrid poplar appear to be in the pulp and paper, laminated strand lumber, and structural panel industries. The mechanical properties of structural lumber cut from hybrid poplar will only compete in the stud market. Structural composites such as OSB and LSL made from indigenous and hybrid poplars can, however, be used effectively in other engineered structural applications (Fig. 5). Secondary manufacturing of clear wood

Fig. 5. Hybrid poplar demonstration center, Broadacres Nursery, Hubbard, OR. Ninety-eight percent of the materials in the center are from hybrid poplar.

cuttings of hybrid poplar may also offer possible uses of this material. A more idealistic goal for poplars is to make use of the rapid growth and deep root structure of these poplar clones for phytoremediation efforts to clean up contaminated sites.

In the past, poplar trees had been regarded as weed trees that needed to be removed from a stand. Changes in resource availability, advances in technology, and imagination have proven that there are many uses for indigenous and hybrid poplar clones that take advantage of their special properties.

References

Bendtsen, B.A., Maeglin, R.R., and Deneke, F. 1981. Comparison of mechanical and anatomical properties of eastern cottonwood and *Populus* hybrid NE-237. Wood Sci. **14**(1): 1–14.

Brashaw, B.K. 1995. Preliminary evaluation of hybrid cottonwood lumber mechanical properties. NRRI/TR-95/46. 4 pp.

Davidson, W.H. 1979. Hybrid poplar pulpwood and lumber from reclaimed strip-mine. USDA For. Serv. Res. Note NE-282. 2 pp.

DeBell, J.D., Gartner, B.L., and DeBell, D.S. 1998. Fiber length in young hybrid *Populus* stems grown at extremely different rates. Can. J. For. Res. **28**: 603–608.

Dickson, R.E., Larson, P.R., and Isebrands J.G. 1974. Differences in cell-wall chemical composition among eighteen three-year-old *Populus* hybrid clones. *In* Proceedings 9th Central States Forest Tree Improvement Conference, Oct. 10–11, 1974, Ames, Iowa. pp. 21–34.

Eckstein, D., Liese, W., and Shigo, A.L. 1979. Relationship of wood structure to compartmentlization of discolored wood in hybrid poplar. Can. J. For. Res. **9**: 205–210.

Forest Products Laboratory. 1999. Wood handbook — wood as an engineering material. USDA For. Serv. FPL-GTR-113.

Geimer, R.L., and Crist, J.B. 1980. Structural flakeboard from short-rotation intensively cultured hybrid *Populus* clones. For. Prod. J. **30**(6): 42–48.

Hall, R.B, Hilton, G.D., and Maynard, C.A. 1982. Construction lumber from hybrid aspen plantations in the central states. J. For. **80**: 291–294.

Hansen, E., Moore, L., Netzer, D., Ostry, M., Phipps, H., and Zavitkovski, J. 1983. Establishing intensively cultured hybrid poplar plantations for fuel and fiber. USDA For. Serv. Gen. Tech. Rep. NC-78. 8 pp.

Hernandez, R.E., Koubaa, A., Beaudoin, M., and Fortin Y. 1998. Selected mechanical properties of fast-growing poplar hybrid clones. Wood Fiber Sci. **30**(2): 138–147.

Holt, D.H., and Murphey, W.K. 1978. Properties of hybrid poplar juvenile wood affected by silvicultural treatments. Wood Sci. **10**(4):198–203.

Isebrands, J.G., and Parham, R.A. 1974. Tension wood anatomy of short-rotation *Populus* spp. before and after kraft pulping. Wood Sci. **6**(3): 256–265.

Isebrands, J.G., Sturos, J.A., and Crist, J.B. 1979. Integrated utilization of biomass, a case study of short-rotation intensively cultured *Populus* raw material. TAPPI. **62**(7): 67–70.

Koubaa, A., Hernandez, R.E, and Beaudoin, M. 1998*a*. Shrinkage of fast-growing hybrid poplar clones. For. Prod. J. **48**: 82–87.

Koubaa, A., Hernandez, R.E., Beaudoin, M., and Poliquin, J. 1998*b*. Interclonal, intraclonal, and within-tree variation in fiber length of poplar hybrid clones. Wood Fiber Sci. **30**(1): 40–47.

Kretschmann, D.E., Isebrands, J.G., Stanosz, G., Dramm, J.R., Olstad, A., Cole, D., and Samsel, J. 1999. Structural lumber properties of hybrid poplar. USDA For. Serv. Res. Pap. FPL-RP-573. 8 pp.

Labosky, R., Jr., Bowersox, T.W., and Blankenhorn, P.R. 1983. Kraft pulp yields and paper properties obtained from first and second rotations of three hybrid poplar clones. Wood Fiber Sci. **15**(1): 81–89.

Law, K.N., and Rioux, S. 1997. Five short-rotation poplar clones grown in Quebec: wood and papermaking properties. Timber management toward wood quality and end product. *In* Proceedings of the CTIA–IUFRO International Wood Quality Workshop, August 18–22, 1997, Quebec City, Canada. VII.19-VII.28.

Lowood, J. 1997. Oriented strand board and waferboard. *In* Engineered wood products, a guide for specifiers, designers, and users. *Edited by* Stephen Smulski. PFS Res. Foundation, Madison, WI. pp. 123–145.

Morley, P.M., and Balatinecz, J.J. 1993. Poplar utilization in Canada: past, present and future. For. Chron. **69**(1): 46–52.

Nelson, N.D., Sturos, J.A., Fritschel, P.R., and Satter, L.D. 1984. Ruminant feedstuff from the commercial foliage of hybrid poplars grown under intensive culture. For. Prod. J. **34**: 37–44.

Nelson, S. 1997. Structural composite lumber. *In* Engineered wood products, a guide for specifiers, designers, and users. *Edited by* Stephen Smulski. PFS Res. Foundation, Madison, WI. pp. 147–172.

Panshin, A.J., and de Zeeuw, C. 1980. Textbook of wood technology. McGraw–Hill, Inc., New York, NY. p. 722.

Parham, R.A., Robinson, K.W., and Isebrands, J.G. 1977. Effect of tension wood on kraft paper from a short rotation hardwood. Wood Sci. Technol. **11**: 291–303.

Phelps, J.E., Isebrands, J.G., and Teclaw, R.M. 1987. Raw material quality of clones grown under short-rotation intensively cultured *Populus* clones II. Wood and bark from first-rotation stems and stems grown from coppice. IAWA Bull. **8**: 182–186.

Riemenschneider, D.E., Netzer, D.A., and Berguson, B. 1996*a*. Intensive culture of hybrid poplars: what's new in Minnesota. *In* Proceedings, 1st Conference, Short Rotation Woody Crops Operations Working Group, September 23–25, 1996, Paducah, Kentucky. *Edited by* B.J. Stokes. Auburn University, Auburn, AL. pp. 53–58.

Riemenschneider, D.E., Stelzer, H.E., and Foster, G.S. 1996*b*. Quantitative genetics of poplars and poplar hybrids. *In* Biology of *Populus* and its implications for management and conservation. *Edited by* R.F. Stettler, H.D. Bradshaw, Jr., P.E. Heilman, and T.M. Hinckley. NRC Research Press, Ottawa, ON. pp. 159–181.

Roos, K., and Brashaw, B. 1993. Effect of hybrid poplar/aspen mixes on mechanical properties of oriented strandboard. NRRI/-96. 8 pp.

Wong, A. 1987. Review of pulping and paper making properties of aspen. Canadian and Alberta Forest Service. 81 pp.

Youngquist, J.A., and Spelter, H. 1990. Aspen wood products utilization: impact of the Lake States composites industry. *In* Proceedings, Aspen Symposium 89, July 25–27, 1989, Duluth, Minnesota. *Edited by* R.D. Adams. USDA For. Serv. NC-GTR-140. pp. 91–102.

Zarges, R.V., and Neuman, R.D. 1980. Kraft pulp and paper properties of *Populus* clones grown under short-rotation intensive culture. TAPPI. **63**: 91–94.

Zhou, D. 1990. A study of oriented structural board made from hybrid poplar. Physical and mechanical properties of OSB. Holz als Roh und Werkstoff. **48**(7–8): 293–296.

CHAPTER 10
The science of poplar culture

Donald I. Dickmann

Introduction

The art and science of poplar culture goes far back into antiquity. This long span of interest in poplars reached a culmination in the 20th century with an acceleration of scientific and developmental research with poplars as the subject. Part A of this book summarizes the state of this scientific knowledge at the 20th–21st century transition. Comparison of the current status with that presented in previous books on poplar culture in North America (e.g., Dickmann and Stuart 1983; DeByle and Winokur 1985; Boysen and Strobl 1991; Navratil and Chapman 1991) shows that we made considerable progress in the last few decades of the 20th century. The purpose of this chapter is to briefly highlight the accomplishments of scientists and practitioners who have worked on *Populus* material and in so doing have advanced poplar culture. I will end with a perspective on future poplar research. Part B of this book, which covers commercial clones and cultivars, also ends with a perspective on future poplar research, but authors Bradshaw and Strauss take a much more futuristic view. For a detailed perspective on the scientific foundation of poplar culture, the companion to this volume — *Biology of Populus* (Stettler et al. 1996), also published by NRC Research Press — is highly recommended.

Poplars in scientific research

In the January 2001 issue of *The Atlantic Monthly*, Jonathan Rauch made this provocative statement in an article on technology and the oil industry:

> Knowledge, not petroleum, is becoming the critical resource in the oil business; and though the supply of oil is fixed, the supply of knowledge is boundless. In every sense except the one that is most literal and least important, the planet's resource base is growing larger, not smaller. Every day the planet becomes less an object and more of an idea.

D.I. Dickmann. Department of Forestry, Michigan State University, East Lansing, MI 48824-1222, U.S.A.
Correct citation: Dickmann, D.I. 2001. The science of poplar culture. *In* Poplar Culture in North America. Part A, Chapter 10. *Edited by* D.I. Dickmann, J.G. Isebrands, J.E. Eckenwalder, and J. Richardson. NRC Research Press, National Research Council of Canada, Ottawa, ON K1A 0R6, Canada. pp. 293–308.

Although Rauch was writing about oil, the same message pertains to the subject of this book. The poplar resource is finite, in a strictly physical sense. But the ingenuity and ideas of scientists and practitioners of poplar culture, combined with the exceedingly tractable biology of the genus, presents limitless possibilities. The demand for poplars to be used for wood or energy, modification of the environment, enhancement of aesthetics, and sequestering carbon certainly will increase. But because we have a remarkable reservoir of practical and scientific knowledge of poplar trees, a reservoir that will continue to grow, current limitations to the application of poplar culture will be surpassed, one by one. Just as the oil industry has been revolutionized by the application of technical knowledge, so poplar culture has evolved in spectacular fashion as its foundation has become less intuitive and more scientific.

To give an indication of the reservoir of scientific and practical knowledge of poplars that currently exists, I searched the TREECD database (2001 CAB International, New York, NY), which includes all forestry, agroforestry, and wood products literature. This search revealed that from 1939 through the early days of 2001 over 20 500 publications on some aspect of poplars appeared (Table 1).

Table 1. Results of a computerized search on forest-tree genus names in the TREECD and *Biological Abstracts* publication databases for the years indicated inclusive.

	Number of publications	
Tree genus	TREECD (1939–2001)	*Biological Abstracts* (1980–2000)
Hardwoods:		
Quercus	29 483	7 295
Populus	**20 589**	4 169
Eucalyptus	17 326	5 222
Betula	13 602	3 541
Acer	10 836	3 341
Acacia	8 829	5 291
Conifers:		
Pinus	85 619	17 258
Picea	40 276	7 867
Abies[a]	12 574	2 753
Pseudotsuga	12 243	2 317
Larix	10 564	1 616

[a]A search on the genus name *Abies* also picks out all publications that reference *Picea abies*. Therefore, the data for *Abies* in this table were calculated by subtracting the number of hits for *Picea abies* from the number of hits for *Abies*.

Approximately 65% of these publications were printed in the last quarter of the 20th century. Note that while *Populus* is a well-studied genus, it comes in well behind *Quercus*, *Picea*, and, especially, *Pinus*. The *Biological Abstracts* database (2001 BIOSIS, Philadelphia, PA) includes the peer-reviewed literature of biological and medical journals, and it shows over 4000 basic science citations on *Populus* from 1980 through 2000. But again, *Populus* lags behind several other hardwood and conifer genera. Nonetheless, these searches indicate how the contributions of countless practitioners and scientists have laid the solid foundation upon which contemporary poplar culture rests.

Members of the genus *Populus* are very compliant subjects for integrated biological research and have served as model plants; they have become to forest biology what the white mouse is to medical research. In this regard, poplars can truly be considered unique. The status of poplars as model plants rests upon certain of their singular characteristics:

- A large reservoir of genetic variability resides in poplars, both at the species and section level.
- The genome of poplars is relatively small, making genetic manipulation and discovery easier than with other tree species.
- Clonal intra- and inter-specific hybrids can be readily produced by conventional breeding, and generation times are short (see Chap. 2). Furthermore, the pliability of poplars to methods of genetic biotechnology allows the engineering of unique or altered gene combinations. Thus, it is possible to engineer genotypes possessing certain predetermined traits for use in a particular experiment or application.
- Propagation of replicate plants using hardwood or softwood cuttings is remarkably simple and inexpensive, allowing genotype to be a constant instead of a variable in both laboratory and field experiments.
- Cell and tissue culture of selected genotypes is relatively easy, making micropropagation as well as basic genetic and physiological studies of selected genotypes possible (Fig. 1).
- Plants grow as well in controlled-environment growth rooms or greenhouses as they do in the field, creating opportunities for studying plants under a multiplicity of environmental conditions.
- The morphology of poplar plants is very uncomplicated and predictable, permitting a wide variety of experimental manipulations and measurement by numerous research instruments to be made.
- The rapid growth rate of poplars allows experiments to be compressed in time, saving money and accelerating accession of new knowledge.

Fig. 1. The ease with which poplar plantlets can be produced by controlled microculture of cells or tissue leads to many opportunities for basic genetic, biochemical, and physiological research. Photo by K.-H. Han.

This suite of characteristics is possessed by no other woody plant genus. Thus, in the future the rate of increase of fundamental knowledge of poplar biology likely will exceed most other woody plants.

It is on the genetic front that the accession of knowledge about poplars has exploded in the past decade, driven largely by the development of new methods in molecular biology. Linkage maps of the *Populus* genome, which identify the chromosomal location of quantitative trait loci (QTL) affecting a variety of traits, have been available for some time (Bradshaw 1996). The number of QTL identified and mapped has been continually growing. Parallel with the acquisition of this basic information has been the development of genetic engineering (transformation) techniques. Poplars are very compliant in this regard, making "molecular breeding" an efficient reality. What still is lacking is the base sequencing of the entire *Populus* genome, as has been done with humans, certain crop plants, and the model herb *Arabidopsis*. The technology of this field — genomics — has advanced spectacularly in recent years, and these advances, combined with the relatively small size of the *Populus* genome, make the goal of understanding the structure and function of the entire set of poplar genes an achievable reality.

Recently, an international group of scientists studying forest genetics proposed "The Poplar Genome Sequencing Project" (<http://poplar2.cfr.washington.edu/popseq/popseqsig.htm>, March 16, 2001). This proposal emphasizes that the sequencing effort should start immediately, that once completed the sequence should reside in the public domain, and that the model species for this effort be quaking aspen (*P. tremuloides*). Because genes among poplar taxa are virtually identical at the level of DNA, the sequence from aspen will be directly transferable to other members of the genus. Current estimates are that the entire aspen genome could be sequenced in approximately 2 years at a cost of approximately $65 million (Canadian). Although expensive, the opportunities for fundamental and practical research created by this effort would be phenomenal, raising the science of poplar culture to a new level.

Critical areas for poplar research

Poplar culture has entered the new millennium on a solid foundation of basic and applied knowledge, but we have no reason to become complacent. Many problems — both biological and socio-economic — remain to be solved or continue to reappear in new variants. Poplar culture can be likened to flying a high-performance aerobatic airplane at low altitude; if you relax your concentration and begin to daydream, you might find the airplane headed straight for the ground with no chance of recovery. Research, therefore, must move forward in a comprehensive and concentrated fashion. In most cases, an interdisciplinary approach will produce the most beneficial results. I will outline some of the critical areas for future poplar research and development below in a general way. Specific research and development objectives or scientific hypotheses in each area will have to be developed by experts in the various fields of inquiry.

Decline of natural populations of poplar

The human species has been altering the environment in which it lives since it first appeared on the earth. These alterations seldom have been beneficial to the totality of the earth's flora and fauna, but in the 19th and 20th centuries biotic upheavals of a magnitude heretofore unknown occurred. Countless species have been driven to extinction by over-exploitation or habitat degradation, while many others teeter on the brink. There are reasons not to be overly optimistic about this trend reversing in the future. Natural populations of several members of the genus *Populus* currently are on the decline due to human activities. In Europe, for example, populations of *P. nigra* are threatened by destruction of their natural riparian habitat and the incursion of genes from exotic poplars. Restoration of *P. nigra* ecosystems, therefore, is a high priority for European poplar ecologists.

North America is not immune from such problems. Although the aspens (*P. tremuloides* and *P. grandidentata*) certainly are in no danger of disappearing, in some parts of their natural ranges populations are on the wane (Fig. 2). For example, in Utah the area occupied by quaking aspen is less than half of what it used to be, and the decline continues (Bartos and Campbell 1998). In Michigan and Wisconsin, the area occupied by quaking and bigtooth aspen has decreased by 36% since its recorded peak in the 1930s (Cleland et al. 2001). These declines

Fig. 2. In some parts of its natural range, the area occupied by aspen is declining because old aspen clones are dying off and being replaced by later successional forest types or grassland.

are directly related to both an excess and a deficiency of disturbances. On the one hand, the disturbance created by burgeoning elk and deer populations and overgrazing by cattle has doomed aspen reproduction, particularly in many parts of the West. In parts of western Canada, land clearing for agriculture also is impacting aspen populations. On the other hand, the absence of a critical disturbance that has been part of the aspen environment for countless millennia — fire — has allowed aspen clones to senesce and die. Succession then proceeds to more tolerant forest types or grassland.

Reintroduction of fire via carefully prepared prescriptions, therefore, is a critical need in the Great Lakes Region and the Rocky Mountain West. But the escalating encroachment of human developments — the urban–wildland interface — places severe logistical and social impediments on this practice. Creative approaches to prescribed burning, therefore, need to be developed. Silvicultural harvesting of aspen (see Chap. 3) could make up some of the fire deficit. In parts of the boreal mixedwood region of Canada, for example, aspen actually is increasing in area because it regenerates more aggressively than its conifer associates following harvesting. But harvesting lags far behind the need in areas where aspen is on the decline. Many private landowners, for example, fail to embrace the concept of timber harvesting or fail to realize its biological and economic benefits. In the West, inaccessibility and lack of strong markets in certain areas provide further obstacles. Furthermore, aspen growing in parks, wilderness set asides, and roadless areas will continue to be off-limits to harvesting unless the controlling agencies adopt a more comprehensive view of ecosystem restoration. But none of these practices aimed at stimulating aspen suckering can be successful unless browsing by ungulate animals is controlled; this is another area where creative approaches are badly needed.

Even more alarming than the decline in aspen is the disappearance of riparian poplar ecosystems throughout western North America. In certain parts of the southwestern U.S. and the Sacramento Valley in California, for example, less than 5% of the original riparian forest remains (Braatne et al. 1996). A similar trend is developing in other parts of western North America. The major culprits in this decline are livestock grazing, water diversion for irrigation, domestic settlement, exotic plant encroachments, onstream reservoirs, channelization of river courses, agricultural clearing, gravel mining, and tree harvesting by humans and beavers. The need for conservation and restoration of riparian poplar ecosystems is easily recognized. The solution to this problem, however, is particularly intractable because it involves the integrated management of land, water, and biological resources as well as the coordination of multiple private, regional, state or provincial, and federal stakeholders (Braatne et al. 1996). As a result, little progress has been made.

Planting aspen clones

Although the technology for planting aspen seedlings has been well established for some time (Fig. 3), planting clonal aspen stands for timber, restoration, or amenity purposes remains an unsolved problem. That is not to say that clonal aspen propagules cannot be produced (Zsuffa et al. 1993). Short segments of aspen roots will produce plantable trees if raised in a conventional nursery. Young aspen suckers excised from the parent root or softwood tip cuttings can be induced to root under mist in a greenhouse or controlled environment room. Aspen plantlets formed under microculture also will root in the appropriate culture medium, and, after a growth period in containers in a greenhouse, can be outplanted. All of these methods, however, are inefficient, cumbersome, and expensive, precluding their use in large-scale planting programs. On the other hand, hardwood cuttings are a simple and cost-efficient method of plantation establishment. But cuttings from most members of section *Populus* simply will not produce adventitious roots because preformed primordia do not form in the bark of young stems as they do in *Aigeiros* and *Tacamahca* poplars. Without hardwood cuttings — or a comparable system — aspen planting will continue to be on the margin of North American poplar culture. Because aspens and white poplars can thrive on coarse-textured, droughty upland soils where their cottonwood and balsam poplar relatives do not, an entire niche for plantation-based poplar culture is currently unexploited.

Genetic biotechnology offers some promise for circumventing this bottleneck. Adventitious root regeneration appears to be under the control of major genes in

Fig. 3. Aspen seedlings can be planted successfully. But until a system of clonal propagation comparable to hardwood cuttings is developed, planting aspen will be on the margin of poplar culture in North America.

Tacamahaca and *Aigeiros* poplars, with a major QTL affecting root regeneration frequency (Han et al. 1994). If these rooting genes could be identified and cloned, could not elite aspen genotypes be transformed with them? Another approach currently being explored is the transformation of aspen genotypes with *rolB* or *aux* genes using the *Agrobacterium rhizogenes* vector (Hamill and Chandler 1994). The *rolB* gene causes an increase in sensitivity to the plant hormone auxin and a release of auxin from conjugates within the plant cells in which it is expressed. The *aux* gene regulates auxin biosynthesis. Because increased levels of endogenous auxin — or more specifically a shift in the auxin/cytokinin hormone ratio towards auxin — has been shown to promote lateral root initiation and emergence, this approach holds some promise for inducing rooting in recalcitrant aspen stem segments.

Diseases and insect pests

If there is one area of poplar culture where unrelenting diligence by the scientific community is absolutely essential, it is this one. The canker caused by *Septoria musiva*, the leaf rusts caused by *Melampsora* spp., and the cottonwood leaf beetle (*Chrysomela scripta*), in particular, continue to cause great grief to poplar growers (Fig. 4). Current approaches to dealing with these pests are summarized in Chaps. 7 and 8 of this book. My main point in this discussion is that there is no "silver bullet" that will eliminate any of them. Because fungi and insects evolve so rapidly in the face of selection pressures caused by an inherent resistance mechanism of the host or a pesticide, clones or control measures that are effective today will be useless in the future. Furthermore, the worldwide trade in consumer goods and exchange of poplar material almost assures that new exotic pest organisms will be introduced into North America. The recent discovery of established populations of the destructive Asian longhorn beetle (poplars are among its many

Fig. 4. Diseases and insect pests are part of the ecology of poplar stands, yet they can cause great economic loss. Continual scientific research is needed to cope with these problems, although no "silver bullet" solutions should be expected.

hosts) in several North American cities is but one example of what can and will happen. Many poplar people are amazed that the destructive bacterial canker so prevalent in European poplar plantations has not made its way to North America, but its introduction is an ever-present threat.

Foremost in the arsenal of options to deal with pest organisms is the rich genetic diversity of the genus *Populus*. Although it may be ephemeral, inherent resistance to most pests exists in any population of poplars. Through conventional breeding or genetic engineering, this resistance can be fixed in a genotype and expanded through cloning. Genetic biotechnology offers the additional option of creating novel gene combinations that are impossible to produce through breeding. For example, the gene from the bacterium *Bacillus thuringiensis* (Bt) that regulates synthesis of the endotoxin that kills many defoliating insects has been transferred to both crop plants and poplar clones. Chemical pesticides will continue to be an option for dealing with insect pests and certain diseases, but this is an expensive silvicultural option. Development of cost-effective and environmentally benign pesticides, therefore, has to be a priority for the future.

The role of genetically modified organisms

Among forest tree genera, *Populus* is the most amenable to the genetic engineering of novel genotypes (Klopfenstein et al. 1997). These transgenic or genetically modified organisms (GMO) have the potential to revolutionize poplar culture because they represent a means for solving intractable biological problems that conventional breeding cannot. Using either an *Agrobacterium* vector or direct gene transfer, poplar genotypes have been transformed with various gene constructs that convey toleration to herbicides, resistance to certain insects and diseases, toleration of air pollution, reproductive sterility, accelerated flowering, and lower lignin content of wood. At present, there seems to be few biological limits to what can be achieved through the practical application of biotechnological methods. And if the proposed *Populus* gene sequencing project discussed above succeeds — and there is every indication that it will — these biotechnological advances will move to an even higher level.

Notwithstanding this spectacular biological potential, serious environmental and social limitations to the use of genetically engineered clones in poplar culture exist. No public release or commercial deployment of a transgenic poplar has occurred at this writing, even though the first successful genetic transformation of *Populus* was reported in 1986 (Fig. 5). Millions of hectares of farmland in North America, however, are currently planted with transgenic agricultural crops. Scientists, politicians, environmental activists, and the public have raised concerns about the long-term effects of commercially deployed GMO on ecosystems (Marvier 2001). The most serious concern is the generation of severe weed problems through the transfer of herbicide-resistance genes from transgenic crops to related or unrelated weed species. Gene escape poses less of a threat for traditional agricultural row crops on this continent because most cultivated species are

Fig. 5. Most poplars are compliant subjects for genetic transformation, and many transgenic poplars are under test. However, no transgenic poplars have been released for commercial production.

exotic imports that have few wild relatives to interbreed with. The majority of poplar genetic improvement programs, however, are only one or two generations away from wild selections, so the risk of gene escape to native populations is very real. Sterility, either genetically engineered or in the form of natural triploids, will be required before transgenic poplars can be used in operational plantings.

Poplars have been transformed with the Bt endotoxin gene, thereby conferring resistance to certain insect defoliators; e.g., the cottonwood leaf beetle (see Chap. 7). Recent laboratory studies, however, have produced evidence that Bt toxin originating from GMO can have serious effects on non-target, beneficial insects and soil organisms. These short-term studies, while useful indicators, beg the question of the long-term ecological effects of GMO. A reflection on the history of exotic plant and insect pest introductions into North America can be particularly unsettling in this regard; many of these organisms did not manifest serious pestilence until decades after they were first introduced. Likewise, the now-hidden dark side of GMO may not show up for many years.

Poplar scientists doing genetic transformations or molecular breeding face an additional concern. Ecological terrorist groups — e.g., the Earth Liberation Front (ELF) — have sabotaged the facilities, records, and field trials of numerous agricultural biotechnologists, causing millions of dollars in damage. The work of several poplar researchers also have been targeted. These attacks will continue in the future, and perhaps intensify, and no poplar biotechnology program is immune from them. Following a recent attack on experimental poplar and other plantings in Rhinelander, Wisconsin (none of which, ironically, were transgenic), this statement appeared in an ELF press release. It inspires no optimism.

> What you don't realize is that we are infiltrating your ranks and gathering our own information. We could be that temporary worker typing away on your disgusting computer. We could even be your secretary or intern. We are everywhere and nowhere and we are watching.

Ecoterrorist groups tend to operate at night and in small, independent, anonymous cells. They are quite dedicated and very cautious, making them difficult to infiltrate or apprehend. Sooner or later, though, law-enforcement authorities will make inroads into these groups, and their cycle of terrorism will be disrupted. In the meantime, the watchword for poplar researchers is — carry on, but be careful!

The economic bottom line

Regardless of the sophistication and reliability of the underlying biological technology, if a commercial venture into poplar culture does not produce the requisite profit margin or is not cost effective, it will not persist. The trick, then, is to convert biological potential into economic viability. Because they grow so rapidly and can be established at a low cost, stands of poplar are considered to be as close to the ideal woody plant system as any that can be grown in the northern latitudes. Natural stands of aspen are the most economically viable of all; simply harvest existing clonal stands and then wait until the resulting sucker regeneration once again reaches commercial size (there are other options for aspen — see Chap. 4). In the case of natural stands, the cost of initial stand establishment — which for aspen may have occurred thousands of years ago — and most tending operations do not have to be paid. This is an economic advantage not to be underestimated.

Contemporary poplar plantations are a different matter. On the plus side, high yields and rotations of 10 years or less are very attractive economically. On the negative side, the costs of site selection and preparation, planting stock (including the cost of the genetic improvement program that produced it), planting, weed and pest control, fertilization, and other tending operations must be carried at compound interest until the end of the rotation. Furthermore, transportation costs of the final product to the mill site are high and limit the "working circle" of a plantation poplar enterprise. At best, the amortization of these costs will produce a return on the investment that meets the grower's expectations or exceeds the return on an alternative investment. For various reasons, however, the best is not always the outcome. In fact, the history of poplar culture in North America is replete with examples of commercial plantation programs that failed to meet the economic bottom line and were killed by company management.

In the past two decades, much attention has been given to the large-scale pulpwood plantings of poplar hybrids by forest industries in Oregon and Washington. These plantations are widely viewed as having passed the economic test. Recent reports from the Pacific Northwest, however, undermine this optimism. Both on the west and east sides of the Cascade Mountains, industries are thinning many of their plantations and extending the rotations to produce large-diameter logs for non-structural solid wood products — primarily plywood, furniture stock, and molding. There are several reasons for this shift in the silvicultural system, but paramount among them is the insufficient rate-of-return that pulpwood plantations have yielded in a soft worldwide pulp market. It remains to be seen whether the solid-wood rotations will do any better.

After an intensive economic analysis of forest plantations on a global scale, Roger Sedjo (1983) claimed that plantations were the highest-value use of marginal agricultural or idle lands (Fig. 6). Sedjo's conclusions, while supported by attractive present net value and internal rate of return data, have not been entirely borne out by the North American record of poplar culture. Poplar plantations indeed may be the highest-value land use, especially if their environmental benefits (see Chap. 6) are taken into account, but in many cases it's still not high enough. Economists, therefore, need to identify ways to cut costs, improve operational efficiencies, alter cultural scenarios, and assign proper economic value to environmental amelioration. Strauss and Grado (1997) identified land rent and taxes, biomass yields, and maintenance operations (site preparation, soil amendments, and herbicide and pesticide sprays) as the dominant factors determining production costs in poplar plantations. The improvement of growth rates — particularly on marginal agricultural land — and building in inherent resistance to herbicides, insect pests, and diseases through conventional breeding and biotechnology, therefore, must continue unabated.

Finally, taking advantage of government subsidies can greatly improve the bottom line for small non-industrial landowners. The Conservation Reserve (CRP) and Forest Stewardship Programs in the U.S.A., and the Prairie Farm Rehabilitation Administration (PFRA) in Canada, which administers the National Soil and Water Conservation Program and shelterbelt efforts, offer cost-sharing and other aids for tree planting. The recent large-scale plantings of poplars on CRP lands in Minnesota and the long-term viability of the poplar program at the PFRA Shelterbelt Centre in Saskatchewan seem to bear this out. But if poplar culture is to have a sustainable economic future, all of the approaches discussed above must be vigorously pursued.

Conclusions

I end this part of the book on a note of optimism. There certainly are problems that remain to be solved, and new problems will arise in the future, but the

Fig. 6. Poplar plantations may represent the highest use for idle or marginal agricultural land, but the economic bottom line for large industrial operations is not always favorable. Photo by L. Zsuffa.

application of scientific research can and will solve them. Thus, poplar culture has a bright future. As we proceed into the 21st century, poplars will continue to be an important part of the landscape mosaic of North America. Natural stands will dominate, as they always have, but plantations probably will increase in areal extent. Projections indicate that the demand for poplar wood fiber and biomass will expand. The amenity uses of poplars discussed in Chap. 6 also will become more prevalent, especially as more proven and tested clones become available. For many reasons, not the least of which is their charismatic quality, many people seem to love poplars and like to have them around. Poplars are, after all, *arbor populi*, the tree of the people. Therefore, those of us who grow and study them will never be looking for work.

Acknowledgements

Thanks to Toby Bradshaw and Andy David for critical reviews of an early draft of this chapter, and to Douglas McRae for information on aspen area in Canada.

References

Bartos, D.L., and Campbell, R.B. 1998. Decline of quaking aspen in the interior West — examples from Utah. Rangelands, **20**(1): 17–24.

Boysen, B., and Strobl, S. (*Editors*). 1991. A grower's guide to hybrid poplar. Ontario Ministry of Natural Resources, Brockville, ON. 148 pp.

Braatne, J.H., Rood, S.B., and Heilman, P.E. 1996. Life history, ecology, and conservation of riparian cottonwoods in North America. *In* Biology of *Populus* and its implications for management and conservation. *Edited by* R.F. Stettler, H.D. Bradshaw, Jr., P.E. Heilman, and T.M. Hinckley. NRC Research Press, Ottawa, Ontario, Canada. pp. 57–85.

Bradshaw, H.D., Jr. 1996. Molecular genetics of *Populus*. *In* Biology of *Populus* and its implications for management and conservation. *Edited by* R.F. Stettler, H.D. Bradshaw, Jr., P.E. Heilman, and T.M. Hinckley. NRC Research Press, Ottawa, ON. pp. 183–199.

Cleland, D.T., Leefers, L.A., and Dickmann, D.I. 2001. Ecology and management of aspen: a Lake States perspective. *In* Sustaining Aspen in Western Landscapes, Symposium Proceedings. *Compiled by* W.B. Shepperd, D. Binkley, D.L. Bartos, T.J. Stohlgren, and L.G. Eskew. USDA For. Serv. RMS-P-18. pp. 81–99.

DeByle, N.V., and Winokur, R.P. (*Editors*). 1985. Aspen: ecology and management in the western United States. USDA For. Serv. Gen. Tech. Rep. RM-119. 283 pp.

Dickmann, D.I., and Stuart, K. 1983. The culture of poplars in eastern North America. Dept. of Forestry, Michigan State University, East Lansing, MI. 168 pp.

Hamill, J.D., and Chandler, S.F. 1994. Use of transformed roots for root development and metabolism studies and progress in characterizing root-specific gene expression. *In* Biology of adventitious root formation. *Edited by* T.D. Davis and B.E. Haissig. Plenum Press, New York, NY. pp. 163–179.

Han, K.-W., Bradshaw, H.D., Jr., and Gordon, M.P. 1994. Adventitious root and shoot regeneration in vitro is under major gene control in an F_2 family of hybrid poplar (*Populus trichocarpa* × *P. deltoides*). For. Gen. **1**: 139–146.

Klopfenstein, N.B., Chun, Y.W., Kim, M.-S., and Ahuja, M.R. 1997. Micropropagation, genetic engineering, and molecular biology of *Populus*. USDA For. Serv. Gen. Tech. Rep. RM-GTR-297. 326 pp.

Marvier, M. 2001. Ecology of transgenic crops. Am. Sci. **89**: 160–167.

Navratil, S., and Chapman, P.B. (*Editors*). 1991. Aspen management for the 21st century. Forestry Canada, Northwest Region and Poplar Council Canada, Edmonton, AB. 174 pp.

Rauch, J. 2001. The new old economy: oil, computers, and the reinvention of the earth. The Atlantic Monthly **287**(1): 35–49.

Sedjo, R.A. 1983. The comparative economics of plantation forestry: a global assessment. Resources for the Future, Washington, DC. 161 pp.

Stettler, R.F., Bradshaw, H.D., Jr., Heilman, P.E., and Hinckley, T.M. (*Editors*). 1996. Biology of *Populus* and its implications for management and conservation. NRC Research Press, Ottawa, ON. 539 pp.

Strauss, C.H., and Grado, S.C. 1997. Economics of producing *Populus* biomass for energy and fiber systems. *In* Micropropagation, genetic engineering, and molecular biology of *Populus*. *Edited by* N.B. Klopfenstein, Y.W. Chun, M.-S. Kim, and M.R. Ahuja. USDA For. Serv. Gen. Tech. Rep. RM-GTR-297. pp. 241–248.

Zsuffa, L., Sennerby-Forsse, L., Weisgerber, H., and Hall, R.B. 1993. Strategies for clonal forestry with poplars, aspens, and willows. *In* Clonal forestry II. Conservation and application. *Edited by* M.R. Ahuja and W.J. Libby. Springer-Verlag, New York, NY. pp. 91–119.

CHAPTER 11
Poplar clones: an introduction and caution

Donald I. Dickmann and J.G. Isebrands

All that glisters is not gold —
Often have you heard that told.

William Shakespeare
The Merchant of Venice, Act II

Introduction

Among the important trees in North American silviculture, none are more closely identified with cloning than poplars (*Populus* spp.). In fact, it is rare that one encounters cultivated poplars that are not clones, whether they be extensively managed natural stands of aspen or intensively managed plantings of a pure species or hybrid. This propensity to clone must rank as one of poplar's most distinctive attributes, especially in the context of forestry (Fig. 1). No other group of trees except willows, which are not yet of great importance in North American forestry, can be cloned with such ease or are so widely managed in clonal stands than poplars. Many foresters working with conifers or high-value hardwoods such as black walnut, sugar maple, or red oak look with envy upon poplar foresters; if they could clone selected phenotypes of their species as readily as poplar people can, their approach to breeding and stand management would be revolutionized. In this chapter, we set the stage for the discussion that follows in Chaps. 12 and 13 of the commercial poplar clones and cultivars that currently are used in commercial plantings. In this context, we refer primarily to clones propagated by hardwood stem cuttings, principally in sections *Aigeiros* and *Tacamahaca* (Fig. 1). Members of section *Populus* — with the exception of certain *P. alba* genotypes and their hybrids — do not produce adventitious roots on hardwood stem cuttings, and they are excluded from this discussion.

D.I. Dickmann. Department of Forestry, Michigan State University, East Lansing, MI 48824-1222, U.S.A.
J.G. Isebrands. USDA Forest Service, North Central Research Station, Rhinelander, WI 54501, U.S.A.
Correct citation: Dickmann, D.I., and Isebrands, J.G. 2001. Poplar clones: an introduction and caution. *In* Poplar Culture in North America. Part B, Chapter 11. *Edited by* D.I. Dickmann, J.G. Isebrands, J.E. Eckenwalder, and J. Richardson. NRC Research Press, National Research Council of Canada, Ottawa, ON K1A 0R6, Canada. pp. 309–324.

Fig. 1. Fast-growing poplar clones in sections *Aigeiros* and *Tacamahaca* are planted widely throughout the world by using dormant hardwood stem cuttings. No other trees except willows can be clonally propagated with such ease.

The advantages of cutting-based clonal forestry or horticulture are two-fold:

- The favorable attributes of selected phenotypes in a wild or breeding population can be fully replicated and rapidly mass produced;
- Plantings of various kinds, from individual specimen trees to 1000-ha industrial plantations, can be easily and economically established.

The fact that a poplar cutting or "stick" planted in the soil will produce a tree identical to the one from which the cutting was taken — and in a hurry — continues to amaze people. In fact, we have been working with poplars for our entire career, and it continues to amaze us!

The good, the bad, and the ugly

Any discussion of poplar clones must begin with an important warning. To restate Shakespeare's phrasing of the ancient adage, every poplar clone that glitters is not necessarily gold. In fact, the ease with which poplar trees can be cloned is both a blessing and a curse. The blessing is obvious — easy establishment of plantings and, because of poplar's inherently rapid growth rate, the subsequent production of large trees in a short period of time. The curse is that any poplar phenotype that superficially looks good or grows rapidly in its early years can quickly be mass propagated and planted widely, regardless of its hidden defects. The most common and serious defect is extreme susceptibility to diseases and insect pests. There are millions of cuttings of these worthless clones available today that continue to be sold to trusting and unsuspecting buyers. Frankly, their elimination would be a boon to poplar culture. If ever the warning *caveat emptor* — let the buyer beware — applied, it's when buying poplar planting stock (Dickmann and Isebrands 1999). This warning applies whether you are a homeowner planting a few landscape trees or a windbreak, a phytoremediation company using poplars to clean up a contaminated site, a tree farmer growing firewood, a forest products company producing fiber for a paper mill, or a nursery establishing propagation beds for commercial cutting production.

Often, the marketing of poplar planting stock is accompanied by eye-catching advertising, aimed principally at small private landowners. Beginning in late winter, the Sunday supplements to newspapers are a good place to find these ads. Readers should not be taken in by the inflated claims and miraculous testimonials imbedded in these misleading advertisements. The more outlandish the ad, the more likely the material being marketed is of questionable worth. These marketers are either intentionally swindling innocent buyers or they are unaware of the problems that plague the clones they are selling. The solution to this problem is to buy poplar cuttings only from an established, reputable nursery, preferably one near your planting location. Insist that they supply only cuttings of clones proven to be hardy and resistant to the diseases and insects prevalent in your geographic area. Ask for proof; a responsible nursery should willingly comply. Forest industries producing cutting stock in their own nurseries must be absolutely sure that their deployment clones are disease resistant and adapted to their location.

If you own a nursery, stock only proven poplar clones (Fig. 2); your reputation will depend on it. These clones will vary from one area of the continent to the next; what is good in the Great Lakes Region, for example, may not be good in the East, the southern U.S., the prairies, or the Pacific Coast. Therefore, a large selection of clones will be required if you market over a large geographic area. In the same way that perennials are targeted for certain plant hardiness zones, individual poplar clones should be marketed only in regions where they have been proven successful. Besides the cultivars recommended in this book, forestry or horticulture departments at local universities, cooperative extension offices, USDA Forest Service and Canadian Forest Service research laboratories, or state and provincial natural resource agencies should be able to supply the necessary information on clones suitable for use in a particular locality.

Standing the test of time

Any clonal recommendation or large-scale deployment of poplar clones must be based on long-term field testing — this requirement is absolute! The early death of individual trees or the breakup of whole stands because poorly adapted clones were planted has been experienced by everyone from small landowners to large

Fig. 2. Nurseries that sell poplar planting stock only should grow clones that have withstood long-term testing, and these clones only should be marketed in localities where they have been proven disease resistant and hardy.

forest industries, bearing out our admonition. Long-term field testing is being carried on in many locations in North America, and the exact methods vary. We use our own experience as an example of how to approach this endeavor, but other approaches are equally valid. There are, however, three immutable attributes of valid field tests.

- Tests must be statistically designed so that reliable data can be recorded;
- Tests should be duplicated on a variety of sites and geographic locations, the extent of this duplication depending on how widely the clones under test will be deployed;
- Each round of field testing should run for 8–10 years, or to within a year or two of the anticipated rotation length.

Although such tests are expensive, they cannot be bypassed; that is just a fact of life with poplars as it is with other crops. Patience also is required. Certain disease problems — e.g., cankers — may not become evident for 5 or more years. Furthermore, the ranking of poplar clones in a test based on their growth rate can be unstable over time. Clones that start growing slowly sometimes move up in the rankings by the time rotation age is reached, whereas some fast early growers may slow down as they get older.

In Michigan, Wisconsin, Minnesota, and Iowa, long-term tests are being conducted by the North Central *Populus* Research Consortium, which is coordinated by the USDA Forest Service North Central Research Station and funded by the U.S. Department of Energy Bioenergy Feedstock Development Program in Oak Ridge, TN (Riemenschneider at al. 2001). The Consortium's objective is to identify promising new poplar clones for the North-Central Region. Although only a few proven clones are currently available in the region, interest in planting poplars is on the rise, as it is throughout North America. So the need for new clones is of critical importance. We began in 1995 with a planting of more than 60 different clones selected by poplar geneticists at Iowa State University, the University of Minnesota, and the USDA Forest Service North Central Research Station. The clones, which included numerous hybrids and native cottonwood (*P. deltoides*), were established in replicated two-tree plots at sites in the four states (Fig. 3). Similar plantings of 90 clones in 1997 and 55 clones in 2000 followed. Growth of these three plantings is measured yearly. Disease and insect incidence, tree form, and breakage from wind, ice, or snow also are scored. After 5–10 years, a small

Fig. 3. A long-term test of *Populus* clones just beginning its sixth growing season. Located in Ames, IA, this plantation is part of the North Central *Populus* Research Consortium test network. Individual clones are planted in two-tree row plots.

number of the "best" all-around clones from these plantings will be selected, and cuttings of them will be replanted in second-generation test plantations that are much larger in size. They will be grown for at least another 10 years in a variety of locations throughout the region. This approach is time consuming but absolutely necessary to identify the most reliable material for widespread release. Of the over 200 clones planted in first-stage tests, less than a dozen may eventually be released for commercial scale-up.

Two case studies amplify our insistence on long-term tests. The first case involves clone NM-6, a hybrid of *P. nigra* and *P. suaveolens* (formerly *P. maximowiczii*). This clonal cultivar has shown impressive early growth and good resistance to the stem canker caused by *Septoria musiva*[1] in initial test plantings. Because of its good early performance, NM-6 recently has been planted widely in the region. But now some problems are beginning to appear. Although the early growth of NM-6 is very rapid, it sometimes declines after trees reach 4 or 5 years of age on poorer quality sites. Other clones that start more slowly than NM-6 may eventually exceed it in size. In the spring of 1998, severe thunderstorms with strong winds swept through the Lake States, causing branch breakage and blow-down of NM-6 trees in some plantations. Other clones resisted the wind better. As a result, we now recommend this clone with less enthusiasm than we used to. Furthermore, we also believe that this clone is being over-planted in our region, a situation that always makes experienced poplar grower nervous.

The second case study reinforcing the importance of long-range field testing comes from the Pacific Northwest. Beginning in the 1960's, silviculture trials by university and government researchers in Washington and British Columbia established the feasibility of poplar culture in that region. The real breakthrough came with the release of exceptionally fast-growing Interamerican (TD) hybrids (*P. trichocarpa* × *P. deltoides* or *P.* ×*generosa*) developed by the poplar genetics and silviculture programs at the University of Washington and Washington State University. In the early 1980's, large-scale industrial plantations of Interamerican clones began to appear in the lower Columbia River Valley. Today, these plantations cover thousands of hectares (Zsuffa et al. 1996). The early Interamerican plantations in the Pacific Northwest were amazingly free of debilitating diseases — no stem cankers and no leaf rusts. But many experienced poplar growers from the East, where diseases are rampant, said, "Just wait!" Then, in 1991 the inevitable happened. An epidemic of defoliating leaf rusts caused by fungi in the genus *Melampsora* (see Chap. 8) devastated many of the Interamerican clones (Newcombe and Chastagner 1993). These rust diseases have since become endemic in the lower Columbia River plantations and have caused growers to completely

---

[1]*Septoria musiva* is an aggressive fungal pathogen common throughout the northeastern and mid-western regions of the U.S.A. and adjacent regions of Canada. Poplars in sections *Aigeiros* and, especially, *Tacamahaca* have been susceptible to *Septoria* in these regions, so resistance to stem cankers is a primary selection criterion (see Chap. 8).

abandon many of the early Interamerican clones. Breeding and selection of new Interamerican and other hybrid clones continues, however, with resistance to Melampsora rust now being the major selection criterion.

Wide-scale testing also can identify the appropriate geographic region for deployment of a particular clone or cultivar. Lombardy poplar (*P. nigra* 'Lombardy'), which is one of the most widely planted cultivars in the world, is a case in point. Experience in North America has shown that when planted in the West or the northern Great Lakes Region, the unique and striking columnar habit of Lombardy can be quite successfully used in landscape or windbreak plantings. In these regions, large, vigorous trees up to 1 m in diameter frequently are seen. In the southern parts of the Midwest, on the other hand, Lombardy succumbs to a canker caused by *Dothichiza populea* and a wetwood bacterium, usually before it reaches 10 years of age. In this region, Lombardy cannot be recommended and should not be sold by nurseries to people in the region.

The northwestern Interamerican hybrids are another striking case of adaptation to the environment of specific geographic regions. In the Northwest, where they were developed, Interamerican hybrids never have shown any incidence of Septoria stem canker, or any other canker diseases for that matter. They do show some leaf spotting caused by *Septoria*, however, although it doesn't infect the stems. But if the same Interamerican hybrids are planted in the Great Lakes Region, they quickly develop multiple Septoria stem cankers, which eventually kill the trees or cause stem breakage (Fig. 4). While their growth rate in the Lake States is just as impressive as it is in the Northwest, the cankers prevent Interamerican hybrids from being a viable option for wide-scale planting. The North Central *Populus* Research Consortium currently is testing clones created by backcrossing Interamerican hybrids to their *P. deltoides* parent in an attempt to capture the fast growth of the Interamericans and the canker resistance of *P. deltoides*.

The need for long-term and geographically diverse testing will continue as poplar culture for wood fiber and other uses becomes more widespread. New hybrids and pure-species clones are continually being created and selected by nurseries, university and government laboratories, genetics cooperatives, and forest industries. Many of these new clones are being developed using traditional breeding and selection protocols, but new molecular biotechnologies also are beginning to be employed (Han et al. 1996). For example, poplar clones have been genetically engineered to be tolerant to certain herbicides (Riemenschneider and Haissig 1991), although these clones have not yet been released for general deployment. Natural barriers to hybridization — e.g., between taxa in section *Populus* and those in other sections — may soon be broken down using advanced genetic methods. Eventually, highly "domesticated" poplar clones may be produced using these technologies (see Chap. 14). But the recent controversy over the use of genetically modified organisms (GMOs) puts their future in doubt and has turned a biological science breakthrough into a volatile political issue. Several poplar

Fig. 4. Incidence of certain poplar diseases is very localized in North America. For example, the canker caused by *Septoria musiva* is absent in plantations of *P. trichocarpa* × *P. deltoides* (TD) hybrids in the Pacific Northwest. But the same hybrids planted in the Great Lakes Region are devastated by the disease.

genetics programs even have been sabotaged by ecological extremist groups. The eventual outcome of this unfortunate situation still is in doubt. Even the "best-case" scenario indicates that the deployment of genetically engineered poplars will be held up for many more years.

Regardless of the glitter of the technology that produced them, new clones will require rigorous, long-term testing. This caveat is true of all human-made hybrids and especially of GMOs, which represent unique genetic combinations not found in nature that can be highly unstable and short-lived. These clones simply haven't been subjected to the millennia of natural selection that eliminates the unfit from native populations.

In our opinion, tree geneticists, breeders, and nurserymen should give as much attention to selecting genotypes of naturally occurring species than they do to developing fast-growing but often unstable hybrids. Eastern cottonwood, for example, has been neglected in the North Central Region and eastern Canada in favor of hybrid clones, which often have been proven unsatisfactory because of diseases. In fact, several of the pure cottonwood clones in the ongoing North Central *Populus* Research Consortium tests — non-cotton producing males — are performing better than the hybrids. The ability of cuttings of eastern cottonwood clones to root, however, is quite variable, and this trait will have to be a primary selection criterion. Many native aspen (*P. grandidentata* and *P. tremuloides*) clones also show superior growth, form, and disease resistance. If barriers to efficient clonal propagation of the aspens by using cuttings can be overcome using biotechnological methods, selected clones would readily find a commercial niche. Not only can proven clones of natural species eventually be released for commercial distribution, but they represent a more stable basis for genetic engineering for disease and pest resistance or other improved traits.

Naming the multitudes

When a particular poplar clone finally is released for commercial scale-up after surviving the long-term field-testing procedure, a whole new set of problems arises. At this stage, the clone can be considered a cultivar (derived from the words "cultivated variety"), and it should be given an epithet or a unique name; e.g., *P. trichocarpa* 'Fritzi Pauley' or *P.* ×*canadensis* 'Gelrica.' Epithets can be used alone, in combination with the Latin binomial (with or without the authority), or attached to the genus name; e.g., *Populus* 'Gelrica.' Epithets *never* are italicized and should be surrounded by single quotation marks — this practice clearly sets them off from the Latin name. (The abbreviation "cv." no longer is required before the cultivar epithet.)

The International Poplar Commission (IPC)[2] has established a formal procedure for registering commercial poplar cultivars and assigning them an epithet. This procedure is based on the "International Code of Botanical Nomenclature" and the "International Code of Nomenclature of Cultivated Plants." The IPC's published registration form includes, besides the requested epithet, information

---

[2]The International Poplar Commission (IPC), founded in 1947, is one of the statutory bodies of the United Nations Food and Agriculture Organization (FAO). The mandate of the Commission includes willows as well as poplars. The functions of the IPC are to study the scientific, technical, social, and economic aspects of poplar and willow cultivation; to promote the exchange of ideas and material among research workers, producers, and users; to arrange joint research programmes; to stimulate the organization of congresses and study tours; to report and make recommendations to the FAO Conference; and to make recommendations to national poplar commissions, through appropriate channels.

about the registrant and their organization, references, origin of the cultivar, morphological descriptions of adult trees and young plants in the nursery, and similar cultivars. According to the IPC, the epithet should be a non-Latinized "fancy name" of not more than 10 syllables or 30 characters, although it can consist of more than one word. Arbitrary sequences of letters and numbers — e.g., NM-6 — should be avoided because they can lead to errors in transcription. Often, epithets consist of a person's name, a geographic location, or some distinctive feature of the clone. Trademarks or company names are not permitted. The cultivar registration form and guidelines for its use, along with a thorough description of the rules for naming cultivars, are found on the IPC website (currently <www.agro.ucl.ac.be/efor/ipc/>).

Currently, the IPC register includes 332 cultivars from sections *Aigeiros*, *Tacamahaca*, and *Populus* (Table 1). Sections *Abaso*, *Turanga*, and *Leucoides* are not represented. The most common registered taxon by far is *P.* ×*canadensis* (Euramerican hybrids), comprising 44% of the total cultivars. *Populus deltoides*, *P. nigra*, *P.* ×*canescens*, *P. trichocarpa*, *P. alba*, and *P.* ×*generosa* also are well represented in the register. Hybrids make up 63% of the total, and pure species 37%.

Oddly, North American poplar breeders and growers are reluctant to assign proper epithets to commercial poplar cultivars and register them with the IPC, although they are by no means alone in that regard. For example, not one of the many Interamerican clones planted widely in the Pacific Northwest has been given a proper epithet and registered with the IPC. In contrast, 12 Interamerican clones have been registered by European workers. One problem is that the IPC registry and registration procedures are not well known among poplar people worldwide. Accessibility of the registry through the Internet will help. Also, overly busy poplar workers may not take the time to fill out the required paperwork, or the formal registration process may somehow offend the free spirit of North Americans and appear to be fettering. Europeans are much more attuned to this process, and the vast majority of currently registered cultivars originated in Europe. Our position is that each commercial cultivar developed in North America should be registered with the IPC and given a proper epithet, and this name should thereafter be consistently used. Because of proprietary issues, clones developed by forest industries probably will continue to be excluded from this process.

Once an epithet has been assigned, great care should be taken to maintain that identity through subsequent propagation. Unfortunately, clonal identifications often are mixed up, so growers must be diligent. To compound the problem, genetically distinct poplar clones of the same taxonomic lineage often show subtle differences in traits such as leaf shape, stem color, bark structure, and crown architecture. Stem cuttings of similar clones are especially difficult to tell apart. As a result, a supposedly pure clonal nursery bed or plantation often will

Table 1. Number of *Populus* cultivars by taxon registered with the International Poplar Commission as of October 2000.[a]

Taxon	Number of cultivars
P. alba	13
P. alba × *P. glandulosa*	1
P. alba × *P. grandidentata*	2
P. balsamifera	1
P. balsamifera × *P. trichocarpa*	1
P. ×*berolinensis*	3
P. ×*canadensis*[b]	145
(*P.* ×*canadensis*) × (*P.* ×*generosa*)	1
(*P.* ×*canadensis*) × *P. yunnanensis*	1
P. ×*canescens*	17
P. deltoides	47
P. deltoides × *P. suaveolens*[c]	2
P. deltoides × *P. yunnanensis*	1
P. ×*generosa*[d]	13
(*P.* ×*jackii*) × (*P.* ×*berolinensis*)	1
P. laurifolia	1
P. suaveolens[c] × *P. trichocarpa*	2
P. suaveolens[c] × (*P.* ×*berolinensis*)	2
P. nigra	40
P. nigra × *P. suaveolens*[c]	6
P. nigra × *P. trichocarpa*	2
P. ×*tomentosa*	6
P. tremula	2
P. tremula × *P. tremuloides*	3
P. tremuloides	1
P. trichocarpa	16
P. yunnanensis	1
Unknown	1
Total	332

[a]A database containing a description of all registered cultivars can be accessed on the International Poplar Commission website <www.agro.ucl.ac.be/efor/ipc/>.

[b]Euramerican hybrids (*P. deltoides* × *P. nigra*), also known as *P.* ×*euramericana*.

[c]Includes the former *P. maximowiczii* (see Chap. 1).

[d]Interamerican hybrids (*P. trichocarpa* × *P. deltoides*), also known as *P.* ×*interamericana*.

contain outliers of a different genotype. These outliers often show up distinctly in the spring because they have a different date of bud flushing.

Another name problem we find especially aggravating: some North American scientists and growers have a seemingly uncontrollable urge to rename clones in the interest of propriety or internal bookkeeping consistency. The cultivar *P.* ×*canadensis* 'Eugenei' is an extreme case in point (Fig. 5). This widely planted clone originated in 1832 at the Simon Louis Frères Nursery in Plantières (Metz), France, and it was imported to North America sometime in the 19th or early 20th century. Since then, it has acquired a plethora of names or cryptic codes including 'Norway,' 'Carolina,' 'Imperial Carolina,' NC-5326, and DN-34. All of these names can be found in the poplar literature, but they all refer to the same clone. This penchant to continually rename has no justification whatsoever; it only leads to confusion, especially among newcomers to poplar culture. As new names

Fig. 5. The *P.* ×*canadensis* cultivar 'Eugenei' originated in France but has been widely planted in eastern North America. This cultivar has been wrongly encumbered with other names such as 'Norway,' 'Carolina,'or 'Imperial Carolina,' which only causes confusion.

proliferate, loss of the true identity of a clone becomes a real possibility. If the IPC system of naming and registration of commercial poplar clones were used in the U.S. and Canada as it is in Europe — and adhered to — this problem would be minimized, although certainly not eliminated.

Adding to the long list of nomenclature problems in *Populus* is the tendency for variety, subspecies, or cultivar names to be converted to the specific name in a Latin binomial. So *P. deltoides* subsp. *monilifera* becomes incorrectly truncated to *P. monilifera*, or *P.* ×*canadensis* 'Robusta' becomes *P. robusta*. An experienced poplar person would recognize these errors, but those not acquainted with the complexities of poplar nomenclature would take the erroneous binomials at face value and — worse still — promulgate them. The times symbol (×) that designates hybrids also is sometimes dropped, and yet another "species" is created. The scientific literature of poplars is rife with such bogus names. The practices that lead to them are at their root simply the result of inattention to detail or a sloppy approach to an important scientific matter, and they should not be tolerated. There certainly are disagreements among scientists about poplar nomenclature and taxonomy (see Chap. 1), and in these cases some latitude must be given. But when Latin names are well established and accepted, a misnomer should be avoided at all costs.

New genetic technologies may provide a partial solution to establishing the identity of unknown or misidentified clones. DNA fingerprinting is becoming common place, and this method works as well in plants as it does in humans. Eventually, a clearing house could be established where DNA fingerprints of a wide range of registered poplar clones could be archived, including those released for commercial deployment. Unknown samples could be sent to the clearinghouse for testing and identification. Clones not recognized as registered cultivars or proven performers could then easily be targeted and discarded.

A clone is a clone — or is it?

Although sometimes viewed as unchanging, a clonal cultivar is not an absolutely stable genetic entity (Fig. 6). Somatic mutations in cells occur at rates that are non-trivial, and these mutations are passed on to subsequent clonal generations (Libby and Ahuja 1993). So clonal plants (ramets) derived from different parts of the mother plant (ortet) or another clonal ramet — e.g., different sections along the length of a stool shoot — may differ from one another at several genetic loci due to localized mutations. Although many of these mutations are not expressed, they will continue to accumulate within and among ramets as donor plants become older and through subsequent generations. At what point the expression of mutations becomes sufficient to warrant the reexamination and renaming of a clonal cultivar becomes a very pertinent issue. Notwithstanding the effects of mutations, genetic differences within a clonal population still are less by orders of magnitude than the differences among progeny produced through sexual reproduction.

Fig. 6. A clone is not necessarily a stable genetic entity because somatic mutations occur continually, sometimes producing a sport or chimera. Propagules of the same clone also may look different because of topophytic (position) effects.

Occasionally, somatic mutations will produce a "sport" (chimera) — usually a distinct shoot — that is markedly different in its phenotypic characteristics from the parent plant. In this case, a significant mutating event has occurred in the dividing cells of an apical meristem. Sports usually are so obvious that they can easily be eliminated from a clonal population, but if they are inadvertently propagated along with normal clonal tissue a confusing situation could result. On the other hand, the sport may have characteristics that warrant its propagation as a separate clonal entity. For example, a putative mutant of Lombardy poplar was grown extensively in Chile at one time because of its rapid height growth. Although it resembled the normal Lombardy in most traits, the mutant had a distinctly semi-evergreen character, retaining some leaves throughout the winter (Pryor and Willing 1965).

Topophytic effects sometimes also are obvious among the ramets of a clone. Topophysis refers to phenotypic effects that are related to the position on a plant where cuttings are taken. These effects may be the result of localized mutations, but more commonly they are due to the turning on or off of certain genes due to age or location on a donor plant. For example, if cuttings are taken in a vertical sequence from the top to the lower branches of a poplar tree, and then established in a plantation together, the bark characteristics among the ramets will vary. Cuttings from the top of the crown will produce trees that have relatively smooth bark, whereas cuttings from lower branches will produce trees with rough bark, mimicking the bark characteristics of the part of tree from which they were taken.

Final thoughts

Poplars and clones are inextricably linked because of the inherent biology of the genus and because cloning is so extremely convenient and useful to breeders and growers. Dormant hardwood stem cuttings will continue to be the major means by which *Aigeiros* and *Tacamahaca* poplars are planted. Perhaps in the near future major biotechnological breakthroughs also will allow plantations of the aspens and other members of section *Populus* to be established in this way. Clonal forestry or horticulture using poplars is a very dynamic enterprise. As old clones become susceptible to new diseases or new variants of old diseases and are surpassed by newly developed clones that have survived screening tests, they will fade from the scene. In the foreseeable future, this cycle will continue unabated. Thus, the majority of clonal cultivars described in Chap. 13 should be viewed only as temporary holders of the commercial label. If a revised edition of this book is published in the future, the list of commercial cultivars likely will be quite different.

Acknowledgements

The authors wish to recognize Don E. Riemenschneider and W.R. Schroeder for critical reviews of an earlier draft of the manuscript for this chapter.

References

Dickmann, D.I., and Isebrands, J.G. 1999. *Caveat emptor.* Am. Nurseryman, **189**(5): 60–65.

Han, K.-H., Gordon, M.P., and Strauss, S.H. 1996. Cellular and molecular biology of *Agrobacterium*-mediated transformation of plants and its application to genetic transformation of *Populus*. *In* Biology of *Populus* and its implications for management and conservation. *Edited by* R.F. Stettler, H.D. Bradshaw, Jr., P.E. Heilman, and T.M. Hinckley. NRC Research Press, Ottawa, ON. pp. 201–222.

Libby, W.J., and Ahuja. M.R. 1993. The genetics of clones. *In* Clonal forestry I: genetics and biotechnology. *Edited by* M.R. Ahuja and W.J. Libby. Springer-Verlag, New York. pp. 5–13.

Newcombe, G., and Chastagner, G.A. 1993. A leaf rust epidemic of hybrid poplar along the Lower Columbia River caused by *Melampsora medusae*. Plant Dis. **77**: 528–531.
Pryor, L.D., and Willing, R.R. 1965. The development of poplar clones suited to the low latitudes. Silvae Genet. **14**: 123–127.
Riemenschneider, D.E., and Haissig, B.E. 1991. Producing herbicide tolerant *Populus* using genetic transformation mediated by *Agrobacterium tumefaciens* C58: a summary of recent research. *In* Woody plant biotechnology. *Edited by* M.R. Ahuja. Plenum Press, New York. pp. 247–263.
Riemenschneider, D.E., Berguson, W.E., Dickmann, D.I., Hall, R.B., Isebrands, J.G., Mohn, C.A., Stanosz, G.R., and Tuskan, G.A. 2001. Poplar breeding and testing strategies in the north central U.S.: demonstration of potential yield and consideration of future research needs. For. Chron. **77**: 245–253.
Zsuffa, L., Giordano, E., Pryor, L.D., and Stettler, R.F. 1996. Trends in poplar culture: some global and regional perspectives. *In* Biology of *Populus* and its implications for management and conservation. *Edited by* R.F. Stettler, H.D. Bradshaw, Jr., P.E. Heilman, and T.M. Hinckley. NRC Research Press, Ottawa, ON. pp. 515–539.

CHAPTER 12
Keys to species and main crosses

James E. Eckenwalder

Simplified key to adults of wild poplar species worldwide, excluding hybrids

1. Leaves similar in color above and beneath 2

1. Leaves conspicuously lighter (often whitened) beneath 7

2. Leaves with a narrow colorless border; terminal buds present; winter buds conspicuously resinous; seed capsules opening with 4 valves; anthers blunt 3

2. Leaf margin green; terminal buds absent; winter buds not resinous; seed capsules opening with 2–3 valves; anthers minutely pointed . . . 4

3. Leaves without glands on the blade at its attachment to the petiole (leaf stalk); shoots often densely hairy; pedicels (flower stalks) less than 3 mm long . . . . *P. fremontii*

3. Leaves with up to 6 glands on the blade at its attachment to the petiole (leaf stalk); shoots usually hairless; at least some pedicels (flower stalks) more than 4 mm long . . . . *P. deltoides*

4. Leaves finely toothed; seed capsules opening with 2 valves. . . *P. mexicana*

4. Leaves coarsely toothed or untoothed; seed capsules opening with 3 valves 5

5. Leaves mostly untoothed, covered with mealy hairs . . . . . . *P. pruinosa*

5. Many leaves coarsely toothed, smooth 6

6. Leaves often broader than long . . . . *P. euphratica*

J.E. Eckenwalder. University of Toronto, Department of Botany, 25 Willcocks Street, Toronto, ON M5S 3B2, Canada.
Correct citation: Eckenwalder, J.E. 2001. Keys to species and main crosses. *In* Poplar Culture in North America. Part B, Chapter 12. *Edited by* D.I. Dickmann, J.G. Isebrands, J.E. Eckenwalder, and J. Richardson. NRC Research Press, National Research Council of Canada, Ottawa, ON K1A 0R6, Canada. pp. 325–329.

6. Leaves usually longer than broad . . . . *P. ilicifolia*

7. Petiole flattened side-to-side at the junction with the blade 8

7. Petiole round or slightly flattened top-to-bottom at the junction with the blade 15

8. Leaves with a narrow colorless border; winter buds sticky with reddish resin . . . . *P. nigra*

8. Leaf margin green; winter buds without sticky resin 9

9. Leaves with persistent white woolly hair underneath and on the petiole. . 10

9. Leaves, if hairy, not persistently white woolly 11

10. Leaves on long shoots weakly 3-lobed to strongly 5-lobed; flower bracts slightly toothed . . . . *P. alba*

10. Leaves on long shoots not lobed; flower bracts not toothed. . . . *P. monticola*

11. Leaf tip drawn out into a long point. . . . . *P. adenopoda*

11. Leaf tip short 12

12. Tips of teeth on leaves rounded. 13

12. Tips of teeth on leaves pointed 14

13. Leaves on spur shoots mostly with more than 15 teeth on each side . . . . *P. tremuloides*

13. Leaves on spur shoots mostly with fewer than 15 teeth on each side . . . . *P. tremula* s. l.

14. Leaves on spur shoots mostly with more than 15 teeth on each side . . . . *P. sieboldii*

14. Leaves on spur shoots mostly with fewer than 15 teeth on each side. *P. grandidentata*

15. Floral bracts fringed with hairs right at the edges, like eyelashes. 16

15. Floral bracts without marginal hairs but sometimes so finely cut as to be hairlike 18

16. Leaves densely woolly beneath. 17

16. Leaves not woolly beneath . . . . *P. gamblei*

17. Deciduous in summer; seed capsules 4–6 mm long, hairy . . . . *P. simaroa*

17. Deciduous in autumn; seed capsules 2–3 mm long, hairless . . . . *P. guzmanantlensis*

18. Leaves woolly beneath, at least at first; winter buds not very resinous . . 19

18. Leaves not woolly beneath; winter buds usually heavily resinous 21

19. Most leaves 20–30 cm long. . . . . *P. lasiocarpa*

19. Most leaves 10–20 cm long 20

20. Leaf tips pointed . . . . *P. glauca* s. l.

20. Leaf tips rounded . . . . *P. heterophylla*

21. Most leaves 3 times as long as wide or more 22

21. Most leaves 2 ½ times as long as wide or less 23

22. Twigs hairy; leaves hairy along the veins beneath . . . . *P. laurifolia*

22. Twigs and leaves hairless . . . . *P. angustifolia*

23. Leaves broadest at or beyond the middle. 24

23. Leaves broadest below the middle 25

24. Leaf base wedge-shaped or narrowly rounded . . . . *P. simonii* s. l.

24. Leaf base broadly rounded or heart-shaped . . . . . . *P. suaveolens* s. l. (including *P. maximowiczii*)[1]

25. Seed capsules opening with 2 valves . . . . *P. balsamifera*

25. Seed capsules opening with 3–4 valves 26

26. Seed capsules densely hairy . . . . *P. trichocarpa*

26. Seed capsules generally hairless 27

27. Leaves mostly about as broad as long . . . . *P. ciliata*

27. Leaves mostly longer than broad 28

28. Leaves with a short-pointed tip . . . . *P. szechuanica*

29. Leaves with a long-pointed tip . . . . *P. yunnanensis*

---

[1]See Chap. 13 (p. 333) in the section on "Other intersectional hybrids" for a discussion of the status of *P. maximowiczii* and *P. suaveolens*.

Simplified key to native poplar species and commonly cultivated species and hybrids in North America north of Mexico

1. Leaves woolly beneath (at least on the veins) and on the petiole near its base and tip 2

1. Leaves not woolly 5

2. Leaves mostly less than 8 cm long 3

2. Leaves mostly more than 8 cm long 4

3. Leaves of long shoots strongly 5-lobed, white-woolly beneath. . . . *P. alba*

3. Leaves of long shoots irregularly toothed, grayish woolly beneath . . . . *P.* × *canescens*

4. Leaves coarsely toothed or wavy-edged, densely white-woolly beneath . . . . *P.* × *tomentosa*

4. Leaves finely toothed, sparsely woolly . . . . *P. heterophylla*

5. Trees with a narrow crown of upright branches 6

5. Trees with a broader crown of spreading branches 8

6. Winter buds scarcely resinous . . . . *P. tremula* (cv. Erecta)

6. Winter buds very resinous. 7

7. Leaves widest at or beyond the middle . . . . . . *P. simonii* (cv. Fastigiata)

7. Leaves widest at or below the middle . . . . *P. nigra* (cv. Italica)

8. Petiole (leaf stalk) strongly flattened side-to-side at its junction with the blade 9

8. Petiole (leaf stalk) round, or flattened top-to-bottom, or a little narrower than high at its junction with the blade 15

9. Winter buds scarcely resinous 10

9. Winter buds conspicuously resinous 12

10. Leaves of short shoots with more than 15 teeth on each side, the largest teeth up to 1 mm deep . . . . *P. tremuloides*

10. Leaves of short shoots with fewer than 15 teeth on each side, the largest teeth more than 2 mm deep 11

11. Leaves of short shoots with sharp-pointed teeth . . . . . . *P. grandidentata*

11. Leaves of short shoots with rounded teeth . . . . *P.* × *wettsteinii*

12. Leaves of short shoots with 2 or more glands at the base near the petiole . . . . *P. deltoides* (subspp. *deltoides* and *monilifera)*

12. Leaves of short shoots with 0–1 glands at the base near the petiole 13

13. Leaves of short shoots with more than 15 teeth on each side, longer than wide, at least some with glands at the base near the petiole . . . . *P.* × *canadensis*

13. Leaves of short shoots with fewer than 15 teeth on each side, about as wide as long or wider, entirely without glands 14

14. Leaves of long shoots wider than long . . . . *P. deltoides* (subsp. *wislizeni)*

14. Leaves of long shoots about as wide as long . . . . *P. fremontii*

15. Most leaves at least 3 times as long as wide . . . . *P. angustifolia*

15. Most leaves less than 3 times as long as wide 16

16. Leaves with a narrow, colorless border 17

16. Leaf margin green. 20

17. Leaves about as broad as long . . . . *P.* × *jackii*

17. Leaves longer than broad 18

18. Leaves mostly less than 8 cm long . . . . . *P.* × *berolinensis* and offspring

18. Leaves mostly more than 10 cm long 19

19. Leaves broadest well below the middle . . . . *P.* × *generosa*

19. Leaves broadest near the middle . . . . *P. nigra* × *P. maximowiczii*

20. Seed capsules opening with 2 valves, hairless . . . . *P. balsamifera*

20. Seed capsules opening with 3–4 valves, at least a little hairy 21

21. Leaves broadest well below the middle, seed capsules densely hairy . . . . *P. trichocarpa*

21. Leaves broadest near the middle, seed capsules sparsely hairy . . . . *P. maximowiczii* × *P. trichocarpa*

CHAPTER 13
Descriptions of clonal characteristics

James E. Eckenwalder

General characteristics of main crosses

Euramerican hybrids, *Populus* × *canadensis* Moench (syn. *P.* × *euramericana* Guinier, nom. illeg.)

Populus × *canadensis* Moench includes all natural and artificial hybrids between any races of *P. nigra* L. and *P. deltoides* Marsh., including backcrosses and advanced generation hybrids. These were the first intercontinental poplar hybrids to be used in plantation culture and among the first tree hybrids of any kind to be so used. They first attained prominence in the 18th century, when *P. canadensis* was named, and many individual combinations of different races of the two parents were given species names in the 19th and early 20th centuries. Botanical nomenclature subsumes all combinations between any two species under a single epithet, the earliest described one that conforms to the rules of nomenclature, in this case *P.* × *canadensis*. The frequently used *P.* × *euramericana* Guinier has no botanical standing and is considered a technically illegitimate renaming of *P.* × *canadensis*, although there is no impediment to referring to these clones informally as Euramerican hybrids.

The numerous Euramerican hybrids as a group share a typical set of characteristics that set them apart from and in between the parent species. They are vigorous and many possess a relatively narrow crown of upwardly angled branches inherited through the Lombardy poplar. Most have grayish mature bark and first-year twigs with an orange cast, contrasting with the darker bark and redder twigs of *P. nigra* and the browner bark and yellowish or tan twigs of *P. deltoides*. Many of the hybrids inherit 5-angled twigs from *P. deltoides*, and such twigs are more prominently expressed in the most vigorous shoots. On the other hand, the reddish resinous winter buds are derived from *P. nigra* and contrast with the browner buds and paler resin of *P. deltoides*. The leaves usually readily distinguish Euramerican hybrids from their parents, but this applies especially to the

J.E. Eckenwalder. University of Toronto, Department of Botany, 25 Willcocks Street, Toronto, ON M5S 3B2, Canada.
Correct citation: Eckenwalder, J.E. 2001. Descriptions of clonal characteristics. *In* Poplar Culture in North America. Part B, Chapter 13. *Edited by* D.I. Dickmann, J.G. Isebrands, J.E. Eckenwalder, and J. Richardson. NRC Research Press, National Research Council of Canada, Ottawa, ON K1A 0R6, Canada. pp. 331–382.

spring flush of leaves that overwinter in the bud (preformed, or early, or short- or spur-shoot leaves) rather than those that develop during the course of the growing season (neoformed, or late or long-shoot leaves), which can be difficult to distinguish from those of *P. deltoides*. Both leaf forms of *P. deltoides* (at least in those races that have contributed to *P.* × *canadensis*) usually have 1–3 pairs of glands at the base of the leaf on the upper side near the attachment to the petiole (basilaminar glands), and these glands are completely lacking in *P. nigra*, while Euramerican hybrids may have a single gland of the pair. The teeth on the preformed leaves of Euramerican hybrids are fairly coarse, intermediate between the very coarse toothing of *P. deltoides*, with the largest teeth up to 7 mm deep, and the much finer toothing in *P. nigra*. The base of the preformed leaves can be particularly diagnostic. Euramerican hybrids, like cv. Eugenei, often have a toothless, broadly wedge-shaped base, with an abrupt shoulder linking it to the toothed portion of the margin.

Flowering characteristics also set Euramerican hybrids apart from the parent species. Female clones have seed capsules opening with 2–3 valves, unlike the 4 valves of *P. deltoides* and uniformly 2 valves of *P. nigra*. Although the two parent species have different modal numbers of stamens in the male flowers, they both can have about 30, as do many hybrids. However, the longer pedicels (flower stalks), at least some of them longer than about 4 mm, distinguish Euramerican hybrids from *P. nigra*, while the bud color distinguishes them from *P. deltoides*.

Interamerican hybrids, *Populus* × *generosa* A. Henry (syn. *P.* × *interamericana* van Broekh.)

Populus × *generosa* A. Henry includes all natural and artificial hybrids between any races of *P. deltoides* Marsh. and *P. trichocarpa* Torr. & A. Gray, including backcrosses and advanced generation hybrids. The increasingly popular interamerican hybrids are the most vigorous of all the intersectional hybrids between balsam poplars in sect. *Tacamahaca* and cottonwoods in sect. *Aigeiros*, at least when the cottonwood parent is one of the southern cottonwood races (*P. deltoides* subsp. *deltoides*). First generation hybrids are fairly typical intersectional hybrids but favor balsam poplar characteristics a bit more than do the other combinations involving eastern cottonwood, *P.* × *acuminata* Rydberg (*P. angustifolia* × *P. deltoides*) and *P.* × *jackii* Sargent (*P. balsamifera* × *P. deltoides*). In particular, although the preformed leaves are a little more coarsely toothed than those of *P. trichocarpa*, the teeth are usually not as large as those of *P.* × *acuminata* or *P.* × *jackii*. The leaves are larger than in the other hybrid combinations, and the neoformed leaves of terminal shoots on saplings may be 15 cm or more long, contributing to the scientific name and rapid growth of *P.* × *generosa*. This rapid growth and good form have promoted interamerican hybrids to a position of pre-eminence in poplar plantation culture in Pacific North America and central and western Europe. Backcrosses and F_2 hybrids vary greatly in vigor and characteristics, so few of these clones are in commercial production compared to F_1 hybrids. As clearly documented by Professor Reinhard Stettler (College of Forest Resources, University of Washington, Seattle) and collaborators in North America and by other workers in Germany and elsewhere, the heterosis of the F_1 hybrids is due, in part, to the combination of the rapid height growth of *P. trichocarpa* with the rapid diameter growth of *P. deltoides*. Perhaps no other tree in temperate climates can match the biomass accumulation capability of interamerican hybrids grown under favorable conditions. Unfortunately, these hybrids cannot be grown commercially in eastern North America because they are devastated by *Septoria* canker.

Other intersectional hybrids

For practical purposes in North America, this group consists exclusively of hybrids between balsam poplars in section *Tacamahaca* and cottonwoods or black poplar of section *Aigeiros*. Other intersectional combinations are possible and have been created in breeding programs, and some are even found naturally, such as *P. deltoides* (sect. *Aigeiros*) × *P. heterophylla* (sect. *Leucoides*), but all of these are rare and none shows much promise for timber production under ordinary conditions. On the other hand, numerous spontaneous hybrids involving both native and cultivated balsam poplars and cottonwoods are found in North America wherever pairs of these species are found growing together. For example, in Quebec, where *P.* × *canadensis* has long been widely cultivated, it hybridizes naturally with *P. balsamifera* to form *P.* × *rollandii*. All of the possible intersectional hybrids among the native species have been found in the wild except for *P. balsamifera* × *P. fremontii*, whose parents do not grow together naturally.

However, even the *Tacamahaca*/*Aigeiros* intersectional hybrids other than *P.* × *generosa*, when cultivated, are found primarily in horticultural or amenity plantings, such as shelterbelts, and have little use in plantation culture. This includes *P.* × *jackii* (*P. deltoides* × *balsamifera*) and its derivatives and *P.* × *berolinensis* (*P. nigra* × *P. laurifolia*) and its derivatives. Clones of these two combinations are readily identified because *P.* × *jackii* has the most heart-shaped leaves, about as wide as long, among the intersectional hybrids, while those of *P.* × *berolinensis* are some of the narrowest, inheriting this trait from the Siberian *P. laurifolia*. Berlin poplar can be distinguished from *P.* × *acuminata* (*P. deltoides* × *P. angustifolia*), widely cultivated for ornament in the Rocky Mountain region, because it has much stronger balsam characteristics, with smaller teeth than those on *P.* × *acuminata* leaves and much stronger color differentiation between the upper and lower leaf surfaces.

The third member of this group included in our list, *P. nigra* × *P. maximowiczii*, is heterotic but has not yet been developed to the same extent as Euramerican or interamerican hybrids, with just a single clone, 'Max 5' (or 'NM6'), commonly cultivated. The same is true of *P. trichocarpa* × *P. nigra*, which occurs spontaneously around cities west of the Cascade Mountains, but has just barely been brought into horticultural or plantation culture there. These hybrids have strong balsam leaf and shoot characteristics and relatively narrow, upright growth. The *P. nigra* × *P. maximowiczii* hybrids inherit the leathery leaf texture of *P. maximowiczii*, a characteristic distinct from all of the other clonal groups considered here but shared with other *P. maximowiczii* derivatives, such as its hybrids with *P. balsamifera*, *P.* × *canadensis*, *P. deltoides*, and *P.* × *generosa*. Except in Quebec, where *P. maximowiczii* has become an important part of the breeding program, all of these intersectional hybrids involving balsam poplars (especially those including *P. maximowiczii*) have limited commercial use in eastern North America because of their susceptibility to *Septoria* canker. On the other hand, *P. deltoides* × *P. maximowiczii* is becoming prominent in the Pacific Northwest because of its rapid growth and resistance to *Septoria*, as well as to *Melampsora* rust and *Venturia* blight. It has become a regular part of regional breeding programs but selections are not yet widely planted and are not described here.

Even with the recent publication of the English version of *Flora of China* (Missouri Botanical Garden, St. Louis, 1998), the Asian balsam poplars as a whole are in need of taxonomic revision. *Populus maximowiczii*, the only one of these used extensively in breeding programs in North America, is ultimately likely to be treated as a maritime subspecies of the eastern Siberian and northern Chinese *P. suaveolens*. No formal botanical name reflecting this status has been published, and it would not be appropriate to simply treat it as a plain synonym of *P. suaveolens*, the earlier name. Therefore, except in the worldwide key to species of poplar, I am retaining the name *P. maximowiczii* in hybrid formulas throughout this text. When a new botanical combination is published, the epithet 'maximowiczii' is still likely to be part of it, and in any event, there is no particular reason to alter such breeding program designations as NM or MT.

Aspen hybrids

This grouping includes a variety of different combinations among European and North American species of section *Populus*. Most are of interest primarily in horticultural and amenity plantings, but some attention has been focused on plantation culture in the Great Lakes region, particularly using triploid clones of *P.* × *wettsteinii* (*P. tremula* × *P. tremuloides*). These two species are the most closely related pair of aspens, and their hybrids are sometimes hard to distinguish from the two parents. Generally speaking, they have coarser toothing on the leaves than *P. tremuloides*, although less prominent than in *P. tremula*, and the winter buds have a touch of hairiness at the base not found in those of *P. tremuloides* (thus making the buds less shiny). Hybrids involving *P. alba* are much easier to recognize, since they all have some of the woolly hair on the buds, twigs, and leaf undersides characteristic of that species. Unlike the white poplar, however, the hybrids usually lack deeply 5-lobed leaves on long shoots, although they are commonly 5-angled or even so jaggedly and coarsely toothed as to appear 5-lobed. Most horticultural plantings of white poplar hybrids use clones of *P.* × *canescens* (*P. alba* × *P. tremula*). These largely replace *P. alba* in the southeastern United

States and differ from it in their somewhat thinner and much grayer (rather than brilliantly white) wool, leading to the common name gray poplar. The spontaneous hybrid between introduced European *P. alba* and the native North American *P. grandidentata*, *P.* × *rouleauiana*, is much more vigorous than *P.* × *canescens*, but is much less frequently cultivated. It typically has larger leaves and hairier buds than *P.* × *canescens*, but is not otherwise particularly easily distinguished from it. Most of the aspen hybrids are harder to root than the other major hybrid groups considered, although there are effective production methods involving root cuttings. On the other hand, once established, the pronounced clonal growth habit of aspens can obviate the need to replant between rotations, suckers from the roots being more satisfactory than coppice shoots for stand regeneration. Susceptibility to hypoxylon canker and other diseases is, perhaps, one reason why most attention for plantation culture has shifted away from aspen hybrids to some of the other hybrid combinations.

Descriptions of some important clones in North America for production and general cultivation[1]

Species selections

Populus alba Linnaeus 'Nivea,' silver poplar

Female clone with a broad crown of long, spreading branches; large trunk cylindrical; bark bone white and smooth on branches, becoming dark gray and deeply furrowed on large trunks; first-year twigs round, reddish brown, initially hidden by dense white wool; winter buds woolly and not sticky. Leaves dark green above, densely white-woolly beneath; preformed (early) leaves of short shots 3–7 cm long, roundly 5-angled; neoformed (late) leaves of long shoots 4–10 cm long, deeply 5-lobed, maple-like. Flower bracts brown, slightly toothed, fringed; flowers numerous, seed capsules narrowly flask-shaped, with 2 valves.

Commonly used in horticulture, but not plantation culture. It root suckers heavily and can persist essentially indefinitely after abandonment of cultivation. It also becomes something of a pest in lawns. The suckers are dominated by 5-lobed neoformed leaves, while the mature trees look rather different with their predominance of unlobed preformed leaves.

[1]An index to the clones described is provided on p. 397.

Populus alba Linnaeus 'Pyramidalis,' Bolleana poplar

Male clone with a narrow, columnar crown of short, upright branches; large trunks buttressed; bark bone white and smooth on branches, becoming grayer and rougher on large trunks; first-year twigs round, reddish brown, variably covered with white wool; winter buds 5–10 mm long, woolly and not sticky. Leaves dark green above, thickly white-woolly beneath; preformed (early) leaves of short shoots 3–5 cm long, roundly 5-angled; neoformed (late) leaves of long shoots 6–12 cm long, deeply 5-lobed, maple-like. Flower bracts brown, slightly toothed, fringed; flowers numerous; stamens about 8.

This clone has no use in plantation culture, but has a modest presence in amenity plantings and as an accent plant in horticulture. It is a little longer-lived than the similarly shaped *P. nigra* 'Italica' in the Great Lakes region. It root suckers to a much lesser extent than *P. alba* 'Nivea' and so becomes less of a pest. The mature trees have a much higher proportion of neoformed leaves than do comparable-sized individuals of *P. alba* 'Nivea.'

Populus deltoides Marshall 'Belleville'

Male clone with a dense, cylindrical crown of moderate, rising branches from a slightly crooked trunk; large trunks cylindrical and furrowed on large trunks; first-year twigs yellowish tan, strongly 5-angled with prominent ribs, not hairy; winter buds 5–13 mm long, scarcely sticky with a yellowish resin, not hairy. Leaves yellowish green above and beneath, 6–9 cm long, roundly triangular, finely toothed along much of the edge, with a modestly pointed tip and a straight base, the petiole 3.5–5 cm long, red, flattened side to side at the junction with the blade. Flower bracts white and deeply cut with brownish tips; flowers numerous but well separated; stamens about 40–50.

Although it grows quickly and has a decent form, this selection from Ontario is not well suited to dense plantation culture where it is outperformed by Euramerican hybrids.

Populus deltoides **Marshall 'S7C8'**

Female clone with an oval crown of stiffly upwardly angled branches; large trunks cylindrical; bark greenish tan on branches, becoming light brown or gray and deeply furrowed on large trunks; first-year twigs yellowish green with sparse, elongate white lenticels, round to strongly 5-angled with sharp, corky, reddish brown ribs, hairless; winter buds 10–20 mm long, sticky with yellowish green, slightly fragrant resin. Leaves bright yellowish green above, a little paler and duller beneath, (6–)9–15 cm long, roundly triangular, with numerous teeth along the upper two thirds, grading from fairly fine near the base to very fine below the narrowly prolonged tip, the base nearly flat across with a shallow indentation bearing a cluster of fingerlike glands at the petiole, which is (3.5–)6–8 cm long, reddish above, and strongly flattened at the blade. Flower bracts deeply cut, white; flowers numerous and loosely packed; seed capsules opening with 4 valves.

This Texas selection is a fairly typical southern cottonwood (subsp. *deltoides*).

Populus deltoides **Marshall 'S7C15'**

Male clone with an oval crown of stiffly ascending branches; large trunks cylindrical; bark orangish tan on branches, becoming grayish brown and deeply furrowed on large trunks; first-year twigs greenish brown with a few, elongate white lenticels, round or 5-angled with sharp but small corky ribs, hairless; winter buds 10–20 mm long, sticky with yellowish, slightly fragrant resin. Leaves bluish green above, a little paler beneath, (7–)10–13(–15) cm long, with roundly hooked, graded teeth along most of the margin, extremely coarse in preformed (early) leaves of short shoots, coarse in neoformed (late) leaves of long shoots, the tip toothless and narrowly triangular, the base flat across to conspicuously heart shaped, often with 2 or more conspicuous, emergent or even leaflike glands on the upperside at the petiole, which is (4.5–)7–10 cm long, yellowish to reddish green, and strongly flattened at the blade. Flower bracts deeply cut, white; flowers numerous and loosely packed; stamens about 30–40.

This Texas selection stands out with its very prominent and unusual glands at the junction of the leaf blade and petiole.

***Populus deltoides* Marshall 'S7C20'**

Female clone with a dense, bushy crown or upwardly angled branches bearing many sylleptic shoots (growing out in the same year as the branch they are borne on); large trunks cylindrical; bark tan with a greenish to orangish cast on branches, becoming light brown to grayish and deeply furrowed on large trunks; first-year twigs greenish brown, with fairly numerous, elongate white lenticels, round to 5-angled, with sharp, shallow ridges about the same color as the twigs, hairless; winter buds 10–20 mm long, sticky with yellowish, slightly fragrant resin. Leaves bluish green above, similar beneath, (5–)9–14 cm long on primary shoots, broadly and roundly triangular to heart-shaped, with numerous, graded, medium to fine teeth along most of the margin, the tip sharply triangular, the base shallowly heart-shaped to straight across, or even rounded, with a few small, fingerlike glands at the junction with the petiole, which is (3–)6–10 cm long, greenish yellow to reddish and strongly flattened at the blade. Flower bracts deeply cut, white; flowers numerous and loosely packed; seed capsules opening with 4 valves.

This Texas selection is somewhat intermediate between southern (subsp. *deltoides*) and northern (subsp. *monilifera*) cottonwoods. Much of central Texas falls into an intergradation zone between these cottonwood subspecies.

***Populus deltoides* Marshall 'ST70'**

Male clone with an egg-shaped crown of gently rising branches; large trunks cylindrical; bark greenish gray on branches, becoming light brown and deeply furrowed on large trunks; first-year twigs green with a few, elongate white lenticels, strongly 5-angled, with sharp, tan, corky ribs, hairless; winter buds 10–20 mm long, sticky with greenish yellow, slightly fragrant resin. Leaves bright green above, a little duller beneath, (6–)10–15(–18) cm long, roundly triangular to elongate heart-shaped, with numerous, graded, coarse, rounded teeth along most of the margin, the tip toothless, prolonged, and narrowly triangular, the base nearly straight across to moderately heart-shaped, with 2 or more finger-shaped glands at the junction with the (3–)5–8 cm long, strongly flattened, yellowish green to red-stained petiole. Flower bracts deeply cut, white; flowers numerous and loosely packed; stamens about 30–40.

This vigorous clone, selected at Stoneville, is a classic southern cottonwood (subsp. *deltoides*) with proportionately very elongate long-shoot (neoformed late) leaves compared to other more northerly or westerly subspecies of eastern cottonwood or Fremont cottonwood.

***Populus deltoides* Marshall 'ST148'**

Female clone with a spreading crown of upcurved branches; large trunks cylindrical; bark greenish gray on branches, becoming light brown and deeply furrowed on large trunks; first-year twigs greenish brown with a few, elongate white lenticels, strongly 5-angled with sharp, tan, corky ribs, hairless; winter buds 10–20 mm long, sticky with greenish yellow, slightly fragrant resin. Leaves rich dark green above, paler beneath, (6–)11–17 cm long, roundly and elongately triangular, with numerous, weakly graded, moderately coarse, rounded teeth along most of the margin, the tip toothless, narrowly triangular, sharply prolonged, the base shallowly wedge-shaped between straight shoulders, with 2 or more finger-like to sac-like glands at the junction with the (5–)7–10(–13) cm long, strongly flattened but moderately slender petiole. Flower bracts deeply cut, white; flowers numerous and loosely packed; seed capsules opening with 4 valves.

This clone, selected at Stoneville, is vigorous and productive but difficult to propagate from hardwood cuttings. It has proportionately longer petioles than most cottonwood clones and so appears relatively sparsely clothed with leaves. The flower buds are unusual in this clone in resembling those of male cottonwoods in size and arrangement, with numerous large buds clustered near the terminal bud. Most female cottonwoods have sparser flower buds that are much smaller than those of male individuals.

Populus nigra Linnaeus 'Italica,' Lombardy poplar

Male clone with a narrow columnar crown of long, upright branches; large trunks heavily buttressed; bark light brown on branches, dark grayish brown and deeply furrowed on large trunks; first-year twigs round, orange, smooth and a little shiny; winter buds 10–20 mm long, sticky with reddish orange resin, not hairy. Leaves light green above and beneath; preformed (early) leaves of the numerous short shoots 5–8 cm long, somewhat diamond-shaped, finely toothed along both sides, with a wedge-shaped base; neoformed (late) leaves of long shoots 3–9 cm long, wider than long, roundly triangular, with an abrupt, short point, finely toothed along the whole length. Flower bracts white, deeply cut, without marginal hairs; flowers numerous; stamens 12–20.

The oldest poplar clone used as an upright accent in North American horticulture, Lombardy poplar still predominates in this function in many areas in addition to its use in windbreaks around western orchards and vineyards. Although old plantings have become senescent or have died out, the tree persists by suckering. In the Great Lakes region, the Lombardy poplar is particularly short-lived, often succumbing to *Dothichiza* canker and other diseases within 10 years. The same malady seems to affect other *P. nigra* clones in this area. Lombardy poplar has no direct role in plantation culture but is a parent or grandparent of many other important clones.

Populus simonii **Carrière 'Fastigiata'**

Male clone with a broadly conical crown made up of short upright branches; large trunks fairly cylindrical; bark light gray on branches, grayish brown on large trunks; first-year twigs angled, greenish gray, hairless, dull; winter buds 6–12 mm long, slightly sticky with reddish resin. Leaves dark green above, white beneath, 3–8 cm long, egg-shaped with the narrow end towards the petiole, with a very short point, very finely toothed along the whole side or just along the outer two thirds, petiole up to 15 mm long. Flower bracts white tipped with brown, deeply cut, not hairy; flowers relatively few; stamens about 8.

This is another pyramidal clone used primarily in horticulture and particularly in the northern United States and southern Canada. Neither it nor the parent species has any role in plantation culture, and it has not contributed to the parentage of any important clones.

Populus tremula **Linnaeus 'Erecta'**

Male clone with a narrow columnar crown of long, upright branches; large trunks slightly buttressed; bark light gray on branches, darker and slightly furrowed on large trunks; first-year twigs round, reddish brown and glossy; winter buds 5–10 mm long, not sticky, inconspicuously hairy on the lower scales. Leaves bright green above, lighter, whitish green beneath; preformed (early) leaves of short shoots 3–4 cm long, wider than long, abruptly narrowed to a short point, with about 5–10 coarse, rounded teeth along the middle of each side; neoformed (late) leaves of long shoots 5–8 cm long, a little longer than wide, finely and evenly toothed along the whole length. Flower bracts brown, deeply cut, fringed; flowers numerous; stamens about 8.

This pyramidal clone is replacing the less hardy *P. alba* 'Pyramidalis' and *P. nigra* 'Italica' in horticulture and amenity plantings in the northern plains and prairies and other cold regions. It has considerable suckering capability.

Hybrid selections

***Populus* × *berolinensis* Dippel 'Berlin poplar' (*P. laurifolia* × *P. nigra*)**

Male clone with a conical crown of long, upright branches; large trunks cylindrical; bark grayish green on branches, becoming darker and shallowly furrowed on large trunks; first-year twigs angled, light grayish green, not hairy, dull; winter buds 10–19 mm long, sticky with reddish resin. Leaves dark green above, white beneath, 3–8 cm long, roundly diamond-shaped, finely toothed along the whole side, wedge-shaped at the base and tip, the petiole 1.5–2 cm long, slightly flattened side to side at the blade. Flower bracts white tipped with brown, deeply cut, hairy on the outer surface but not around the edge; flowers relatively few; stamens 15–20.

An old cross between *P. laurifolia* and *P. nigra* that is declining in use in horticulture and amenity plantings because of susceptibility to diseases and late spring frosts. It has no direct role in plantation culture except as a part of the pedigree of some more recent clones.

***Populus × berolinensis* Dippel 'Petrowskyana,' Russian poplar (*P. laurifolia × P. nigra*)**

Female clone with a narrowly cylindrical crown of long, upright branches from a crooked, forked trunk; large trunks cylindrical; bark greenish tan with darker blotches on branches, long remaining smooth but finally becoming gray and furrowed at the base of large trunks; first-year twigs dark greenish brown, round, slightly shiny, sometimes finely hairy; winter buds 10–20 mm long, very sticky with red resin, hairless. Leaves dark green above, white beneath, about 7–8 cm long, egg-shaped with a fairly long point, rounded base usually without glands, and fine, even teeth along at least three quarters of the edge, the petiole about 2.5–4 cm long, finely hairy. Flower bracts white with just the long teeth dark brown; flowers numerous; seed capsules opening with 2–3 valves.

The Russian poplar is a common sight in horticultural and amenity plantings on the northern prairies but its moderate growth rate and frequent susceptibility to diseases and pests preclude serious use in plantation culture. Its main virtues are its hardiness and rich green, dense foliage.

***Populus* 'Brooks'**
(*P. deltoides* × *P.* × *berolinensis*)

Male clone with a dense, cylindrical crown of short, slightly rising branches from a crooked trunk; large trunks cylindrical; bark grayish green to pale gray, long remaining smooth, but finally becoming dark gray and shallowly furrowed between broad, flat ridges at the base of large trunks; first-year twigs dark brown, round, hairless and dull; winter buds 12–22 mm long, very sticky with dark red resin, hairless. Leaves darker green above, paler beneath, about 6–7 cm long, roundly triangular with a fairly long tip and broadly wedge-shaped to heart-shaped base, usually with inconspicuous glands or none; the petiole about 2.5–3 cm long, a little flattened at the blade, pink, fuzzy. Flower bracts white with numerous dark brown, long, narrow teeth; flowers numerous; stamens 25–35.

This clone from the prairies, originally called Brooks #2, is a sibling to 'Griffin,' originally called Brooks #1. The dense, narrow crown of dark foliage make it a handsome and effective choice for windbreaks and, to a lesser extent, for horticulture on the prairies. Unfortunately, it is susceptible to various ills and is frequently disfigured by bud-gall mites and frost damage. Its modest growth rate also restricts its potential for plantation culture.

***Populus* 'Griffin'**
(*P. deltoides* × *P.* × *berolinensis*)

Male clone with a dense, narrow cylindrical crown of short, rising branches from a crooked and forked trunk; large trunks round; bark greenish gray on branches, gray and furrowed on large trunks; first-year twigs dark brown, weakly 5-angled, variably fuzzy, dull; winter buds 8–15 mm long, sticky with orange resin, hairless. Leaves dark green above, paler beneath, 5–8 cm long, egg-shaped, finely toothed along most of the edge, the base broadly wedge-shaped with inconspicuous glands, the tip a bit extended, the petiole 2–3.5 cm long, a little flattened, yellowish green, fuzzy or smooth. Flower bracts white with numerous brown, long, slender teeth; flowers numerous; stamens 25–35.

The most widely planted of E. Griffin's "Brooks" selections, this one (originally Brooks #1) was renamed for Griffin when distributed for shelterbelt planting by Indian Head forest nursery in Saskatchewan. Unfortunately, it appears to be one of the least suitable Brooks clones for the prairies, frequently suffering diseases, pests, and growing season frost damage.

Populus × canadensis Moench 'Allenstein' [DN1]
(_P. deltoides_ × _P. nigra_)

Female clone with a narrowly egg-shaped crown of long, rising branches; large trunks cylindrical; bark greenish gray on branches, dark brown and strongly furrowed on large trunks; first-year twigs light grayish brown, weakly 5-angled and slightly ridged, hairless, dull; winter buds 8–18 mm long, sticky with reddish resin, not hairy. Leaves bright green above, paler beneath, 5–7 cm long, heart-shaped or with a broadly rounded base, finely toothed along the whole edge, the petiole 3–4 cm long, flattened side to side at the blade, hairless. Flower bracts white with brown edges, deeply cut; flowers many, but not tightly spaced; seed capsules opening with 2–3 valves.

This clone from Germany is a fast grower at about 12–15 dm in height per year and puts much of its biomass into the straight trunk rather than the relatively slender branches. Substantial rust susceptibility can lead to early defoliation as can early frosts.

Populus × canadensis Moench 'Baden 431' [DN2] (P. deltoides × P. nigra)

Male clone with a dense, narrow crown of mid-length sharply rising branches; large trunks round; bark greenish gray on branches, long remaining smooth, darker gray and finally furrowed on large trunks; first-year twigs brown, weakly 5-angled and slightly ridged, hairless, a little shiny; winter buds 6–15 mm long, sticky with reddish resin, not hairy. Leaves dark green above with a reddish midrib, paler beneath, 5–7 cm long, generally heart-shaped but often with a broadly rounded base, conspicuously and a little irregularly toothed along the whole margin with rounded teeth, the petiole 3–5 cm long, red, flattened side-to-side at the blade, hairless. Flower bracts white with brown tips, deeply cut; flowers many; stamens about 20–30.

This clone, known as 'Rintheim' in Germany where it originated, is tolerant of close spacing because of its narrow crown. It is reasonably free from diseases and grows at a rate of about 14–18 dm in height per year without particularly rapid diameter growth.

***Populus* × *canadensis* Moench 'Blanc du Poitou' [DN19] (*P. deltoides* × *P. nigra*)**

Male clone with a broad, oval crown of long, gently rising branches; large trunks cylindrical; bark light brown, becoming dark grayish brown and shallowly furrowed on large trunks; first-year twigs greenish brown, distinctly 5-angled with prominent ridges, hairless, slightly shiny; winter buds 6–12 mm long, sticky with pale resin, not conspicuously hairy. Leaves dull bluish green on both sides, 5–8 cm long, heart-shaped or with a straight-across base, with a short, narrow tip, finely toothed along the whole edge, with small, elongate glands at the base near the petiole, which is 3.5–7 cm long, red, and flattened side-to-side at the blade. Flower bracts white with the deeply cut edges brown; flowers numerous; stamens about 20–30.

Another old clone (ca. 1870) from France that is still used in plantation culture, it is tolerant of late frosts due to its late flushing, but also retains its leaves late and so has a reasonable growth rate. Its susceptibility to diseases varies greatly with site.

***Populus* × *canadensis* Moench 'Canada Blanc' [DN30]**
(*P. deltoides* × *P. nigra*)

Male clone with a sparse, broadly cylindrical crown of long, gently rising branches; large trunks cylindrical; bark grayish brown on branches, long remaining smooth, finally becoming dark grayish brown and furrowed at the base of large trunks; first-year twigs purplish brown, strongly angled with distinct ridges, shiny, hairless; winter buds 5–12 mm long, very sticky with reddish resin, hairless. Leaves bluish green above and beneath, 5–8 cm long, roundly triangular, finely toothed around most of the edge, with a short, narrow point and a broadly wedge-shaped base with small round glands at the petiole, which is 2–4 cm long, red, flattened side-to-side at the blade, and hairless. Flower bracts white with deeply cut, brown tips; flowers numerous; stamens about 25–30.

This clone is a fairly typical Euramerican hybrid, although with a somewhat more open, cylindrical crown of branches with browner bark than usual. It seems to have characteristics closer to those of *P. deltoides* than many other clones.

***Populus* × *canadensis* Moench 'DN21'**
(*P. deltoides* × *P. nigra*)

Male clone with a narrow crown of sharply rising branches; large trunks cylindrical or fluted; bark tan with an orange cast on branches, becoming dark grayish brown and deeply furrowed on large trunks; first-year twigs orange–brown, round, hairless; winter buds 8–15 mm long, sticky with orange resin. Leaves dark green above, grayish green beneath, 7–10 cm long, triangularly egg-shaped, with numerous, even, moderately coarse, rounded teeth along almost the whole margin, the tip abruptly triangular, moderately prolonged, the base broadly wedge-shaped to broadly rounded, mostly without conspicuous glands at the junction with the 4.5–7(–8.5) cm long petiole, which is flattened and yellowish green to reddish. Flower bracts deeply cut, white with brown just at the tips; flowers numerous and loosely packed; stamens about 20–30.

This is one of the most frequently used commercial Euramerican clones in the Great Lakes region.

Populus × canadensis Moench 'DN70'
(_P. deltoides × P. nigra_)

Clone of uncertain sex as of this writing, with a narrow crown of upright branches; large trunks cylindrical or irregular; bark tan with an orange cast on branches, becoming grayish brown and deeply furrowed on large trunks; first-year twigs reddish orange, round, hairless; winter buds 8–15 mm long, sticky with reddish orange resin. Leaves dark green above, much paler beneath, about 6–8 cm long, roundly triangular, with numerous, moderately fine, rounded, and hooked teeth, grading slightly from near the base to just below the protruding, narrowly triangular tip, the base flat across to broadly rounded, with 0–2 round or fingerlike glands at the junction with the 3.5–5 cm long petiole, which is greenish yellow and strongly flattened, although narrow. Flower bracts deeply cut, white with brown tips; flowers numerous and loosely packed; expect about 20–30 stamens *or* seed capsules opening with 3 valves.

Another of the commercial Euramerican clones in production in the Great Lakes region.

***Populus* × *canadensis* Moench 'Eugenei,' Carolina poplar [DN34] (*P. deltoides* × *P. nigra*)**

Male clone with a broadly conical crown of long, rising branches; large trunks cylindrical to somewhat fluted; bark light tan with an orange cast on branches, becoming grayish tan and deeply furrowed on large trunks; first-year twigs round, yellowish orange, hairless, slightly shiny; winter buds 1–2.5 cm long, very sticky with reddish orange resin. Leaves bright green above, paler and a little duller beneath, the petiole (3–)6–8 cm long, strongly flattened side-to-side at the blade; preformed (early) leaves of short shoots 4–10 cm long, angularly egg-shaped, with about 15–25 graded teeth along the middle of each side, the base broadly wedge-shaped and shouldered below the teeth, sometimes with a single round gland at the petiole, the tip drawn out in a long point; neoformed (late) leaves of long shoots 6–11(–14) cm long, broadly egg-shaped, with about 30–45 teeth along most of the side, the base broadly rounded, the tip with a short point. Flower bracts white with brown tips, deeply cut, hairless; flowers numerous; stamens 20–30.

This is, by far, the most prevalent Euramerican clone in general cultivation in North America. It has been planted for so long and is so pervasive that land managers, naturalists, foresters, and botanists across the continent often mistake it for the native *P. deltoides* (one of its parents) or *P. fremontii*, depending on the region. Although not itself widely used in plantation culture, it has been used in breeding programs and has passed on its relatively narrow crown, inherited from its Lombardy poplar parent. The Carolina poplar is relatively free from pests and diseases and, in the Great Lakes region, in contrast to other regions, it is an important commercial clone, one of the few recommended in the region for plantations and windbreaks around orchards.

***Populus* × *canadensis* Moench 'Gelrica' [DN5]**
(*P. deltoides* × *P. nigra*)

Male clone with an egg-shaped to rounded crown of sharply rising to nearly horizontal branches on a crooked, leaning, and often forked trunk; large trunks cylindrical; bark very pale gray to white on branches, becoming darker with age, long remaining smooth, but finally becoming ridged and furrowed at the base of large trunks; first-year twigs greenish brown, slightly 5-angled with variable ribs depending on vigor of shoot, densely covered with short hairs; winter buds 5–15 mm long. Leaves yellowish green above and beneath, 5–7 cm long, roundly triangular, tip short-pointed, base broadly wedge-shaped to shallowly indented with small glands, shallowly toothed along most of the edge with rounded teeth, the petiole 2.5–4 cm long, red, flattened side-to-side at the blade, hairless. Flower bracts white with yellowish brown edges, deeply cut; flowers many; stamens 20–30.

This old clone (ca. 1850) from the Netherlands is one of the slower growing Euramerican hybrids, adding about 10–12 dm in height per year. Its susceptibility to pests and diseases varies greatly in different regions. It has been replaced by newer, more vigorous clones for plantation culture but is still useful for horticultural and amenity plantings.

***Populus* × *canadensis* Moench 'I-45/51'**
(*P. deltoides* × *P. nigra*)

Male clone with a broad conical crown of long, rising branches from a straight trunk; large trunks round and strongly tapering; bark tan on branches, quickly becoming grayish brown and furrowed on large trunks; first-year twigs pale greenish brown, strongly angled with prominent ridges, glossy, hairless; winter buds 5–18 mm long, only slightly sticky with reddish resin, hairless. Leaves slightly bluish green above and beneath, 5–8 cm long, heart-shaped but with a flat base, tip short-pointed, finely toothed along almost the whole margin, the petiole about 2.5–4 cm long, pale red, flattened side-to-side at the blade. Flower bracts white with a dark brown, deeply cut edge; flowers numerous; stamens about 20–30.

This Italian clone has good growth on heavy soils as well as on better ones. It is fairly susceptible to pests and diseases but has more use in plantation culture than in horticultural or amenity plantings.

***Populus* × *canadensis* Moench 'NE367'**
(*P. deltoides* × *P. nigra*)

Male clone with an oval crown of upwardly arching branches, large trunks cylindrical or somewhat fluted; bark tan with an orange cast on the branches, becoming dark grayish brown and deeply furrowed on large trunks; first-year twigs reddish orange, round, hairless; winter buds 9–18 mm long, sticky with reddish orange resin. Leaves bluish green above, paler and duller beneath, (7–)10–14 cm long, roundly triangular, with numerous, coarse, rounded teeth grading from near the base to just below the prolonged, triangular, toothless tip, the base straight across or slightly wedge-shaped, often with 2–3 domed glands at the junction with the 6–7 cm long, strongly flattened, red petiole. Flower bracts deeply cut, white with brown tips; flowers numerous and loosely packed; stamens 20–30.

This Euramerican hybrid clone from New England is grown in production plantations east of the Cascade Mountains in Oregon and Washington. It was bred by Ernst Schreiner for the Oxford Paper Company (Frye, Maine) program. The extra glands at the junction of the leaf blade and petiole are unusual among Euramerican clones.

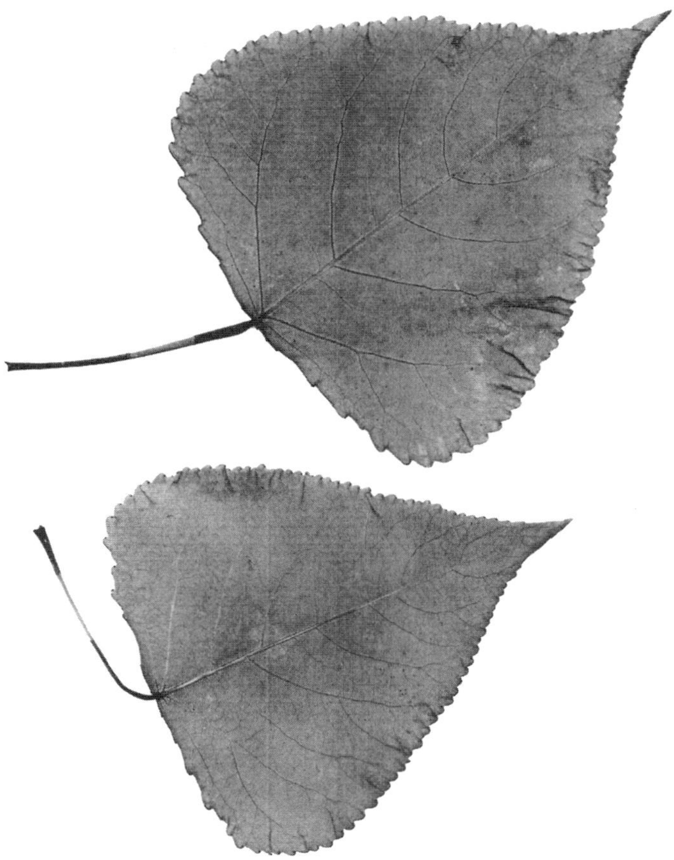

***Populus* × *canadensis* Moench 'Regenerata' [DN16]**
(*P. deltoides* × *P. nigra*)

Male clone initially with a narrowly cylindrical crown of short rising branches but ultimately broadly rounded and irregular, large trunks cylindrical; bark brown on branches becoming dark gray and furrowed on large trunks; first-year twigs dark brown, 5-angled and prominently ribbed, hairless and somewhat shiny; winter buds 5–16 mm long, very sticky with reddish resin, not hairy. Flower bracts white with brown edges, sharply cut; flowers numerous; stamens about 20–30.

This very old clone (ca. 1814) from France has a modest growth rate for a Euramerican hybrid of about 10–12 dm in height per year. In compensation, perhaps, it is relatively resistant to diseases. Like other older clones, it is now largely replaced by more recent and faster growing clones, but still retains some use in horticultural and amenity plantings.

***Populus* × *canadensis* Moench 'Robusta' [DN17]**
(*P. deltoides* × *P. nigra*)

Male clone with a broadly conical crown of long, rising branches; large trunks cylindrical; bark greenish gray on the branches, quickly becoming dark gray and furrowed on trunks; first-year twigs light reddish brown, round, except slightly angular on especially robust shoots, hairless; winter buds 5–18 mm long, only slightly sticky with reddish resin, hairless. Leaves shiny yellowish green on both sides, 6–10(–12) cm long, roundly triangular with short triangular tip, the base fairly flat across and with small glands at the petiole, toothed along almost the whole edge, the petiole 3.5–5 cm long, red, flattened side-to-side at the blade, hairless. Flower bracts white with light brown tips, deeply cut; flowers many; stamens about 20–25.

This old (ca. 1895) clone from France has good growth at 12–19 dm in height per year, which is achieved partly through early flushing and late leaf fall, making it susceptible to damage by late spring frosts. The newly flushing foliage is bright red, while other clones of Euramerican hybrids range from green through brown and dull red to dark purple. It is prone to develop stem cankers at wounds. Still used to some extent in plantations in Europe, it is mostly used in horticultural and amenity plantings in North America.

***Populus* × *canadensis* Moench 'Stormont' [DN74]**
(*P. deltoides* × *P. nigra*)

Male clone with a narrow, irregular crown of long, sharply rising branches from a crooked and forking trunk; large trunks cylindrical; bark grayish brown on branches, becoming furrowed on large trunks; first-year twigs brown, strongly angled with prominent ridges, shiny, hairless; winter buds 5–10 mm long, sticky with red resin, not hairy. Leaves triangularly egg-shaped, (5–)7–11 cm long, finely toothed along most of the edge, with a narrowly triangular, prolonged tip and very broadly wedge-shaped base bearing small round glands at the petiole, which is 4.5–8 cm long, yellowish green to light red, and flattened side-to-side at the blade. Flower bracts white with light brown, deeply cut edges; flowers numerous; stamens about 20–30.

This clone was developed in Ontario and is most suitable for short-rotation intensive culture or amenity plantings rather than for production of timber or horticultural uses. It is one of the clones used commercially in the Great Lakes region.

***Populus* × *canadensis* Moench 'Walker'**
(*P. deltoides* × *P. nigra*)

Female clone with a sparse crown of short branches from a trunk that bears a few spreading major branches; large trunks cylindrical or somewhat irregular; bark greenish gray, soon becoming grayish tan and strongly furrowed on the trunk; first-year twigs greenish tan, variably 5-angled with prominent ridges, fuzzy; winter buds 10–18 mm long, a little sticky with orange resin, fuzzy. Leaves yellowish green above, a little paler beneath, 5–7 cm long, somewhat diamond-shaped, a little coarsely toothed along the middle portion of the edge, with a long pointed tip and wedge-shaped base without prominent glands, the petiole 2–3.5 cm long, somewhat flattened at the blade, light red, hairless. Flower bracts white with a dark brown margin cut into long, narrow branching teeth; flowers numerous but not dense; seed capsules opening with 3 valves.

Although this clone from the Indian Head nursery in Saskatchewan has been considered a selection of *P. deltoides*, the leaf shape and lack of glands at the blade/petiole junction show that it originated through hybridization with *P. nigra*, perhaps via another Euramerican hybrid. It is fairly susceptible to pests and diseases but is widely used for plantation, amenity, and horticultural plantings in the southern Canadian prairies. It is the mother of 'Assiniboine,' 'CanAm,' and 'Manitou,' additional prairie selections that are probably backcrosses to *P. deltoides*.

***Populus* × *canescens* J. E. Smith 'Gray Poplar'**
(*P. alba* × *P. tremula*)

Female clone with a broad, rounded crown of long, spreading branches; large trunks cylindrical; bark greenish white on branches, dark gray and deeply furrowed on large trunks; first-year twigs round, reddish brown, woolly at first, becoming smooth with age; winter buds 3–8 mm long, not resinous, thinly woolly. Leaves dark green above, light green and thinly grayish woolly beneath, becoming smooth with age, 3–7 cm long, with 2 prominent round glands at the base, petioles 2–6 cm long, slightly flattened side-to-side at the blade; preformed (early) leaves of short shoots egg-shaped with about 4–11 coarse, rounded teeth on each side; neoformed (late) leaves of long shoots very broadly egg-shaped, irregularly toothed with numerous, uneven, angular teeth. Flower bracts coarsely toothed, brown, fringed; flowers very numerous; seed capsules narrowly flask-shaped, with 2 valves.

This clone largely replaces *P. alba* 'Nivea' in horticulture in the southeastern United States and is found less commonly elsewhere. It is an old clone and, like the white poplar, persists through root suckers long after abandonment. Although some selections of *P.* × *canescens* are suitable for plantation culture, this one, either selected for its ornamental qualities or just a clone of convenience, is not.

Populus × canescens J. E. Smith 'Tower'
(_P. alba × P. tremula_)

Male clone with a narrow columnar crown of short upright branches; large trunks somewhat irregular; bark grayish green on branches, gray and slightly furrowed on large trunks; first-year twigs reddish brown, sparingly woolly; winter buds 4–10 mm long, not sticky, somewhat woolly. Leaves dark bright green above, light green and variably woolly beneath, 4–8 cm long, the petioles 2–6 cm long, slightly flattened side-to-side at the junction with the blade; preformed (early) leaves of short shoots egg-shaped, coarsely toothed along the middle with rounded teeth; neoformed (late) leaves of long shoots very broadly egg-shaped, irregularly toothed with uneven, pointed teeth. Flower bracts coarsely toothed, brown, fringed; flowers numerous; stamens about 8.

This pyramidal clone is sometimes used in place of the Bolleana poplar in horticulture plantings and the two may be confused. The neoformed leaves of long shoots on the Bolleana poplar are distinctly 5-lobed, while those of 'Tower' are much more shallowly marked by irregular angles.

***Populus* × *generosa* A. Henry '11-11'**
(*P. trichocarpa* × *P. deltoides*)

Vigorous male clone with an irregular spreading crown of upwardly angled branches; large trunks cylindrical; bark light gray on branches, becoming dark gray and deeply furrowed on large trunks; first-year twigs reddish brown, cylindrical to 5-angled, hairy; winter buds 10–20 mm long, very sticky with greenish orange, strongly fragrant resin. Leaves dark green above, somewhat whitened beneath, 10–18(–22) cm long, elongate egg-shaped to broadly egg-shaped, with numerous, fine, fairly even teeth along practically the whole margin, the tip sparsely toothed, short and bluntly triangular to somewhat prolonged, the base broadly and shallowly heart-shaped, with 2 or more slightly protruding glands at the junction with the 8–10 cm long, yellowish green, slightly flattened petiole. Flower bracts deeply cut, with a white centre shading to brown tips; flowers numerous and densely packed; stamens about 30–40.

Bred by Reinhard Stettler, this is one of the older modern interamerican hybrid clones that is still in production. Although once more widely planted, it has been superseded by more consistent and better adapted newer clones. It is still used for phytoremediation where its vigor and strong uptake of soil materials are more important than superior form or fiber production. It is one of the main clones used by R. Ceulemans in his book, *Genetic variation in functional and structural productivity determinants in poplar* (Thesis Publishers, Amsterdam, 1990), on the effects of form and physiology on productivity in short-rotation intensive poplar culture.

***Populus* × *generosa* A. Henry '15-29'**
(*P. trichocarpa* × *P. deltoides*)

Female clone with a narrow crown of steeply upwardly angled branches; large trunks cylindrical but markedly sinuous when young; bark light gray on branches, becoming dark gray to tannish brown and deeply furrowed with flat-topped ridges on large trunks; first-year twigs reddish brown, round, hairy; winter buds 10–20 mm long, very sticky with red, fragrant resin. Leaves dark green above, waxy white beneath, 20–25 cm long, heart-shaped or elongate heart-shaped, with numerous, moderately coarse teeth grading slightly from the base to the abrupt, short, pointed tip, the base shallowly heart-shaped to broadly rounded, often with 2 or more inconspicuously protruding glands at the junction with the ca. 9–10 cm long petiole, which is reddened and strongly flattened. Flower bracts deeply cut, with a white center shading to brown tips; flowers numerous and crowded; seed capsules opening with 3–4 valves.

A typical example of the current production interamerican hybrid clones grown west of the mountains in British Columbia, Washington, and Oregon, '15-29' has the enormous leaves that accompany the fastest growing clones — but makes them susceptible to drought stress, frost, and low-temperature damage. In the milder climate of California, this clone outgrows other interamerican hybrids. It was bred by Reinhard Stettler.

***Populus* × *generosa* A. Henry '20-88-183'**
(*P. trichocarpa* × *P. deltoides*)

Female clone with a spreading crown of slender horizontal branches; large trunks cylindrical and straight; bark greenish gray on branches, light brown and scaly on young trunks, becoming deeply furrowed with age; first-year twigs dark reddish brown, cylindrical to 5-angled, hairy; winter buds 10–20 mm long, very sticky with reddish, fragrant resin. Leaves dark green above, strongly whitened between the prominent veins beneath, up to 20–23 cm long, heart-shaped, with numerous coarse, rounded teeth along the whole margin, but shallower near the triangular tip and shouldered and heart-shaped base marked by 2 or more domed to fingerlike glands at the junction with the 9–10 cm long, reddish green, flat-topped round petiole. Flower bracts deeply cut, white with brown tips; flowers numerous and densely packed; seed capsules opening with 3–4 valves.

The enormous leaves of this clone that was bred by Brian Stanton (James River Corp., Camas, WA) have yellow to red veins on the upper surface. The toothing is remarkably uniform as it follows the wavy margin, which is a distinctive feature among commercial clones of *P.* × *generosa*. The straight stem is due in part to strong apical dominance, preventing the crown from breaking up into major branches.

Populus × _generosa_ A. Henry '23-91'
(_P. trichocarpa_ × _P. deltoides_)

Male clone with a narrow crown of short branches, many produced in the same year as the parent stem; bark greenish gray on branches, becoming rich brown with flaking scales and finally deeply furrowed on large trunks; first-year twigs dark reddish brown, cylindrical to 5-angled, hairy; winter buds 10–20 mm long, very sticky with reddish, fragrant resin. Leaves dark green above, white between the prominent dark veins beneath, up to 25–30 cm long, and 23–24 cm wide, broadly egg-shaped, with numerous, coarse, wave-shaped teeth of mixed sizes along the whole margin from the narrowly triangular, moderately prolonged tip to the broadly rounded to shallowly heart-shaped base, the latter with 2 or more fingerlike but inconspicuous glands at the junction with the 7–10 cm long, moderately flattened red petiole. Flower bracts deeply cut, white with brown tips; flowers numerous and densely packed; stamens about 30–40.

This triploid clone, bred by Reinhard Stettler, has more cottonwood characteristics in its leaves than some of the other interamerican hybrids. Because the leaves are broad as well as long, they have an unusually large surface area. The strongly hooked teeth, narrowly prolonged tip, and somewhat flattened petiole are all indicative of their departure from the predominant balsam characteristics of most of the common clones. On the other hand, the undersides are whiter than in many of the other clones, a balsam characteristic.

***Populus* × *generosa* A. Henry '24-305'**
(*P. trichocarpa* × *P. deltoides*)

Male clone with a narrow crown of short branches, including many sylleptic branches on the current year's growth; trunks somewhat sinuous at first, becoming cylindrical; bark greenish gray on branches, becoming rich brown with flaking scales and finally deeply furrowed on large trunks; first-year twigs dark reddish brown, cylindrical to 5-angled, hairy; winter buds 10–20 mm long, very sticky with reddish orange, fragrant resin. Leaves bright green above, light gray to stark white between the reddish purple veins beneath, up to 25–30 cm long and 24–27 cm wide, roundly heart-shaped, with numerous, fairly fine and even rounded teeth between the narrowly triangular, moderately prolonged tip and the straight across or shallowly heart-shaped base bearing 2 or more, inconspicuous, slightly protruding glands at the junction with the 9–11 cm long, strongly flattened dark red petiole. Flower bracts deeply cut, white with brown tips; flowers numerous and densely packed; stamens about 30–40.

Although predominantly balsam-like in general characteristics, the enormous leaves of this triploid clone bred by Reinhard Stettler stand out for being almost as wide as long, a trait inherited from its cottonwood parent, as is the strongly flattened petiole. The narrow, prolonged leaf tip is also a *P. deltoides* trait, but the white underside can be as prominent as in any balsam poplar. The leaves have a relatively thin texture for their expanse, which makes them somewhat susceptible to wind damage. However, this is one of the last interamerican clones to flush in the Pacific Northwest and hence is tolerant of spring flooding.

***Populus* × *generosa* A. Henry '49-177'**
(*P. trichocarpa* × *P. deltoides*)

Female clone with a narrow crown of moderately rising branches; large trunks straight and cylindrical; bark greenish gray on branches, becoming gray and smooth to warty, finally deeply furrowed on large trunks; first-year twigs reddish brown, cylindrical to 5-angled, hairy; winter buds 8–15 mm long, very sticky with reddish orange, fragrant resin. Leaves dark green above, whitish green between the dark veins beneath, 20–27 cm long and 15–18 cm wide, elongately egg-shaped, with numerous, irregularly coarse, rounded teeth along the whole margin, from the short, triangular tip to the broadly rounded base with a shallow indentation bearing 1–3, inconspicuous, protruding glands at the junction with the 8–11 cm long, modestly to strongly flattened petiole. Flower bracts deeply cut, white with brown tips; flowers numerous and densely packed; seed capsules opening with 3–4 valves.

The leaf shape of this clone bred by Reinhard Stettler is a more rounded version of that of *P. trichocarpa*. The clone grows well because of its strong apical dominance, but susceptibility to leaf rust, and corresponding growth reductions, have become more evident since its release.

***Populus* × *generosa* A. Henry '52-225'**
(*P. trichocarpa* × *P. deltoides*)

Female clone with a dense, narrow crown of steeply rising branches including many proleptic branches; large trunks cylindrical but spiraling at first and tending to fork; bark greenish gray on branches, becoming gray, relatively smooth during a production cycle but deeply furrowed on large trunks; first-year twigs reddish brown, cylindrical, hairy, with prominent, persistent lenticels; winter buds 8–15 mm long, very sticky with reddish, fragrant resin. Leaves dull green above, somewhat whitened beneath between conspicuous, yellowish green, major veins, but moderately inconspicuous minor veins, 16–19 cm long and 11–14 cm wide, narrowly and triangularly egg-shaped, the numerous, fairly even, moderately fine, rounded teeth along almost the whole margin, gradually tapering to the prolonged tip, the base straight across or rounded, with 2 or more slightly protruding glands at the junction with the 7–8 cm long, strongly flattened petiole. Flower bracts deeply cut, white with brown tips; flowers numerous and densely packed; seed capsules opening with 3–4 valves.

The leaves of this clone are very like those of *P. trichocarpa* in shape, but the toothing is much more prominent and the petiole is flattened like that of *P. deltoides*. The veins are yellow with a tinge of red near the leaf base. The relatively small leaves are advantageous for the dry atmosphere east of the Cascades, where this clone is usually grown. When grown rapidly, however, growth outstrips lignification, resulting in a crooked trunk. The poor growth form makes it ill-suited to timber production, so it is mainly grown for chips.

***Populus* × *generosa* '184-411'**
(*P. trichocarpa* × *P. deltoides*)

Male clone with a narrow crown of long, slender branches; large trunks cylindrical or lobed; bark grayish brown on branches, becoming dark grayish brown and deeply furrowed on large trunks; first-year twigs reddish brown, cylindrical, hairy; winter buds 15–20 mm long, sticky with red, fragrant resin. Preformed (early) leaves of short shoots roundly triangular, (5–)8–11 cm long and (4–)7–10 cm wide, neoformed (late) leaves of long shoots heart-shaped, 12–14 cm long and 11–12 cm wide, both bright green to dark green above, grayish green and often streaked with resin beneath, rather finely toothed along the whole margin, with pointed to rounded teeth becoming finer towards the narrowly to broadly triangular, moderately prolonged tip, the base broadly wedge-shaped or rounded, straight across, or even slightly indented, without glands or with 1 or 2 inconspicuous round or raised glands at the junction with the (2–)4.5–9 cm long, modestly flattened petiole. Flower bracts deeply cut, brown except towards the white center; flowers numerous and densely packed; stamens 30–40.

This clone, used primarily east of the Cascades, was developed by Reinhard Stettler. It has much smaller leaves than most of the *P.* × *generosa* clones grown on the rainier west side.

***Populus* 'Androscoggin'**
(*P. maximowiczii* × *P. trichocarpa*)

Male clone with a broad, egg-shaped crown of long, rising branches; large trunks cylindrical; bark gray on branches, becoming dark grayish brown on large trunks; first-year twigs greenish brown, sharply 5-angled with prominent, corky ribs, densely fuzzy; winter buds 8–14 mm long, sticky with reddish brown, fragrant resin. Leaves dark green above, white beneath, 7–10 cm long, egg-shaped, finely and evenly toothed along the whole edge, the tip short, the base heart-shaped, the petiole 2.5–3.5 cm long, red, round at the blade. Flower bracts white with light brown edges, deeply cut; flowers numerous and fairly densely packed; stamens about 30–40.

This fast-growing clone from the Oxford Paper Company breeding program in the 1930s is entirely balsam poplar in parentage and characteristics. It has a long growing season, with early leaf flush and late leaf drop, although leaf loss can be accelerated by rust infection to which this clone has moderate susceptibility. It is otherwise relatively disease and pest resistant and is suitable for plantation culture in less continental areas of northeastern North America. However, in the more continental climate of the Great Lakes region and even in the St. Lawrence valley of Quebec, it is one of the worst available commercial clones because of its extreme susceptibility to Septoria canker.

***Populus* '282-189'**
(*P. trichocarpa* × *P. maximowiczii*)

Clone of uncertain sex as of this writing, with an oval crown of gently rising branches; large trunks cylindrical; bark greenish gray on branches, becoming dark gray and deeply furrowed on large trunks; first-year twigs reddish brown, round or 5-angled, densely fuzzy; winter buds 8–16(–20) mm long, sticky with red, fragrant resin. Leaves dark green above, white beneath, 15–25 cm long, broadly egg-shaped, moderately finely and evenly toothed along the whole edge with hairy, gland-tipped teeth, the tip bluntly triangular, the base broadly rounded, with 2–4 inconspicuous, round glands at the junction with the 5–7 cm long, channeled, round, reddish brown petiole. Flower bracts deeply cut, white with brown lobes, flowers numerous and densely packed; expect about 30–40 stamens *or* capsules with 3–4 sparsely hairy valves.

This vigorous clone, hardy only in the northwest coastal region, has enormous leaves that are wider than those of *P. trichocarpa* and more purely balsam-like than the *P.* × *generosa* clones. The hairiness of the plant extends to the margins of the leaves and the veins on the lower surface, which are densely woolly.

***Populus* 'NM6' [Max5]**
(*P. nigra* × *P. maximowiczii*)

Female clone with a large, narrowly egg-shaped crown of long, gently to sharply rising branches; large trunks cylindrical to fluted; bark tan on branches, becoming dark grayish brown and deeply furrowed on large trunks; first-year twigs greenish brown to dark reddish brown, round, variably sparsely hairy to fuzzy; winter buds 8–16 mm long, sticky with fragrant red resin. Leaves dull dark green above, white beneath, 7–10(–14) cm long, elongately heart-shaped to almost round, finely and evenly toothed along the whole edge, with a short, pointed tip and deeply indented base without glands, the petiole 1.5–3.5(–4.5) cm long, round and yellowish green. Flower bracts white with deeply cut, brown tips; flowers numerous and fairly densely packed; seed capsules opening with 2–3 valves.

This clone from Germany is more resistant to diseases than its Asian balsam poplar parent, *P. maximowiczii*. It grows quite rapidly, at about 12–22 dm in height per year, and so is being used extensively in plantation culture in the Great Lakes region. The brittle branches and large leaves make it too susceptible to wind damage for use in horticultural and amenity plantings.

Populus × _jackii_ Sargent 'Balm of Gilead'
(_P. balsamifera_ × _P. deltoides_)

Female clone with a broad rounded crown of long, spreading branches; large trunks cylindrical; bark light greenish gray on branches, becoming darker and roughly furrowed on large trunks; first-year twigs round, dark reddish brown, shiny, often covered with a fuzz of very short hairs; winter buds 9–15 mm long, very sticky with reddish resin, covered with a fuzz of very short hairs. Leaves very dark green above, white beneath, 4.5–8 cm long, heart-shaped, finely and evenly toothed along most of the length, the tip abruptly short-pointed, the petiole 2–5 cm long. Flower bracts deeply cut, white with brown tips; flowers numerous; seed capsules nearly spherical, with 2–3 valves, rarely maturing.

This very old clone was long cultivated in eastern North America for medicinal use of the bud resin and may still be found, as if wild, persisting by root suckers long after abandonment of cultivation. It has no role in plantation culture today, and its sterility has kept it out of breeding programs. Its origin is still disputed, with some authors considering it a clone of *P. balsamifera*.

***Populus × jackii* Sargent 'Jackii 4'**
(*P. balsamifera × P. deltoides*)

Male clone with a broad low crown of long spreading branches; large trunks cylindrical but uncommon because of shrubby forking; bark light grayish brown on branches, becoming gray and roughly furrowed on large trunks; first-year twigs angled, purplish brown, somewhat shiny, not conspicuously hairy; winter buds 6–12 mm long, sticky with reddish resin. Leaves dark green above, white beneath, 5–8 cm long, broadly egg-shaped, with numerous, very fine, even teeth along the whole length, the base broadly rounded, the tip with an abrupt, short point, the petiole about 3–4 cm long. Flower bracts deeply cut, white with brown tips; flowers numerous; stamens about 30–40.

This shrubby clone has had some use in plantation culture in Ontario, when the emphasis was on very short rotation via coppicing. It has little use otherwise in plantation culture and is not suitable for horticultural or amenity plantings.

***Populus* × *jackii* Sargent 'Northwest'**
(*P. balsamifera* × *P. deltoides*)

Male clone with a broadly egg-shaped crown of long, gently rising branches; large trunks cylindrical; bark grayish green to light gray on branches, dark gray and roughly furrowed on large trunks; first-year twigs round, dark reddish brown, slightly shiny, covered with a fuzz of short hairs; winter buds 10–18 mm long, sticky with reddish resin, fuzzy with short hairs. Leaves bright green above, whitened beneath, triangularly egg-shaped, 5–9 cm long, finely but conspicuously and regularly toothed along the whole margin; the base broadly rounded, the tip with an abrupt, short point, the petiole 3–4 cm long, fuzzy with very short hairs. Flower bracts deeply cut, light brown with dark brown tips, hairless; flowers in moderate numbers; stamens about 60.

This is one of the more important clones in horticulture, amenity plantings, and plantation culture on the northern prairies where it originated by hybridization between locally adapted *P. balsamifera* and *P. deltoides*. It is very hardy and reasonably fast growing, despite its short growing season that contributes to its frost tolerance. Although it is resistant to some diseases and pests, it is highly susceptible to canker, leaf rust, and poplar bud gall mites.

***Populus* × *rouleauiana* B. Boivin 'Pyramidalis'**
(*P. alba* × *P. grandidentata*)

Male clone with a conical crown of long, upswept branches; large trunks somewhat irregular; bark greenish gray with an orange cast on branches, dark gray and deeply furrowed on large trunks; first-year twigs reddish brown, slightly woolly at first; winter buds 6–12 mm long, not sticky, woolly at first. Leaves dark green above, grayish green and variably woolly beneath, 5–10 cm long, with 2 obvious cup-shaped glands at the base near the petiole; the petioles 4–6 cm long, flattened side-to-side at the blade; preformed (early) leaves of short shoots broadly egg-shaped, with few, large, rounded, wavy teeth; neoformed (late) leaves of long shoots very broadly egg-shaped, somewhat 5-angled, very coarsely and irregularly toothed with sharp, angular teeth. Flower bracts coarsely toothed, brown, fringed; flowers numerous; stamens about 10.

While wild clones of this hybrid combination are common in temperate northeastern North America, this relatively recently derived offspring of the Bolleana poplar is infrequently encountered. It is hardier and more disease resistant than the Bolleana poplar but does not have as narrowly columnar a crown.

Some additional important commercial clones not described here

Clone name	Sex	Regions of use
Populus balsamifera × *P. maximowiczii* clone 915004	Female	Quebec
Populus balsamifera × *P. maximowiczii* clone 915005		Quebec
Populus × *canadensis* 'DN33'		North Central region
Populus × *canadensis* 'Assiniboine'	Male	Prairies
Populus × *canadensis* 'CanAm'	Female	Prairies
Populus × *canadensis* 'Drapal Wulst' (DN181)		Great Lakes
Populus × *canadensis* 'Manitou'	Male	Prairies
Populus × *canadensis* 'NE237'		Rocky Mountains
Populus × *canadensis* 'Spijk' (DN177)		Great Lakes
Populus × *canadensis* × *P. maximowiczii* clone 915508	Female	Quebec
Populus × *canadensis* × *P. maximowiczii* clone 916401	Male	Quebec
Populus deltoides 'Fitler 2'		Mississippi delta
Populus deltoides 'Hill'	Female	Prairies
Populus deltoides 'Prairie Sky'		Prairies
Populus × *generosa* '50-197'	Male	Pacific Northwest
Populus × *generosa* '52-226'	Male	Pacific Northwest
Populus × *generosa* 'Beaupre'		Northeast
Populus × *generosa* 'Boelare'		Northeast
Populus × *generosa* 'Unal' (DTAC-7)	Male	Northeast, Pacific Northwest
Populus × *hastata* [BT] clone 747210		Quebec
Populus × *hastata* [BT] clone 747215p		Quebec
Populus maximowiczii × *P. balsamifera* clone 915311	Female	Quebec
Populus maximowiczii × *P. balsamifera* clone 915319	Female	Quebec
Populus maximowiczii × *P.* × *generosa* clone 505227	Male	Quebec
Populus maximowiczii × *P.* × *generosa* clone 505326		Quebec
Populus nigra '14-272'	Female	Rocky Mountain
P. × *rouleauiana* 'Crandon'	Female and male	Midwest, South

CHAPTER 14
Breeding strategies for the 21st Century: domestication of poplar

H.D. Bradshaw, Jr., and Steven H. Strauss

Introduction

The goal of plantation culture is the economically and ecologically sustainable production of wood to meet a growing worldwide demand. At the end of the 20th Century, poplar culture has reached a fork in the road to the future, and both paths must be followed in order to achieve the goal of sustainability. One branch of the road leads to *poplar forestry*, where natural or extensively managed stands have multiple functions in addition to wood and fiber production, such as watershed protection, maintenance of biodiversity, and recreation. Aspen in the boreal ecosystem of North America is an example of a species for which a poplar forestry management strategy seems appropriate. This form of forestry strongly supplements the ecological preservation functions of biological and riparian reservations that include poplars as key species.

The other branch leads to *poplar agriculture*, where "tree farms" are intensively managed with the dominant goal of producing the maximum volume and quality of wood, fiber, and biomass for energy on the smallest possible land base. The best agricultural practices will be used to increase economic yields, including planting on the best available sites; fertilization and irrigation; control of weeds, pests, and diseases; and genetic improvement through conventional breeding, interspecific hybridization, vegetative propagation (cloning), and biotechnology. Examples of poplar agriculture in North America are the large-scale hybrid poplar fiber and energy farms established in the past 20 years, primarily by the forest products industry and government agencies. Increasing the productivity of plantation forests could have many benefits, including providing an alternative crop for farmers and sparing native forests from overharvesting. In this chapter, we

H.D. Bradshaw, Jr. College of Forest Resources, University of Washington, Seattle, WA 98195, U.S.A.
Steven H. Strauss. Department of Forest Science, Oregon State University, Corvallis, OR 97331, U.S.A.
Correct citation: Bradshaw, H.D., Jr., and Strauss, S.H. 2001. Breeding strategies for the 21st Century: domestication of poplar. *In* Poplar Culture in North America. Part B, Chapter 14. *Edited by* D.I. Dickmann, J.G. Isebrands, J.E. Eckenwalder, and J. Richardson. NRC Research Press, National Research Council of Canada, Ottawa, ON K1A 0R6, Canada. pp. 383–394.

will focus upon breeding strategies designed for poplar agriculture, where the greater potential for yield increases justifies the expense and effort required to implement state-of-the-art breeding technology.

Predicting the future course of technology has led to more embarrassments than successes. Nevertheless, we venture the forecast that the 21st Century will be remembered by the forest products industry as the Era of Tree Domestication, and that poplar will lead the way. Just as poplar farmers have adopted an agricultural paradigm for cultural practices such as fertilization, irrigation, and weed control, poplar breeders will follow the agricultural example in genetics, beginning with the most fundamental process in crop development: *domestication*.

Plant domestication

Crop domestication is among the greatest technological advances in human history (Diamond 1997), enabling the efficient, reliable production of food on a small land area. The domestication of agricultural crops from their wild ancestors was begun by prehistoric farmers 5000–10 000 years ago, and continues today as a result of ongoing breeding programs. Few people realize how different domesticated crop plants are from their wild relatives. Whereas a wild plant struggles for existence in a complex ecosystem, expending much of its energy sensing its environment and competing with neighboring plants for access to light, water, and nutrients, the environment of a cultivated crop is simplified and optimized by the farmer to allow the domesticated plant to devote its energy almost exclusively to the production of useful structures such as seeds, tubers, or fruits. The process of domestication greatly improves the yield and quality of the crop for human use, and simultaneously reduces or eliminates the ability of the plant to survive in its original wild state. The crop depends upon the farmer for survival as much as the farmer depends upon the crop for food and income.

The conversion of wild plants into domesticated crops involved radical changes (mutations) in the genes of crop ancestors. It is only in the past 15 years, with the development of modern molecular genetic methods to examine the detailed structure of plant DNA, that we have begun to discover the number and identity of genes that were mutated during the course of plant domestication.

For example, the conversion of wild mustard flowering stalks into the striking "flower-heads" of cauliflower appear to have involved mutations in just two master genes that control flower and fruit development (Smyth 1995). Dwarf and semi-dwarf rice and wheat strains, the foundation of the Green Revolution which doubled grain yield between 1960 and 2000, are the result of a single mutation (Peng et al. 1999). Based upon the work of George Beadle (Beadle 1972) and John Doebley (Doebley and Stec 1991) we know that mutations in just five major genes transformed the inedible seed-head of teosinte (the wild ancestor of corn) into an "ear" of corn (Fig. 1). Maize (corn) has at least 50 000 different genes, but only five of them had to be altered by mutation to transform teosinte from a

Fig. 1. The transition from inedible wild teosinte (A) to primitive maize (B–D) required mutations in just five major domestication genes. Modern maize is shown on the right (E). Photo courtesy of John Doebley.

weedy grass growing exclusively in the central highlands of Mexico into the most important agricultural crop across North America.

A genetically similar, though less visibly dramatic, domestication history has occurred in many other crops. Interestingly, although a few trees used in horticulture (e.g., apples, olives, cherries) have been domesticated to a significant degree, none have included forest trees grown for wood. There are several reasons for the lack of forest tree domestication. First, until the 20th Century forest trees were grown mostly under natural or semi-natural conditions, so that "domestication genes" would be a disadvantage to them. It is only under the benign conditions of a farm that domesticating mutations have value. In contrast, crops have been grown in an agricultural setting for the past 100 centuries, allowing a long time period for domestication genes to be recognized and perpetuated through selective breeding. Second, the generation time of forest trees grown under near-wild conditions is very long, commonly 10–30 years from planting a seed until the tree produces its own seeds, so that progress by conventional breeding is slow when compared with annual crops. Wheat, rice, corn, soybeans, and other crop plants have been bred by humans for many thousands of generations. A major facet of domestication of fruit trees was selection for earlier and more intense reproduction, which then would feed back to speed further domestication. In contrast, the most advanced forest tree breeding programs are only in their third or fourth breeding cycle. Third, many “domestication” mutations are genetically recessive, and so can only be observed by careful inbreeding. Most tree species are strictly outbreeding (cross-pollinating), and so recessive domestication genes remain hidden. Many crop plants are naturally self-pollinated, or can be easily inbred, making it simple to discover beneficial recessive mutations.

If we accept the idea that trees can be domesticated to dramatically improve their yield and quality under the intensive agricultural systems typical of modern farming, and that domestication may require mutations in just a handful of critical genes, we are left with four important questions:

1. What will a domesticated tree look like?
2. How will we identify domestication genes?
3. How will domestication genes be deployed in breeding programs and production plantations?
4. What consequences will domestication genes have for poplar agriculture?

Characteristics of a domesticated tree

Before attempting to describe the (future) appearance of a domesticated poplar, it is advisable first to ask why wild forest trees have the appearance that they do. Wild trees have been shaped by the force of Darwinian natural selection, not by the hand of humans as crops have. The primary commercial interest in forest trees is for the wood they produce. Why do wild trees produce wood? The answer is: “So that they can grow tall.” Without the strength and stiffness that wood provides to the stem, herbaceous plants can never grow as tall as a tree because they would collapse under their own weight. Why do wild trees grow tall? They do so because there is competition among plants for sunlight to power photosynthesis and thus provide resources for sexual reproduction. Trees are Nature’s ultimate solution to the problem of competition for light. Only the dominant trees in the forest canopy are able to produce the maximum number of seeds and (or) pollen, and thus pass on their genes to future forests. Reproductive output, and not wood production, is the currency of natural selection in wild trees. Wood is just a tree’s way of making seeds, and a wild tree makes as little wood as is necessary to achieve its reproductive goals.

Domesticated trees, on the other hand, will be expected to channel their photosynthetic effort into the production of wood, and not into reproduction. Domesticated trees growing in a plantation do not need to compete with each other or with weeds for access to light, since the farmer will plant the trees with enough space between them, and will control weeds. In a clonal plantation, each neighboring tree is genetically identical, so it makes no sense for the trees to compete with copies of themselves.

Domesticated trees will therefore have a greater stem diameter in proportion to their height than wild trees do. In physiological terms, the sink strength of the vascular cambium will become greater relative to the sink strength of the apical meristem. Thicker, stiffer stems will result in increased wood quality (less reaction wood), larger piece sizes, less waste during sawing, and a greatly increased harvest index (ratio of harvestable product to total biomass) because of the reduced need for photosynthate allocation to structural roots. Because of their inherently stiffer stems and reduced need for longevity compared to wild trees, it will be possible to reduce significantly the proportion of lignin without causing stem form or bending strength to be compromised. The reduction in lignin amount, which is energetically very costly to synthesize, is likely to increase cellulose yield and tree growth rate (Hu et al. 1999). Changes in lignin structure can also reduce the intensity, and thus the cost, of chemical and mechanical treatment during pulping (Baucher et al. 1996).

In a domesticated tree, the number, length, and diameter of branches will be reduced, while maintaining useful dimensions (4–8 feet) of internodes (clear wood) between whorls of branches. This will improve wood quality by producing fewer knots, and may improve productivity by increasing water flow to the leaves (Tyree et al. 1983; Tyree 1988). The domesticated tree will carry the minimum number of leaves necessary to support rapid growth. Crown geometry will become optimized for efficient use of space. Changes in branching and canopy geometry are a common feature of domesticated plants, including maize and the other cereal crops where increased leaf angles have accompanied the shift to very high planting densities (Duvick 1992).

Domesticated forest trees would flower late or not at all, making available additional resources for vegetative growth. A majority of the increased productivity during domestication of crop plants has come from increased allocation of photosynthate to reproductive tissues (Evans 1980). In forest trees, we desire the opposite result: increased vegetative stem mass at the expense of reproductive organs and other vegetative parts. Many of the tree species grown under intensive culture, including poplars, are early successional species that are adapted to grow rapidly in full sunlight while colonizing a new site. After a short time, these pioneer species begin to reproduce vigorously, and continue dispersing copious quantities of seeds and pollen for the remainder of their lives. The combination of rapid growth potential and heavy flowering should enable a significant increase in tree growth rate if flowering could be postponed until after harvest, or prevented entirely (Strauss et al. 1995). Because poplars grown in tree farms are vegetatively propagated, and often composed of hybrids that are not used for further breeding, elimination of flowering in plantations would not pose a difficulty for poplar agriculture. However, preventing flowering in a controlled, specific manner will be very difficult to achieve without genetic engineering (discussed below).

Many species of poplars, particularly the aspens and related taxa, are difficult to propagate vegetatively via rooting or other means. The loss of competence for regeneration with age and tissue differentiation is almost universally observed in forest trees, greatly limiting the economic use of cloning for tree improvement. However, the amenability of trees to vegetative regeneration usually shows great genetic diversity within and among species, suggesting that genes can be isolated that strongly modify competence for regeneration. The addition of such genes via transgenesis, preferably subject to regulation by an inducible promoter (one that causes expression in response to an exogenously applied substance), could greatly increase the rate of genetic gain via clonal propagation.

Discovery of domestication genes for poplar

Three experimental elements may be integrated to discover domestication genes. First, thousands or even millions of poplar seedlings could be tested for rare domestication traits arising as the result of a mutation in a major domestication gene. This random, hit-or-miss approach was used successfully to domesticate every modern crop plant, and so might be considered the "traditional" method of domestication. The principal disadvantage of this method is that progress is very slow and sporadic. Because of the long generation interval of trees, and their cross-pollinating mating systems, rare recessive mutations are extremely difficult to find. A directed program of inbreeding is likely to be required to expose recessive mutations; however, inbreeding will result in plants of poor form and low vigor due to the high genetic load of trees, making domestication alleles difficult to discern. Domestication by this traditional route could take many decades or centuries, even with the rapidly growing knowledge of plant genetics and the amenability of poplar to biotechnology.

The other two approaches to poplar domestication depend entirely upon the use of genetic transformation (i.e., asexual gene transfer), also known as transgenic technology, to overcome the severe limitations of more traditional domestication methods. The second technique for identifying domestication genes in poplars is to use the rapidly growing genetic information database surrounding model herbaceous plants such as the tiny mustard weed *Arabidopsis*. *Arabidopsis* is the first plant to have all of the DNA in its chromosomes completely sequenced, giving us an unabridged list of all the genes in a higher plant. Its small genome, small size, rapid life cycle, and amenability to transformation have also made *Arabidopsis* the plant species of choice for cloning genes that regulate plant development. There is a large and growing collection of genes known to affect such traits as the response to competition, flowering, stem growth, disease resistance, and tolerance to abiotic stresses (e.g., cold and drought).

This wealth of "candidate genes" for domestication from *Arabidopsis* and crops can be used directly to inform efforts to isolate similar kinds of genes (homologs) from poplar and other forest trees (Table 1). This will be most efficient once a large set of expressed sequence tag (EST) sequences are available (Sterky et al. 1998), enabling direct identification of the desired genes via computer searches. However, it is also feasible to isolate new genes directly from poplar DNA based upon sections of the genes that are highly conserved among species. After isolation of each of the candidate domestication genes, their function must be tested in poplars. In trees this can only be done by testing for production of a "mutant" phenotype via transformation. Typically, the effects of hyper-expression and suppression of the gene are studied to understand its function. The gene can also be transformed under the control of its own, natural promoter to assess its subtler effects. Because transgenes are inserted randomly into the genome, study of a population of transgenics is akin to studying a range of alleles (gene variations). A gene that is deleterious when completely knocked-out or highly expressed could still be useful when a less dramatic change in its expression is engineered.

A third strategy for finding domestication genes in trees requires direct, random mutagenesis. Because of the difficulty of inbreeding and the delayed flowering in trees, methods that cause dominant mutations are desirable, as they can be discerned directly after transformation. Activation tagging (reviewed by Weigel et al. 2000) is a method whereby a strong promoter or enhancer of gene expression is randomly inserted into a genome to hyper-activate nearby genes, causing dominant mutations (Fig. 2). Because the frequency of gene activation is low, it requires that a very large population of transgenic trees be produced and screened for domestication traits in a field trial (e.g., tens of thousands to hundreds of thousands of unique transgene insertions). Once traits of interest are observed in the mutated trees, the responsible gene can be readily isolated because the promoter and associated DNA act as a tag to facilitate recovery from the genome. The recovered DNA is then sequenced and re-transformed to verify that it is the cause of the domestication trait. In *Arabidopsis*, activation tagging resulted in the isolation of a number of regulatory genes that were missed with other mutagenesis methods. Although most domestication alleles identified in crops have been

Table 1. Genes with potential for forest tree domestication.

Trait	Gene function	Gene identified in:	Example
Height growth	Hormone synthesis	*Arabidopsis*	Lester et al. 1997
Height growth	Transcription factors	Rice, wheat, maize	Peng et al. 1999
Stem thickness	Unknown	*Arabidopsis*	Talbert et al. 1995
Lateral branching	Transcription factors	Tomato, maize	Doebley et al. 1997
Competition sensing	Phytochrome	*Arabidopsis*, tobacco	Smith 1995
Flowering	MADS box	*Arabidopsis*	Michaels and Amasino 1999

Fig. 2. A schematic diagram of how activation tagging works. A gene enhancer lacking a transcription start site is randomly inserted into the genome. When it inserts sufficiently near to a gene to enhance expression over wild-type levels, it can lead to a novel phenotype of biological or economic interest. The enhancer element and associated DNA then provides a molecular "tag" with which to isolate the gene.

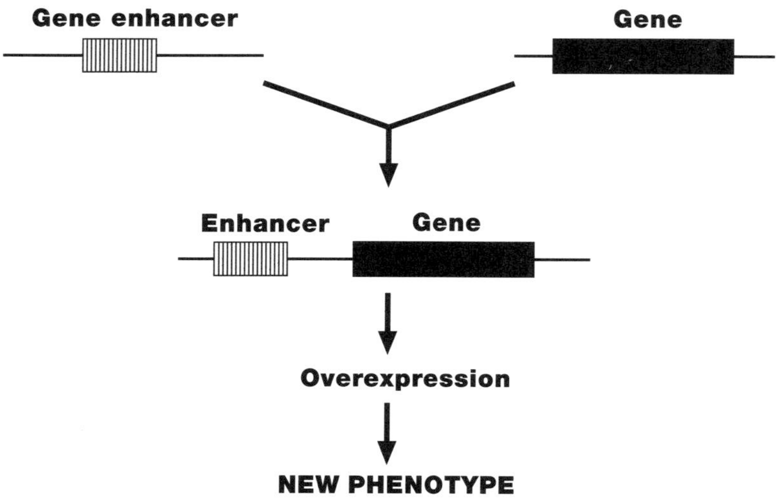

recessive in gene action, the identification of dominant alleles via activation tagging would be especially valuable for trees because of the difficulties of inbreeding-based genetic improvement systems.

Genetic engineering as a core technology for the "Gene Revolution" in poplar culture

Domestication genes are expected to be alleles of wild-type genes. Because these alleles would have deleterious effects on tree fitness in the wild, they should be very rare in germplasm collections from natural populations. However, even if they could be discovered in wild collections and indexed using genetic markers, it would take numerous generations to introgress them into the number of breeding lines required for deployment in production populations. With the delayed flowering of trees, and the extreme genetic diversity within and among regional, national, and international breeding programs, a broad domestication program using traditional breeding would require centuries to achieve, even in poplars. Unless flowering can be greatly and routinely accelerated (see below), genetic transformation appears to be the only option for the use of domestication genes in forestry.

A key advantage of transformation is that new genetic functions can be introduced into highly productive genotypes with minimal disruption of its genetic qualities. In contrast, the introduction of new genes via breeding requires a complete shuffling of the genotype, and is limited to those genes available in the gene pool of the species and its interfertile relatives. Because of mutations that are introduced during transformation and associated *in vitro* culture (somaclonal variation), it is necessary to screen a number of "progeny" (lines) from transformation experiments to identify those with normal growth and stable, desirable expression of the new trait. This can take several growing seasons. The commercial successes of genetic engineering, and the production of many normally growing transgenic poplars in field trials around the world (Meilan et al. 2000), suggest that finding normal, stably-performing transgenic trees is not a substantial obstacle.

The deployment of domestication transgenes in trees will raise very different concerns than have the first generation of transgenes. The first wave of transgenic trees largely consisted of herbicide- and insect-resistant varieties. These two traits raise concerns about the potential for increasing the "weediness" of transgenic poplars. In contrast, domestication genes should handicap trees if they spread to a natural environment, reducing concerns that transgenic poplars might disrupt ecosystems due to their increased vigor, or reduce options for weed control by farmers.

Domestication transgenes could still harm wild populations, however, if not deployed wisely. If domestication transgenes do not include those which prevent sexual reproduction, the use of large populations of transgenic poplars near small, fragmented populations of wild poplars could result in a loss of viability in the wild stands over time. Genetically unfit pollen and seeds from plantations might numerically overwhelm the reproduction of small wild populations, driving them toward local extinction. The combination over the landscape of healthy wild or extensively managed populations, with highly productive agriculture-style plantations that do not produce sexual progeny, is therefore highly desirable. This kind of landscape integration is also likely to be most effective at providing habitat and other ecosystem services while allowing highly efficient wood production.

Goals for poplar domestication

What are the targets for domestication in forest trees that we can foresee in the next century? The goals would be highly diverse, depending on management objectives. However, we suspect that transgenic plantations would embody several of the trait improvements described below:

1. ***Increased yield.*** The ultimate goal of domestication is to produce more wood fiber per unit area and time at less expense. This should be possible by allocation of more photosynthate to stems versus roots, branches, and reproductive tissues; less investment in costly biochemicals such as lignin above the levels needed for structural stability and herbivore protection in tree farms; and improved physiological efficiency due to changes in

height and crown architecture that maximizes photosynthesis per unit area. If fundamental aspects of development can be controlled, such as the timing of flowering and the juvenile–mature transition, it will be possible to speed genetic gain by allowing more rapid breeding and less costly clonal propagation. By combining domestication transgenes with genes for agronomic traits such as herbicide and pest resistance, the yield and environmental benefits of these genes (e.g., reduced water usage, reduced use of undesirable pesticides and herbicides) should be more readily achieved with fewer undesired side effects — and thus, hopefully, with greater public acceptance.

2. ***Tailoring of wood properties to specific end uses.*** Wood that is used for structural purposes, pulp, or energy has very different chemical and physical requirements. For example, by changing the amount and chemistry of lignin in wood, its energy content, digestibility, and amenability to processing into biofuels such as ethanol can be altered markedly. The rapidly expanding catalog of genes expressed in xylem and cambium as a result of genomic sequencing projects should make it possible to alter traits such as fiber length, cell wall thickness, and microfibril angle, enabling major changes in wood and pulp strength, density, and dimensional stability.

3. ***New products and ecological functions.*** Breeding and hybridization has on many occasions given rise to crops with novel properties that determine their use (e.g., the various forms of wheat, rice, and oil crops). Transformation can speed this process; however, it can also enable entirely new biochemical products, or metabolic processes, to be introduced. For example, production of cellulase by trees in bioenergy plantations, or xylanase for trees in pulp plantations, could be a co-product that increases the value of the trees and aids their eventual processing (Morrison et al. 1999). These products could be produced continuously, or synthesized on demand via induced gene expression systems. By engineering genes into trees that facilitate the removal of heavy metals and other pollutants from soils, their value for bioremediation of polluted soils (e.g., Rugh et al. 1998) and biofiltration of farm runoff could be greatly enhanced.

Conclusion

The combination of new knowledge from genomics and molecular biology, with the established feasibility of genetic engineering in poplar, make it an ideal genus with which to demonstrate the power of biotechnology for domesticating trees to meet human needs. Because of the recalcitrance of trees to traditional genetic manipulation, without genetic engineering domestication would proceed extremely slowly and many goals would be effectively unobtainable. Whereas the domestication of the major crops required thousands of years, we expect that genetic engineering will allow the domestication of trees to be accomplished within a few decades. This Gene Revolution may be essential to meeting the wood, fiber, energy, and environmental needs of the rapidly growing world.

Acknowledgements

We are grateful to Don Dickmann, Jerry Tuskan, and John Davis for their critical review of an earlier draft of this manuscript, and to Reinhard Stettler, Richard Waring, Barbara Bond, and Tom Hinckley for helpful discussions. We wish to thank the members of the Poplar Molecular Genetics Cooperative and the Tree Genetic Engineering Research Cooperative for their support. This work was carried out in part with funds from the U.S. Department of Energy Bioenergy Feedstock Development Program via UT-Batelle subcontracts 4000003344 and Lockheed Martin Energy Systems/DOE contract 72X-ST80-7V.

References

Beadle, G.W. 1972. The mystery of maize. Field Mus. Natl. Hist. Bull. **43**: 2–11.

Baucher, M., Chabbert, B., Pilate, G., van Doorsselaere, J., Tollier, M.-T., Petit-Conil, M., Cornu, D., Monties, B., van Montagu, M., Inzé, D., Jouanin, L., and Boerjan, W. 1996. Red xylem and higher lignin extractability by down-regulating a cinnamyl alcohol dehydrogenase in poplar. Plant Physiol. 112: 1479–1490.

Diamond, J. 1997. Guns, germs, and steel: the fates of human societies. W.W. Norton, New York. 480 pp.

Doebley, J., and Stec, A. 1991. Genetic analysis of the morphological differences between maize and teosinte. Genetics, **129**: 285–295.

Doebley, J., Stec, A., and Hubbard, L. 1997. The evolution of apical dominance in maize. Nature (London), **386**: 485–488.

Duvick, D.N. 1992. Genetic contributions to advances in yield of U.S. maize. Maydica, **37**: 69–79.

Evans, L.C. 1980. The natural history of crop yield. Am. Sci. **68**: 388–397.

Hu, W.-J., Harding, S.A., Lung, J., Popko, J.L., Ralph, J., Stokke, D.D., Tsai, C.-J., and Chiang, V.L. 1999. Repression of lignin biosynthesis promotes cellulose accumulation and growth in transgenic trees. Nature Biotech. **178**: 808–812.

Lester, D.R., Ross, J.J., Davies, P.J., and Reid, J.B. 1997. Mendel's stem length gene *Le.* encodes a gibberellin 3-beta-hydroxylase. Plant Cell, **98**: 1435–1443.

Meilan, R., Ma, C., Cheng, S., Eaton, J.A., Miller, L.K., Crockett, R.P., DiFazio, S.P., and Strauss, S.H. 2000. High levels of Roundup and leaf-beetle resistance in genetically engineered hybrid cottonwoods. *In* Hybrid poplars in the Pacific Northwest: Culture, Commerce and Capability. *Edited by* K.A. Blatner and J.J. Johnson. Washington State University Cooperative Extension, Pullman. In press.

Michaels, S.D., and Amasino, R.M. 1999. FLOWERING LOCUS C encodes a novel MADS domain protein that acts as a repressor of flowering. Plant Cell, **11**: 949–956.

Morrison, D.L., Coleman, G., Dale, B.E., Finizza, A.J., Hall, R., Johnson, D., Nichols, R., Sperling, D., and Strauss, S.H.. 1999. Review of the research strategy for biomass-derived transportation fuels. National Research Council (U.S.A.), Board on Energy and Environmental Systems. National Academy Press, Washington, D.C. 48 pp.

Peng, J., Richards, D.E., Hartley, N.M., Murphy, G.P., Devos, K.M., Flintham, J.E., Beales, J., Fish, L.J., Worland, A.J., Pelica, F., Sudhakar, D., Christou, P., Snape, J.W., Gale, M.D., and

Harberd, N.P. 1999. “Green revolution” genes encode mutant gibberellin response modulators. Nature (London), **400**: 256–261.
Rugh, C.L., Senecoff, J.F., Meagher, R.B., and Merkle, S.A. 1998. Development of transgenic yellow poplar for mercury phytoremediation. Nature Biotechnol. **16**: 925–928.
Smith, H. 1995. Physiological and ecological function within the phytochrome family. Ann. Rev. Plant Physiol. Plant Mol. Biol. **46**: 289–315.
Smyth, D.R. 1995. Origin of the cauliflower. Curr. Biol. **5**: 361–363.
Sterky, F., Regan, S., Karlsson, J., Hertzberg, M., Rohde, A., Holmberg, A., Amini, B., Bhalerao, R., Larsson, M., Villarroel, R., Van Montagu, M., Sandberg, G., Olsson, O., Teeri, T.T., Boerjan, W., Gustafsson, P., Uhlen, M., Sundberg, B., and Lundeberg, J. 1998. Gene discovery in the wood-forming tissues of poplar: Analysis of 5,692 expressed sequence tags. Proc. Natl. Acad. Sci. U.S.A. **95**: 13330–13335.
Strauss, S.H., Rottmann, W.H., Brunner, A.M., and Sheppard, L.A. 1995. Genetic engineering of reproductive sterility in forest trees. Molec. Breed. **1**: 5–26.
Talbert, P.B., Adler, H.T., Parks, D.W., and Comai, L. 1995. The *REVOLUTA* gene is necessary for apical meristem development and for limiting cell divisions in the leaves and stems of *Arabidopsis thaliana*. Development, **121**: 2723–2735.
Tyree, M.T. 1988. A dynamic model for water flow in a single tree: evidence that models must account for hydraulic architecture. Tree Physiol. **4**: 195–217.
Tyree, M.T., Graham, M.E.-D., Cooper, K.E., and Bazos, L.J. 1983. The hydraulic architecture of *Thuja occidentalis*. Can. J. Bot. **61**: 2105–2111.

APPENDIX
English to metric conversions

Table A-1. Some English to metric conversions for use in poplar culture.[a]

English unit	Metric unit
Dimensional, weight, volume, and energy	
1 foot	0.304 meters (m)
1 square foot (ft^2)	0.093 square meters (m^2)
1 acre	0.405 hectares (ha)
1 ton	0.907 megagrams (Mg) or metric tons (tonnes)
1 pound	0.454 kilograms (kg)
1 ounce	28.35 grams (g)
1 fluid ounce	28.41 cubic centimeters (cm^3)
1 pint	0.568 cubic decimeters (dm^3)
1 quart	1.14 cubic decimeters (dm^3)
1 British thermal unit (Btu)	1054.4 joules (J)
1 Btu	0.25 kilogram-calories (kcal)
Forestry	
1 cord	3.62 cubic meters (m^3)
100 cubic feet (ft^3) or 1 cunit	2.83 m^3
1 thousand board feet	2.36 m^3
Commodity yield	
1 square foot per acre	0.23 m^2 per ha
1 cord per acre	8.96 m^3 per ha
1 cunit per acre	7.0 m^3 per ha
1 thousand board foot per acre	5.83 m^3 per ha
1 ton per acre	2.24 Mg per ha
1 Btu per acre	2605 joules per ha
1 Btu per acre	0.62 kCal per ha

[a]The numerical factor to convert from metric to English units (i.e., right to left in this table) is the reciprocal of the factor to convert from English to metric. For example, 1 cord = 3.62 m^3, so 1 m^3 = 0.276 cords (1 ÷ 3.62).

Clone index

Index to clones described in Chap. 13.